ACS SYMPOSIUM SERIES 246

Geochemical Behavior of Disposed Radioactive Waste

G. Scott Barney, EDITOR
Rockwell Hanford Operations

James D. Navratil, EDITOR
Rockwell International Rocky Flats Plant

Wallace W. Schulz, EDITOR
Rockwell Hanford Operations

Based on a symposium
jointly sponsored by the Divisions of
Nuclear Chemistry and Technology,
Industrial and Engineering Chemistry,
and Geochemistry at the 185th Meeting
of the American Chemical Society,
Seattle, Washington,
March 20–25, 1983

American Chemical Society, Washington, D.C. 1984

Library of Congress Cataloging in Publication Data

Geochemical behavior of disposed radioactive waste.
(ACS symposium series, ISSN 0097-6156; 246)

"Based on a symposium jointly sponsored by the Divisions of Nuclear Chemistry and Technology, Industrial and Engineering Chemistry, and Geochemistry at the 185th Meeting of the American Chemical Society, Seattle, Washington, March 20-25, 1983."

Bibliography: p.
Includes index.

1. Radioactive waste disposal in the ground—Congresses. 2. Radioisotopes—Congresses. 3. Geochemistry, Analytic—Congresses.

I. Barney, G. Scott. II. Navratil, James D., 1941– . III. Schulz, Wallace W. IV. American Chemical Society. Division of Nuclear Chemistry and Technology. V. American Chemical Society. Division of Industrial and Engineering Chemistry. VI. American Chemical Society. Division of Geochemistry. VII. Series.

TD898.G35 1984 621.48'38 83-3106
ISBN 0-8412-0827-1

PRINTED IN THE UNITED STATES OF AMERICA

ACS Symposium Series

M. Joan Comstock, *Series Editor*

FOREWORD

The ACS SYMPOSIUM SERIES was founded in 1974 to provide a medium for publishing symposia quickly in book form. The format of the Series parallels that of the continuing ADVANCES IN CHEMISTRY SERIES except that in order to save time the papers are not typeset but are reproduced as they are submitted by the authors in camera-ready form. Papers are reviewed under the supervision of the Editors with the assistance of the Series Advisory Board and are selected to maintain the integrity of the symposia; however, verbatim reproductions of previously published papers are not accepted. Both reviews and reports of research are acceptable since symposia may embrace both types of presentation.

CONTENTS

PREFACE

For a complex technology to reach a productive maturity, the disparate scientific disciplines underlying that technology must be gathered into a coherent whole. It is difficult to imagine a technology for which this is better exemplified than the geological disposal of radioactive wastes. The underlying scientific areas include surface chemistry (sorption–desorption, dissolution, ion exchange, corrosion), solution chemistry (hydrolysis, complexation, oxidation–reduction, precipitation), colloid chemistry, ceramics, metallurgy, hydrology, rock mechanics, and geology. These and other fields of study must be synthesized into a useful, practical whole through modelling and rational thought. Finally, the rationalized body of information and conclusions drawn must be substantiated by tests performed in the actual waste disposal environment.

This volume makes an important contribution to the information needed for disposal of wastes in geological media, demonstrating the advanced state of knowledge in many of the above fields of research. It represents a major part of what must be known before high-level radioactive waste disposal may become a reality.

Governmental agencies in the United States and other countries have sponsored a large amount of research on the behavior of radioactive wastes in various environmental settings. The overall objective of this research has been to protect the health and safety of the public by assessing the potential hazard of radionuclides in disposed wastes over periods of time when these radionuclides are significantly active. Making such an assessment requires an understanding of radionuclide distribution and inventory in or near the disposal site, and of the transport processes (chemical, physical, and biotic) that control the movement of radionuclides. The geochemistry of radionuclides in disposal environments is clearly one of the most important aspects of safety assessment because radionuclide release from the disposal site is controlled by complex chemical processes. The goal of this volume is to provide the reader with a single source of the most recent and significant findings of research on the geochemical behavior of disposed radioactive wastes.

Radioactive wastes of concern include wastes that result from operation of the nuclear fuel cycle (mining, fuel fabrication, reactor operation, spent fuel reprocessing, and waste storage), from nuclear weapons testing, and from medical and research activities. In recent years, the emphasis has been on predicting the behavior of disposed high-level wastes in deep geologic repositories. The many chapters on high-level waste reflect this emphasis. However, the chemical behavior of the individual radionuclides described will apply to many types of waste in geologic environments.

The chapters of this volume are organized into sections that cover the chemical aspects that are important to understanding the behavior of disposed radioactive wastes. These aspects include radionuclide sorption and desorption, solubility of radionuclide compounds, chemical species of radionuclides in natural waters, hydrothermal geochemical reactions, measurements of radionuclide migration, solid state chemistry of wastes, and waste-form leaching behavior. The information in each of these sections is necessary to predict the transport of radionuclides from wastes via natural waters and thus to predict the safety of the disposed waste.

Radionuclide transport in natural waters is strongly dependent on sorption, desorption, dissolution, and precipitation processes. The first two sections discuss laboratory investigations of these processes. Descriptions of sorption and desorption behavior of important radionuclides under a wide range of environmental conditions are presented in the first section. Among the sorbents studied are basalt interbed solids, granites, clays, sediments, hydrous oxides, and pure minerals. Effects of redox conditions, groundwater composition and pH on sorption reactions are described.

Solubility constraints define the maximum concentrations of radionuclides at the point of release from the waste. In the second section, radionuclide solubilities in natural waters are reported as measured values and estimated values from thermodynamic data. In addition, information is given concerning the chemical species of radionuclides that could be present in natural waters.

If the heat generated from the waste by radioactive decay is great enough (as in the case of high-level waste disposed of in deep geologic repositories), hydrothermal reactions will occur between the groundwater, host rocks, and waste. The resulting alteration of these solids and groundwaters will affect the behavior of radionuclides in these systems. In the third section, the effects of these hydrothermal reactions are described.

Field measurements of radionuclide migration can be used to help substantiate laboratory measurements of sorption, solubility, and identification of important chemical species. The fourth section describes three field investigations that provide information on the effects of organics, colloids and environmental conditions (Eh, pH, and temperature) on radionuclide transport. The chemical species of radionuclides that are mobile under specific field conditions are identified.

Solid state chemistry of potentially important waste forms is covered in the fifth section. Solid state reactions can determine the oxidation state and physical and chemical stability of radionuclides in various host waste forms. This information can be used to evaluate the utility of crystalline materials as potential hosts for radioactive wastes.

Groundwater leaching of radionuclides from waste forms is the first step in radionuclide transport from a disposal site. The release rate of radionuclides from the waste form is dependent on the waste form's leaching behavior. The sixth section describes the factors that affect the leaching behavior of several potential waste forms and radionuclides.

Finally, Mike McCormack, former Washington state Congressman, discusses the Federal legislation affecting nuclear waste disposal in the United States and the impact of several new laws passed by the Congress—the Nuclear Waste Policy Act of 1982 and the Low-Level Radioactive Waste Policy Act of 1980.

This volume covers ongoing research and, thus, leaves many questions unanswered and many problems unsolved. The geochemistry of disposed radioactive wastes involves many complex issues that will require years of additional research to resolve. High-priority problems include: integration of geochemical data with computer models of chemical interaction and transport, definition of environmental conditions that affect the behavior of radionuclides at specific disposal sites, evaluation of complex formation of dissolved radionuclides with inorganic and organic complexants, and determination of radionuclide solubilities in natural waters.

The editors would like to express their deep appreciation and admiration to Teresa Bess of Rockwell Hanford Operations whose editorial assistance greatly speeded the publication of this volume.

G. SCOTT BARNEY
Rockwell Hanford Operations
Richland, Washington

JAMES D. NAVRATIL
Rockwell International Rocky Flats Plant
Golden, Colorado

WALLACE W. SCHULZ
Rockwell Hanford Operations
Richland, Washington

RAYMOND G. WYMER
Oak Ridge National Laboratory
Oak Ridge, Tennessee

December 1983

LABORATORY STUDIES OF RADIONUCLIDE SORPTION

Radionuclide Sorption and Desorption Reactions with Interbed Materials from the Columbia River Basalt Formation

G. S. BARNEY

Rockwell International, Energy Systems Group, Richland, WA 99352

The sorption and desorption behavior of radionuclides in groundwater-interbed systems of the Columbia River basalt formation was investigated. Radionuclides chosen for study were those of concern in assessing the safety of a high-level radioactive waste repository in basalt (isotopes of technetium, neptunium, plutonium, uranium, americium, cesium, strontium, and radium). Sandstone and tuff materials from selected interbed layers between basalt flows were used in these experiments. Effects of groundwater composition and redox potential (Eh) on radionuclide sorption and desorption on the geologic solids were studied. Sodium, potassium, and calcium in the groundwater decrease sorption of cesium, strontium, and radium by ion exchange reactions. Groundwater Eh strongly affects sorption of technetium, neptunium, plutonium, and uranium since chemical species of these elements containing the lower oxidation states are more extensively sorbed by chemisorption than those containing higher oxidation states. Effects of radionuclide complexation by groundwater anions on sorption were not observed except for neptunium carbonate (or bicarbonate) complexes and plutonium sufate complexes.

Sorption and desorption isotherms were obtained for sorption of radionuclides under oxidizing and reducing conditions. The Freundlich equation accurately describes most of these isotherms. Most radionuclides are apparently irreversibly sorbed on each of the geologic solids since the slopes of sorption and desorption isotherms for a given radionuclide are different. This hysteresis effect is very large and will cause a significant delay in radionuclide transport. It, therefore, should be included in modeling radionuclide transport to accurately assess the isolation capabilities of a repository in basalt.

0097–6156/84/0246–0003$06.00/0

The groundwater transport of radionuclides through water-bearing interbed layers in the Columbia River basalt formation will be controlled by reactions of the radionuclides with groundwater and interbed solids. These interactions must be understood to predict possible migration of radionuclides from a proposed radioactive waste repository in basalt. Precipitation and sorption on interbed solids are the principle reactions that retard radionuclide movement in the interbeds. The objective of the work described herein was to determine the sorption and desorption behavior of radionuclides important to safety assessment of a high-level radioactive waste repository in Columbia River basalt. The effects of groundwater composition, redox potential, radionuclide concentration, and temperature on these reactions were determined.

Geochemical models of sorption and desorption must be developed from this work and incorporated into transport models that predict radionuclide migration. A frequently used, simple sorption (or desorption) model is the empirical distribution coefficient, K_d. This quantity is simply the equilibrium concentration of sorbed radionuclide divided by the equilibrium concentration of radionuclide in solution. Values of K_d can be used to calculate a retardation factor, R, which is used in solute transport equations to predict radionuclide migration in groundwater. The calculations assume instantaneous sorption, a linear sorption isotherm, and single-valued adsorption-desorption isotherms. These assumptions have been shown to be erroneous for solute sorption in several groundwater-soil systems (1-2). A more accurate description of radionuclide sorption is an isothermal equation such as the Freundlich equation:

$$S = KC^N \tag{1}$$

where

S = the equilibrium concentration of sorbed radionuclide in moles/g
C = the equilibrium concentration of radionuclide in solution in moles/L
K and N = empirical constants.

This equation has been successfully applied to many sorption and desorption reactions of dissolved metals and organic compounds. In the case of irreversible sorption (hysteresis), sorption and desorption isotherms are not identical. However, both sorption and desorption Freundlich isotherm equations can be substituted into the transport equation(2):

$$\frac{\rho}{\phi}\frac{S}{t} + \frac{\partial C}{\partial t} = \frac{D\partial^2 C}{\partial x} - v\frac{\partial C}{\partial x} \tag{2}$$

where

D = dispersion coefficient
v = average pore water velocity
ρ = bulk density
ϕ = saturated water content
x = distance in the direction of flow
t = time.

This equation can be used to describe one-dimensional transport of radionuclides through porous media (e.g. radionuclide elution curves from laboratory columns packed with interbed solids) assuming instantaneous sorption and desorption. Van Genuchten and coworkers have demonstrated the importance of using both sorption and desorption isotherms in this equation when hysteresis is significant. Isotherm data for sorption and desorption reactions of radionuclides with interbed materials are presented in this paper which can be used to predict radionuclide transport.

Experimental

Materials. The groundwater compositions of waters in the major water bearing zones of the Columbia River basalts at the Hanford Site have been determined (3). There are two distinct groundwaters present in the basalts: a sodium-bicarbonate buffered groundwater (pH 8 at 25°C) characteristic of the Saddle Mountains and Upper Wanapum basalts and a sodium-silicic acid buffered groundwater (pH 10 at 25°C) characteristic of the Lower Wanapum and Grande Ronde Basalts. Synthetic groundwater compositions have been established that simulate these two groundwater types. The compositions of the synthetic groundwaters used in the sorption experiments are given in Table I. The GR-1A groundwater simulates the groundwater composition of the Mabton Interbed in the Saddle Mountain Basalts. The GR-2 and GR-2A groundwaters simulate the dominant groundwaters in the Lower Wanapum and Grande Basalts. Synthetic groundwaters are used rather than actual groundwaters in order to ensure the availability of a stable, compositionally consistent groundwater for the sorption experiments.

Three interbed materials from the Columbia River Basalt Group have been investigated in the radionuclide sorption experiments. Interbeds are porous sedimentary layers located between many of the basalt flows in the Columbia River Basalt Group and comprise a potential preferential pathway for groundwater and, therefore, radionuclide transport.

Two interbed samples, a sandstone and a tuff, were taken from as outcrop of the Rattlesnake Ridge Interbed above the

Table I. Synthetic Groundwater Compositions

Constituents	Concentration (mg/L)		
	GR-1A	GR-2	GR-2A
Na^+	107	225	246
K^+	11.2	2.5	2.5
Ca^{2+}	2.0	1.06	1.01
Mg^{2+}	0.4	0.07	0
Cl^-	50*	131	152
CO_3^{2-}	0	59	28.8
HCO_3^-	215	75	36.6
F^-	1.3	29	37
SO_4^{2-}	2.4	72	108
SiO_2	30	108	82
pH	8.5	10.0	10.0
Ionic strength	–	0.014	–

*In experiments having reducing conditions, Cl^- content increased to 479 mg/L due to addition of HCl to neutralize N_2H_4 to obtain pH = 8.5.

Pomona basalt flow. A third interbed sample was taken from drilling cores of the Mabton Interbed, located between the Saddle Mountains and Wanapum Basalts. The Mabton Interbed is the first continuous, major interbed above the candidate repository horizons in the Grande Ronde Basalts. Mineralogical characteristics of the interbed materials are summarized in Table II. A more complete discussion of the characteristics of the interbed materials may be found in Reference (4).

Procedures. Batch equilibrations of interbed solids (Mabton Interbed, Rattlesnake Ridge sandstone, or tuff), tracers, and groundwaters were used to measure radionuclide distributions between solid and liquid phases. Triplicate measurements were made for each combination of temperature, redox condition, tracer concentration, tracer type, groundwater composition, and interbed sample. Constant temperatures were maintained by placing the

Table II. Sources and Characteristics of Interbed Materials

Source/Characteristic	Interbed sandstone	Interbed sandstone/ claystone	Interbed tuff
Source	Rattlesnake Ridge Interbed	Mabton Interbed	Rattlesnake Ridge Interbed
Size fraction	0.25 to 0.5 mm	16% = <250 mesh 60% = 40 to 250 24% = <40	0.5 to 1.0 mm
Surface area[a]	50.9 ± 0.8 m^2/g	not available	188 ± 3 m^2/g
Cation exchange capacity[b]	9.9 ± 0.2 meq/100 g	not available	52.2 ± 4.2 meq/100 g
Minerals present[c]	Quartz	Quartz	Smectite Clay
	Plagioclase	Albite	Glass
	Orthoclase	Anorthite	
	Hornblende	Nontronite	
	Muscovite	Biotite	
	Biotite	Orthoclase	
	Clinopyroxene (Also basalt fragments)	Illite	

[a] Ethylene glycol method (5).
[b] Strontium cation exchange (6).
[c] By X-ray diffraction and electron microprobe.

mixtures in environmental shaker air baths (New Brunswick Scientific Company) set at 23°C, 60°C, or 85°C. For experiments under reducing conditions, 0.05M hydrazine was added to the groundwater to establish an Eh of -0.8 V at the pH values used. This Eh value was calculated from the standard potential for hydrazine oxidation in basic solution (+1.16 V) assuming a pH of 9.0, N_2 pressure of 1 atmosphere and a hydrazine concentration of 0.05M.

Details of the procedure are as follows: 2.00 ± 0.02 g of disaggregated interbed material was added to a 1-oz polyethylene bottle. These interbed materials were soft and easily broken up using gentle action with mortar and pestle. Twenty milliliters of unspiked groundwater (reducing or oxidizing, as appropriate) were added to the bottle and the mixture was gently shaken overnight to preequilibrate the system. The bottle was centrifuged and the supernate discarded. This preequilibration was repeated and the bottle + interbed + solution weighed to estimate the volume of residual solution. Spiked groundwater solutions were prepared by dissolving a measured amount of solid tracer (obtained by evaporating stock tracer solution) in the appropriate groundwater solution. The spiked solution was filtered through a 0.3-μm Millipore filter to remove any undissolved solids and analyzed. The spiked solution was added to the bottle from the preequilibration step, and the bottle was capped and shaken gently for 14 days at the appropriate temperature. The contents were then filtered through a 30 Angstrom Amicon 50A ultra-filter. The filtrates were analyzed for tracer concentrations using standard counting techniques.

For desorption measurements, the above procedure was used to load tracer radionuclides onto the interbed solid sample. The solution-solid mixture was centrifuged and the supernate carefully removed to avoid removing any solids with it; it was then filtered through a 30 Angstrom Amicon 50A filter before analysis. Fresh groundwater solution was added to the separated solid and this mixture was equilibrated for 1 week at the same temperature as during the loading of the tracer. The solution was separated as before and the concentration of desorbed tracer determined. The tracer-loaded solid was equilibrated with fresh groundwater 10 to 13 consecutive times. Each time the resulting solution was analyzed for tracer concentration.

Changes in groundwater composition (major ion concentrations) during equilibration of the synthetic groundwaters with interbed solids were measured as follows. Either 20 or 40 g of interbed solids were added to 16-oz polyethylene bottles along with 400 mL of reducing or oxidizing groundwater. The bottles were placed in the environmental shakers set at either 23°C or 60°C. Each bottle was sampled weekly by turning off the shaker, allowing the solids to settle for several hours, and then decanting approximately 20 mL into 30 Angstrom Amicon 50A ultra-filter cones for filtration. The filtrates were analyzed for cations by an inductively coupled plasma spectrometer (ICP; Applied Research Laboratory) and anions by anion chromatography (Dionex, Inc.). It was found that groundwater-interbed solid reactions were fast, reaching steady-state concentrations after less than 1 week. Concentrations of major cations increased over this time; however, interbed solids control the groundwater composition to such a great extent that initial groundwater composition is not very significant.

Results and Discussion

Sorption and Desorption Isotherms. To model radionuclide transport in groundwater through geologic media, it is necessary to mathematically describe sorption and desorption in terms of isotherms. The Freundlich isotherm was found to accurately describe sorption and desorption of all radionuclides studied in the interbed-groundwater systems, except when precipitation of the radionuclide occurred.

In addition to describing sorption and desorption, isotherms can be used to estimate the solubilities of radionuclides in the groundwater-radionuclide-geologic solid system. For radionuclides that form slightly soluble compounds (e.g., $SrCO_3$, $PuO_2 \cdot H_2O$) in these systems, isotherms can define the approximate concentrations above which precipitation, rather than sorption, dominates removal from solution.

Sorption of cesium, strontium, selenium, technetium, radium, uranium, neptunium, and americium on the standard Rattlesnake Ridge sandstone interbed material was measured using the GR-2 groundwater composition. Reducing conditions (0.05$\underline{M}$ N_2H_4) were used for selenium, technetium, uranium, and neptunium. Freundlich plots for cesium and uranium at 23°C, 60°C, and 85°C are shown as examples in Figure 1. Precipitation was observed in the cases of selenium and technetium at 23°C. Selenium begins to precipitate at $10^{-5}\underline{M}$ and technetium precipitates at concentrations above $10^{-7}\underline{M}$. At the higher temperatures, technetium solubility increases and selenium sorption increases so that precipitation is not observed.

Freundlich constants and ranges of K_d values for radionuclide sorption on the Rattlesnake Ridge sandstone are given in Table III. The constants K and N were calculated using linear regression. Linear sorption isotherms (N = 1.0) are observed only for strontium, selenium, and radium.

Sorption isotherms were also measured for sorption of selenium, technetium, tin, radium, uranium, neptunium, plutonium, and americium on the reference Mabton Interbed solids. The GR-1A groundwater composition was used in these experiments. Two temperatures (23°C ± 2°C and 60°C ± 1°C) were used, and both oxidizing and reducing conditions were used for each radionuclide.

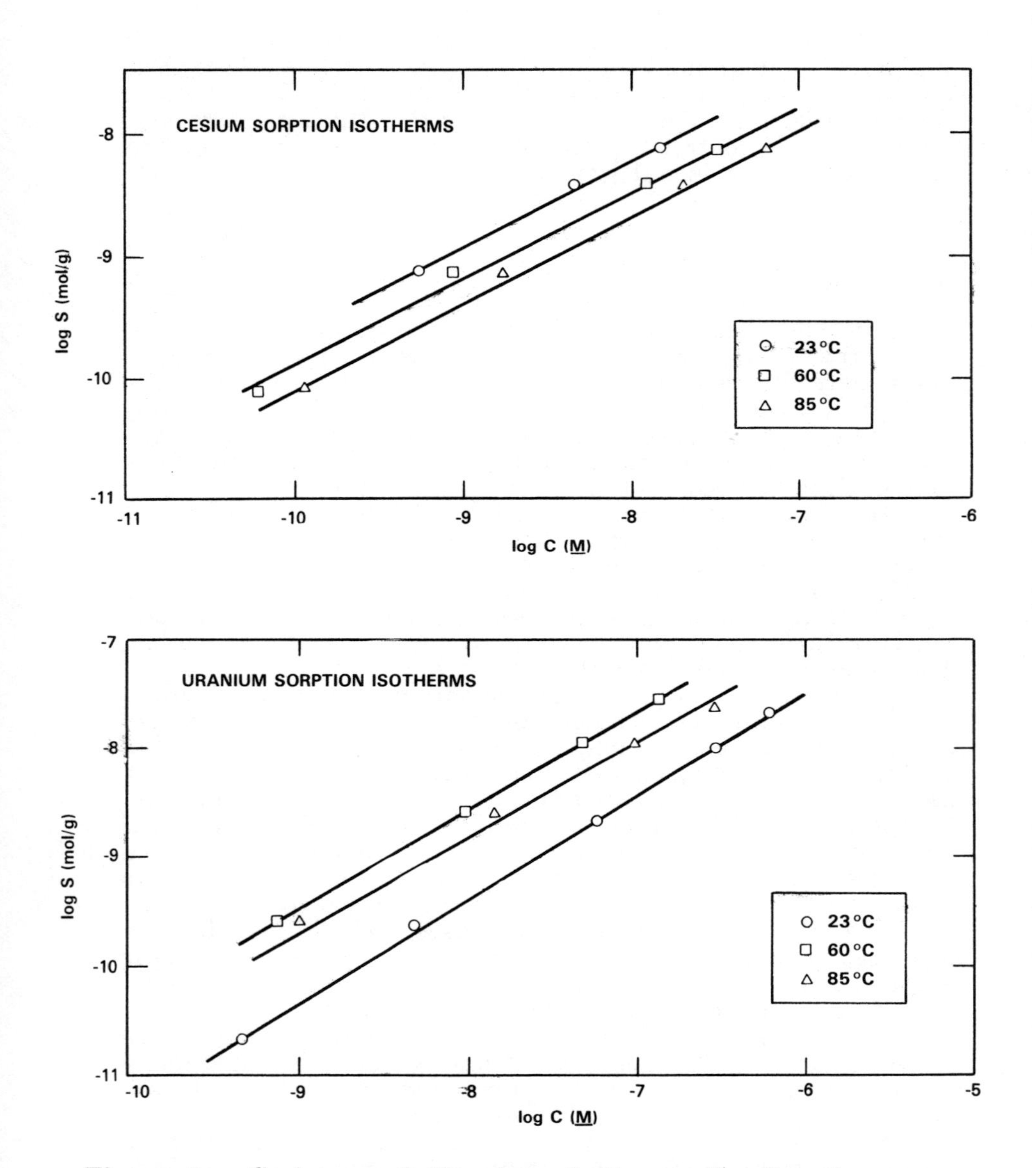

Figure 1. Cesium and Uranium Isotherms for Sorption on Sandstone at 23(o), 60(□), and 85°C(Δ). Oxidizing Conditions Were Used for Cesium and Reducing Conditions for Uranium.

Table III. Freundlich Constants for Radionuclide Sorption (Molar Basis) on Sandstone

Radionuclide	Redox	Temperature (°C)	K	N	r^2*
Cesium	Oxidizing	23	0.0023	0.70	0.99
		60	0.0029	0.72	0.96
		85	0.0013	0.72	0.99
Strontium	Oxidizing	23	0.78	1.02	0.99
		60	1.17	1.03	0.99
		85	12.2	1.18	0.99
Selenium	Reducing	23	0.010	1.03	0.98
		60	0.053	0.97	0.99
		85	0.70	1.06	0.99
Technetium	Reducing	23	1,056	1.50	0.92
		60	5.6	1.30	0.99
		85	0.2	1.09	0.99
Radium	Oxidizing	23	0.62	0.94	0.98
		60	0.48	0.95	0.98
		85	83.1	1.16	0.98
Uranium	Reducing	23	0.013	0.94	0.99
		60	0.046	0.91	0.99
		85	0.0037	0.79	0.99
Neptunium	Reducing	23	3.5	1.03	0.99
		60	–	–	–
		85	1,743	1.36	0.99
Americium	Oxidizing	23	0.030	0.89	0.99
		60	2.89	1.03	0.99
		85	–	–	–

*Coefficient of determination.

Tin and americium were so extensively sorbed under all conditions that isotherm data could not be obtained. These elements are not significantly mobile in the Mabton Interbed aquifer. Values of Freundlich constants for technetium, radium, uranium, neptunium, and plutonium are given in Table IV. The Freundlich equation did not fit the selenium sorption data very well probably because of slow sorption kinetics or precipitation. Precipitation was also observed for technetium at 23°C for concentrations above 10^{-7}M. This is about the same solubility observed for technetium in the sandstone isotherm measurements. Linear isotherms were observed only in the case of radium sorption. In general, sorption on the Mabton Interbed was greater than on the Rattlesnake Ridge sandstone. This is probably due to the greater clay content of the Mabton standard.

The Freundlich equation requires the assumption that sorption reactions are reversible. However, several studies (7) have recently shown that K and N depend on sorption direction, i.e., whether sorption or desorption occurred. In each case, N was less and K was greater for desorption than sorption.

This chemical hysteresis will, of course, affect radionuclide transport. For example, if hysteresis occurs during a column experiment in which a pulse of tracer is added to the influent, the effluent curve will show heavy tailing and a reduction in peak concentration. Ignoring hysteresis effects could cause serious errors in predicting radionuclide movement.

Desorption isotherms for selenium, technetium, neptunium, uranium, and radium have been measured for the Mabton Interbed materials under both oxidizing and reducing conditions at 60°C using the Grande Ronde groundwater composition, GR-1A. The reason for measuring desorption isotherms is to determine whether or not the sorption reactions are reversible (i.e., exhibit hysteresis). An example of the results of desorption isotherm measurements is shown in Figure 2. These curves are Freundlich plots of the sorption and desorption data for neptunium at 60°C under oxidizing and reducing conditions. The two desorption curves were obtained using different tracer loadings (initial S values). Both plots show hysteresis since the slopes of the desorption curves are less than the sorption curves. Hysteresis is a much greater effect for neptunium sorption under reducing conditions, however. The ratio of Freundlich exponents N_s/N_d, where N_s and N_d are the measured exponents for sorption and desorption, respectively, is a measure of the magnitude of hysteresis. Larger values for N_s/N_d indicate greater hysteresis effects. For neptunium sorption under oxidizing conditions $N_s/N_d = 2$ and for reducing conditions $N_s/N_d = 435$.

The sorption and desorption behavior of uranium is similar to neptunium. Figure 3 shows that hysteresis is more important for uranium sorption under reducing conditions than under oxidizing conditions. Values of N_s/N_d are 10 and approximately 200 for oxidizing and reducing conditions, respectively.

Table IV. Freundlich Constants (Molar Basis) for Radionuclide Sorption on Mabton Interbed Solids

Radionuclide	Redox	Temperature (°C)	K	N	r2*
Technetium	Oxidizing	23	2.96×10^{-4}	0.91	0.978
		60	9.40×10^{-4}	0.99	0.979
	Reducing	23	4.36×10^{-1}	1.09	0.978
		60	1.14×10^{-4}	0.58	0.974
Radium	Oxidizing	23	1.83×10^{2}	1.18	0.971
		60	9.64	0.99	0.919
	Reducing	23	5.56×10^{-2}	0.91	0.994
		60	8.40×10^{-2}	0.98	0.990
Uranium	Oxidizing	23	3.97×10^{-4}	0.74	0.993
		60	2.52×10^{-1}	0.96	0.992
	Reducing	23	1.30×10^{-2}	0.88	0.997
		60	1.51×10^{-2}	0.79	0.990
Neptunium	Oxidizing	23	2.82×10^{-4}	0.69	0.994
		60	5.50×10^{-4}	0.68	0.997
	Reducing	23	3.35×10^{-2}	0.80	0.996
		60	1.64×10^{-2}	0.87	0.995
Plutonium	Oxidizing	23	5.06×10^{-1}	0.95	0.967
		60	1.44×10^{-2}	0.78	0.874

*Coefficient of determination.

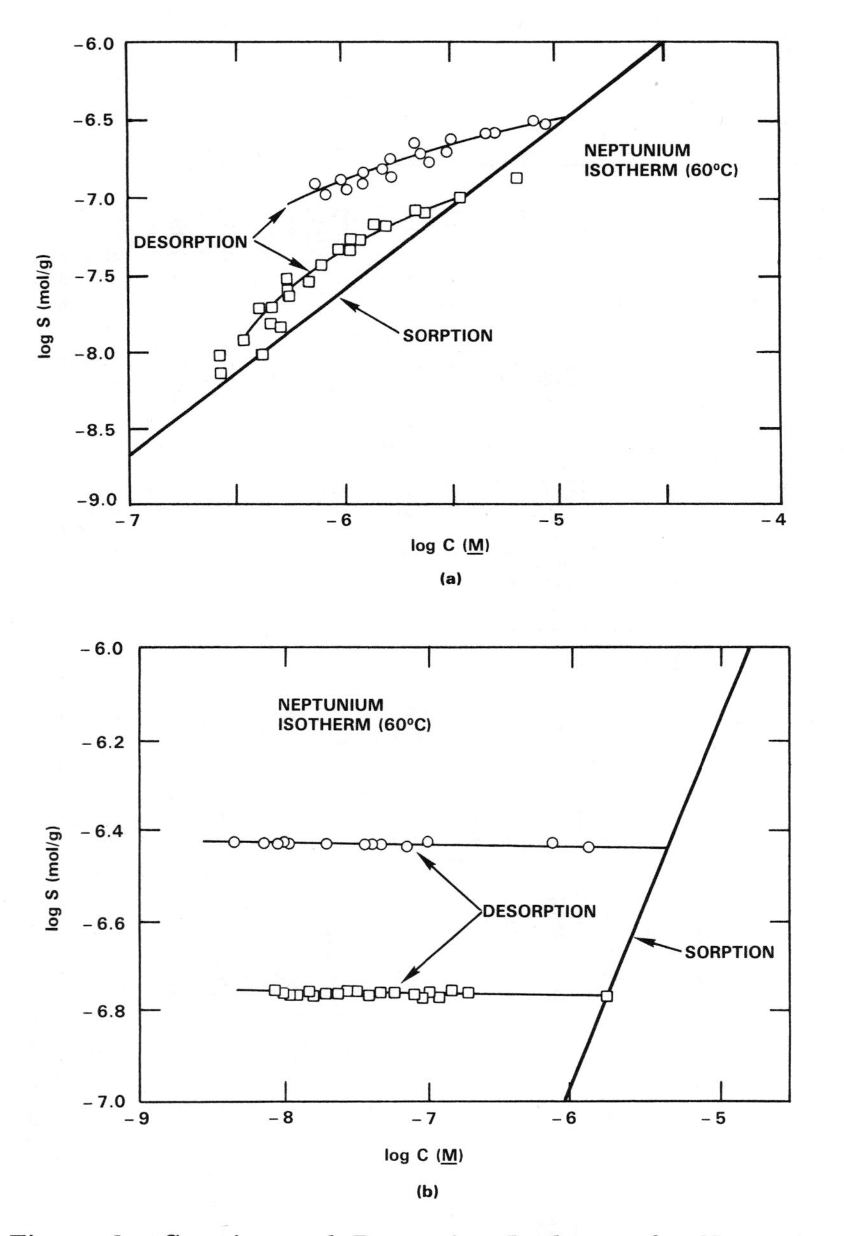

Figure 2. Sorption and Desorption Isotherms for Neptunium Sorption on Mabton Interbed Solids. (a) Oxidizing Conditions. (b) Reducing Conditions.

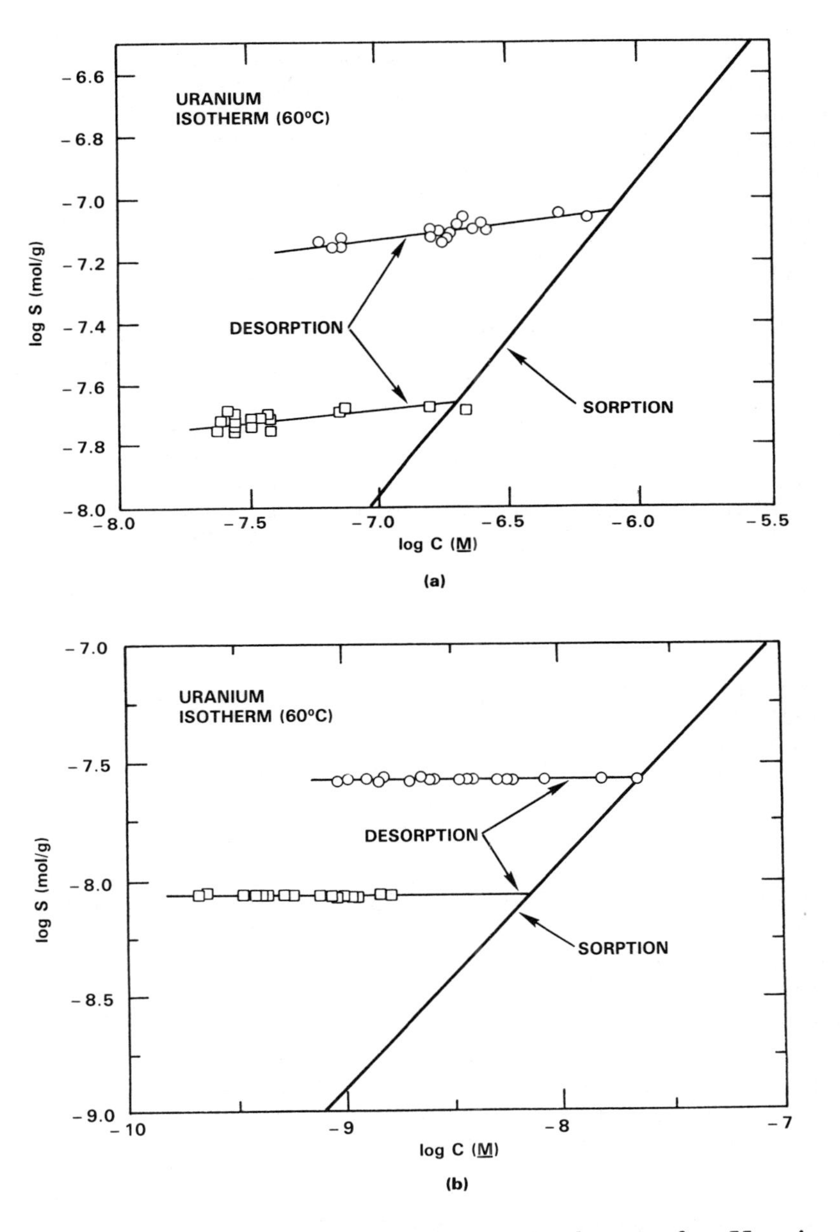

Figure 3. Sorption and Desorption Isotherms for Uranium Sorption on Mabton Interbed Solids. (a) Oxidizing Conditions. (b) Reducing Conditions.

Results for selenium and technetium under oxidizing conditions show that these reactions are fully reversible. However, under reducing conditions, sorption reactions of these radionuclides show significant hysteresis (Figure 4). Sorption of radium also shows a large hysteresis effect. Table V gives a summary of N_s/N_d values for each radionuclide sorption-desorption reaction with Mabton Interbed material at 60°C. These data show that the reduced species of selenium, technetium, neptunium, and uranium are more irreversibly sorbed on Mabton Interbed material than the oxidized species. The reducing environment expected in the interbed layers found in basalt formations at depth should significantly decrease the mobility of these radionuclides and thus provides an additional safety factor for waste storage.

Effects of Groundwater Composition and Eh. Radionuclide sorption on geologic solids is dependent on the chemical composition of the groundwater solution and the redox potential (Eh) of the solid-groundwater system. Aquifers at various depths in the Columbia Plateau formation have been observed to have significant differences in composition. To accurately model radionuclide migration, it is necessary to understand the effects of chemical components and Eh on sorption and solubility of key radionuclides. An additional benefit of this work is to better understand the mechanisms of sorption and desorption of the radionuclides.

The objectives of this work were to determine effects of the major groundwater components found in the Grande Ronde formation (Na^+, Ca^{2+}, K^+, Mg^{2+}, Cl^-, F^-, CO_3^{2-}, HCO_3^-, and SO_4^{2-}) on radionuclide sorption on geologic solids expected in the flow path. Interbed materials lying between basalt layers are of particular importance because of their porous nature. Interbed sandstone and tuff were studied because of their abundance in Columbia River basalt interbeds. The effect of Eh was examined by adding a chemical redox buffer, hydrazine, as a variable. Hydrazine lowered the Eh from about +0.6 V (air saturated solution) to about -0.8 V at pH 9. Because of the large number of variables to be studied, it was necessary to use statistical methods to design the experiments. An efficient design for screening the nine solution variables is the 20-run Plackett-Burman design (8). This is fractional, 2-level, factorial design that is used to identify significant variables and determine their effects. The application of this design is discussed in detail by Barney (4). The design requires a high and low value for each parameter (groundwater component concentration or Eh). Values were chosen to cover the range of concentrations found in the Grande Ronde formation.

Table VI compares the values used in these experiments to those actually found in the Grande Ronde aquifers. The 20-run experimental design requires preparation of 20 solutions, each with a different combination of groundwater parameters and having the high and low values given in Table VI. Silicate

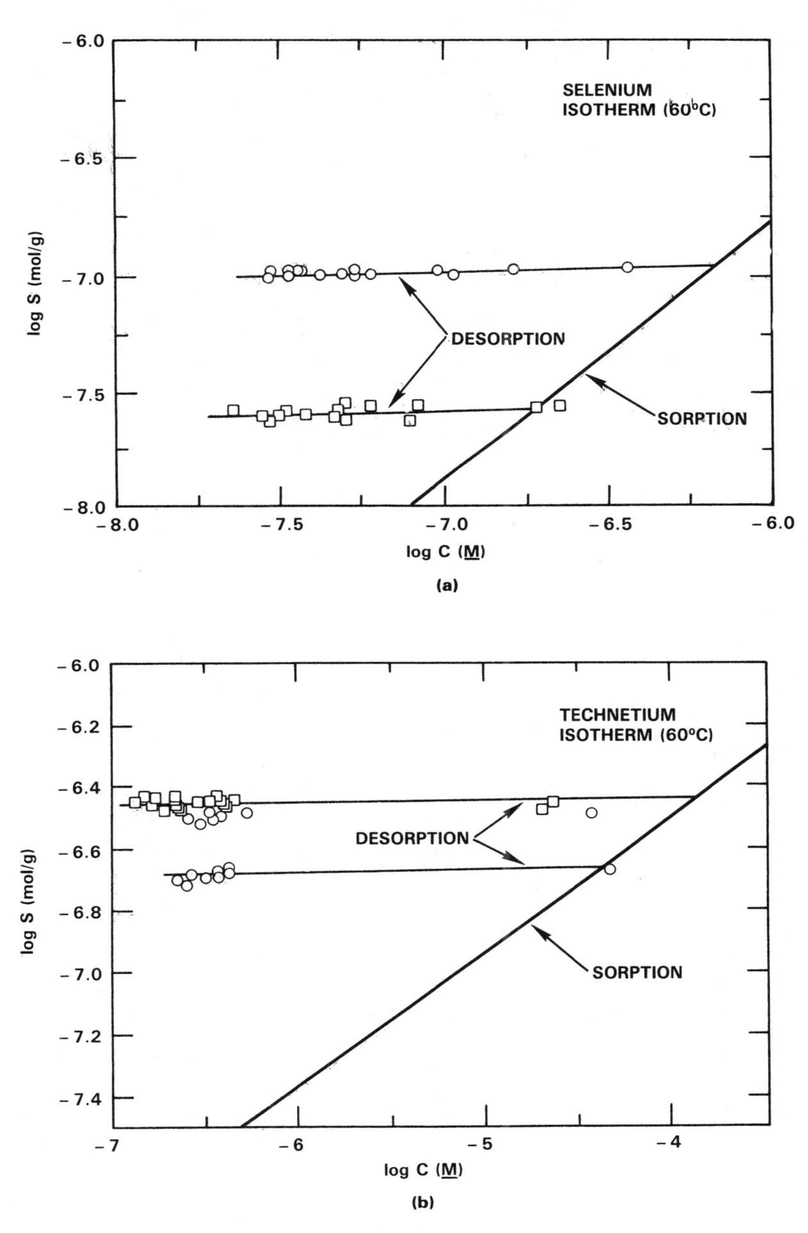

Figure 4. Sorption and Desorption Isotherms for (a) Selenium and (b) Technetium Sorption on Mabton Interbed Solids under Reducing Conditions.

Table V. Freundlich Constants for Radionuclide Sorption-Desorption on Mabon Interbed at 60°C

Isotope	Redox	S_o (mol/g)	N_s	N_d	N_s/N_d
Selenium	Reducing	1.11×10^{-7}	1.0	0.018	56
		2.76×10^{-8}		0.027	37
Technetium	Reducing	1.88×10^{-7}	0.58	0.018	32
		3.50×10^{-7}		0.003	193
Neptunium	Oxidizing	3.13×10^{-7}	1.0	0.45	2.2
		1.38×10^{-7}		0.91	1.1
Neptunium	Reducing	3.84×10^{-7}	0.87	0.002	435
		1.81×10^{-7}		0.002	435
Uranium	Oxidizing	8.88×10^{-8}	0.96	0.093	10
		2.11×10^{-8}		0.074	13
Uranium	Reducing	2.70×10^{-8}	0.79	0.008	100
		8.58×10^{-7}		0.003	263
Radium	Reducing	7.65×10^{-10}	0.98	0.016	61
		1.81×10^{-10}		0.087	11

Table VI. Comparison of Experimental and Measured Parameters

Concentration parameter	Experimental range	Measured range*
Na^+	0.002 to 0.022M	0.009 to 0.016M
K^+	0.0 to 0.0005M	0.00001 to 0.0004M
Ca^{2+}	0.0 to 0.0001M	0.00002 to 0.0001M
Mg^{2+}	0.0 to 0.00002M	$<$0.0000002 to 0.00001M
Cl^-	0.001 to 0.011M	0.001 to 0.008M
F^-	0.0 to 0.002M	0.001 to 0.002M
SO_4^{2-}	0.0 to 0.003M	0.0008 to 0.002M
CO_3^{2-}	0.0 to 0.001M	0.0 to 0.001M
HCO_3^-	0.0 to 0.001M	0.0 to 0.001M
SiO_3^{2-}	0.001M	0.001 to 0.002M

*Based on unpublished analyses of Grande Ronde groundwater by T. E. Jones.

(0.001M) was added to each solution so that steady-state silicate concentrations would be approached more rapidly during equilibration. Since the interbed solids will control silicate concentration, it was not possible to study it as a variable. The final pH of interbed-equilibrated solutions ranged from 8.6 to 9.8 for sandstone and 8.5 to 9.2 for tuff.

Variables found to be significant for sorption of cesium, strontium, technetium, selenium, neptunium, plutonium, americium, and radium on sandstone and tuff at 23°C are given in Table VII. They are ranked in order of significance where more than one variable was found to be significant. The (+) and (–) signs indicate whether sorption is increased or decreased.

Table VII shows that for cesium sorption, both KCl and N_2H_4 are significant for the two geologic solids studied. The negative values indicate that the presence of either KCl or N_2H_4 lowers sorption. Both appear to be competing with Cs^+ ion for sorption sites. Competition between K^+ and Cs^+ ions for sorption sites on mica-like minerals is well known. However, displacement of Cs^+ by hydrazine was surprising since N_2H_4 should exist mainly as a neutral species at pH 9-10. A small amount (0.0005M to 0.005M) will be protonated and apparently competes with Cs^+. Ammonium ion is known to effectively compete with Cs^+ for mineral sorption sites. Hydrazinium ion with a similar molecular structure should also displace Cs^+. Since hydrazine will not reduce or complex Cs^+, the only possible effects on cesium sorption is to compete for sorption sites or to alter the surface of the solid minerals. No evidence of surface alteration (change in color or texture) was observed. Therefore, it appears that an Eh buffer is not required for Cs^+ sorption studies and hydrazine only interferes with the sorption reaction.

Table VII shows that for strontium sorption, sodium salts lower K_d values significantly. Those sodium salts that contribute the largest concentrations of Na^+ (NaCl and Na_2SO_4) are the most significant. There is a linear correlation between the ln of total Na^+ concentration and ln strontium K_d values in these experiments.

Calcium concentration can also affect strontium sorption on sandstone and tuff even through concentrations of Ca^{2+} are 30 to 220 times lower than Na^+. This is due to similar chemical behavior, ion charge, and size of Sr^{2+} and Ca^{2+}. Both ions are sorbed onto similar sites and effectively compete for these. Hydrazinium ions also exchange with Sr^{2+}. Ions of similar size effectively compete with both Cs^+ and Sr^{2+} for sorption sites.

Table VII shows that hydrazine is the only important variable for technetium sorption on each of the geologic solids. Hydrazine causes technetium to be removed from solution either by sorption or by precipitation of the reduced technetium species. Hydrazine is a powerful reducing agent and should reduce TcO_4^- to technetium(IV) according to standard half-cell potentials. No TcO_2 was observed; however, since the technetium passed through

Table VII. Significant Variables for Radionuclide Sorption at 23°C*

Radio-nuclide	Solid	Variables								
		NaCl	NaF	Na_2SO_4	Na_2CO_3	$NaHCO_3$	$CaCl_2$	MgCl	KCl	N_2H_4
^{137}Cs	sandstone	–	–	–	–	–	–	–	-1	-2
	tuff	-3	–	–	–	–	–	–	-2	-1
^{85}Sr	sandstone	-1	-3	-2	–	–	-4	–	–	-5
	tuff	-1	-5	-2	–	–	-4	–	–	-3
^{95m}Tc	sandstone	–	–	–	–	–	–	–	–	+1
	tuff	–	–	–	–	–	–	–	–	+1
^{75}Se	sandstone	–	–	–	-1	–	+2	–	+3	–
	tuff	-2	–	–	-3	–	+4	–	+1	–
^{237}Np	sandstone	–	–	–	-2	-3	–	–	–	+1
	tuff	–	–	–	–	-2	–	–	–	+1
^{238}Pu	sandstone	–	–	–	–	–	+2	-3	–	+1
	tuff	–	–	-1	–	–	+3	–	–	+2
^{241}Am	sandstone	+3	–	–	+1	-2	–	–	–	–
	tuff	+1	–	–	–	–	–	–	–	–
^{226}Ra	sandstone	-4	–	-3	–	–	–	–	-2	-1
	tuff	–	–	-1	–	–	–	–	-2	-3

*The number indicates ranking of importance (1 is the most important). A plus sign indicates that the presence of the variable increases sorption, a minus sign that its presence decreases sorption.

0.45-m filters with the spiked solutions containing N_2H_4, it appears that technetium(IV) exists in solution as a positively charged species, possibly TcO^{2+} as proposed by Gorski and Koch (9).

Selenium is weakly sorbed on sandstone and tuff. Although selenium exists as an anion in these solutions, it is not sorbed by anion exchange on the solids. Their anion exchange capacity is very low at the pH of these experiments. Selenium must, therefore, be sorbed by chemisorption or precipitation. Calcium increases sorption while CO_3^{2-} decreases sorption. A slightly soluble calcium-selenium compound is apparently forming on the surface of solids. The effect of CO_3^{2-} is to reduce the concentration of Ca^{2+} in solution by formation of $CaHCO_3^-$ and $CaCO_3^\circ$ making the calcium unavailable for formation of the compound. Eh-pH diagrams for selenium predict reduction of SeO_3^{2-} to HSe^- by hydrazine under the conditions of these experiments. These anions apparently behave similarly since hydrazine is not a significant variable for selenium sorption.

Hydrazine is by far the most important variable controlling sorption/ precipitation of neptunium. Hydrazine greatly increases removal of neptunium from solution. This is due to reduction of neptunium(V), which exists in oxygenated solutions, to neptunium(IV). Standard half-cell potentials for hydrazine and neptunium(V) indicate that neptunium(V) should be easily reduced. As with other actinides, the (IV) oxidation state should be sorbed more strongly than the (V) because of the greater complex-forming ability of the (IV) state. The effect of CO_3^{2-} and HCO_3^- is to decrease sorption due to formation of carbonate complexes by both neptunium(V) and neptunium(IV). Possible carbonated complexes are $NpO_2HCO_3^\circ$ for neptunium(V) and $Np(CO_3)_4^{4-}$ for neptunium(IV)(10). These neutral anionic species would not be expected to sorb as strongly as positively charged, noncomplexed neptunium species.

Sorption/precipitation of plutonium is greatly affected by the presence of hydrazine. Since hydrazine increases sorption, it appears that at least some of the plutonium is present initially as plutonium(V) or plutonium(VI) and is reduced to plutonium(IV) by hydrazine. According to standard half-cell potentials, both plutonium(VI) and plutonium(V) should be reduced to plutonium(IV) or plutonium(III) under the conditions of the experiments, assuming that the hydroxyl complexes are important at the pHs of the experiments.

Parameters affecting sorption of americium on sandstone were Na_2CO_3, $NaHCO_3$, and NaCl. It seems unlikely that a carbonate complex or solid is formed since Na_2CO_3 increases sorption and $NaHCO_3$ decreases sorption. If carbonate or bicarbonate take part in the reaction, both salts should affect sorption in the same way. However, they do affect the pH differently--Na_2CO_3 slightly increasing pH and $NaHCO_3$ slightly lowering the pH. It appears that a slightly soluble americium compound is formed that is more soluble at low pH.

Significant parameters for radium sorption are Na_2SO_4, KCl, and N_2H_4. Each of these variables decrease sorption of radium by ion exchange competition. The effective ion diameter of Ra^{2+} is near those of Na^+ and $N_2H_5^+$ so that they compete for similar sorption sites.

Conclusions

Most of the sorption reactions of radionuclides with interbed materials were found to be irreversible (i.e., hysteresis was significant). Only selenium and technetium sorption under oxidizing conditions was found to be fully reversible. Sorption hysteresis was much greater for reduced species of selenium, technetium, neptunium, and uranium than for oxidized species. The Freundlich equation fit both sorption and desorption data quite well for most of the radionuclides studied. The ratio of Freundlich exponents for sorption and desorption, N_s/N_d, is a measure of the magnitude of the hysteresis effect. Values of this ratio have been calculated for selenium, technetium, neptunium, uranium, and radium. The large values obtained for each radionuclide under reducing conditions suggest that radionuclide transport will be significantly delayed due to sorption hysteresis in the reducing environment of the basalt interbeds. This effect is an important safety factor that should be considered in radionuclide transport modeling to accurately assess the safety of a nuclear waste repository in basalt.

Statistically designed experiments have identified groundwater components that affect sorption of key radionuclides on basalt interbed materials. Sodium, potassium, and calcium in the groundwater decrease sorption of cesium, strontium, and radium by competing with these radionuclides for sorption sites on the solids. These radiouclides are at least partially sorbed by ion exchange reactions. Groundwater Eh greatly affects sorption of technetium, neptunium, plutonium, and uranium. The reducing Eh values produced by hydrazine (-0.8 V) increased sorption of these elements by reducing them to lower oxidation states--technetium(IV), neptunium(IV), plutonium(III), and uranium(IV). The sorption mechanism for these elements (and for americium) is chemisorption. Over the range of groundwater component concentrations studied, metal complex formation with groundwater anions does not greatly affect sorption. The only evidence of this effect is for formation of neptunium carbonate (or bicarbonate) complexes and plutonium sulfate complexes. Hydrazine reduces SeO_3^{2-} to HSe^- in sorption experiments and these anions have similar sorption behavior. Since selenium should exist as elemental Se in basalt aquifers, a weaker reducing agent (e.g., thiosulfate) must be used to control the Eh in selenium sorption experiments.

Literature Cited

1. DiToro, D. M.; Horzempa, L. M. "Reversible and Resistant Components of PCB Adsorption-Desorption: Isotherms," Environ. Sci. Technol. 1982, 16, 594.
2. Van Genuchten, M. Th.; Davidson, J. M.; Wierenga, P. J. "An Evaluation of Kinetic and Equilibrium Equations for the Prediction of Pesticide Movement Through Porous Media," Soil Sci. Soc. Amer. Proc. 1974, 38, 29.
3. Gephart, R. E.; Arnett, R. C.; Baca, R. G.; Leonhart, L. S.; Spane, F. A. "Hydrolic Studies within the Columbia Plateau, Washington: An Integration of Current Knowledge," RHO-BWI-LD-1, Rockwell Hanford Operations, Richland, Washington, 1979.
4. Barney, G. S. "FY 1981 Annual Report: Radionuclide Sorption on Basalt Interbed Materials," RHO-BW-ST-35 P, Rockwell Hanford Operations, Richland, Washington, 1981.
5. Rai, D.; Franklin, W. T. "Effect of Moisture Content on Ethylene Glycol Retention by Clay Minerals," Geoderma. 1978, 21, 75.
6. Routson, R. C.; Wildung, R. E.; Serne, R. J. "A Column Cation-Exchange-Capacity Procedure for Low-Exchange-Capacity Sands," Soil Sci. 1973, 115, 107.
7. Van Genuchten, M. Th.; Cleary, R. W. "Movement of Solutes in Soil: Computer-Simulated and Laboratory Results," in "Soil Chemistry, B. Physico-Chemical Models," Bolt, G. H., Ed., Elsevier Scientific Publishing Company, New York, 1979.
8. Plackett, R. L.; Burman, J. P. "The Design of Optimum Multifactorial Experiments," Biometrika 1946, 33, 305.
9. Gorski, B.; Koch, H. "The Chemistry of Technetium in Aqueous Solution," J. Am. Chem. Soc. 1969, 31, 3565.
10. Rai, D.; Serne, R. J. "Solid Phases and Solution Species of Different Elements in Geologic Environments," PNL-2651, Pacific Northwest Laboratory, Richland, Washington, 1978.

RECEIVED October 28, 1983

Reactions Between Technetium in Solution and Iron-Containing Minerals Under Oxic and Anoxic Conditions

T. T. VANDERGRAAF, K. V. TICKNOR, and I. M. GEORGE

Atomic Energy of Canada Limited, Whiteshell Nuclear Research Establishment, Pinawa, Manitoba R0E 1L0 Canada

The behaviour of technetium in the geosphere is of particular importance in nuclear fuel waste management studies because this man-made element has a long half-life and, under ambient conditions in the laboratory, is not readily sorbed on geologic materials.

Autoradiographic analyses of rock and mineral thin sections contacted with $^{95m}TcO_4^-$-containing solutions, under oxic and anoxic conditions, have confirmed that virtually no sorption takes place in the presence of oxygen. However, under anoxic conditions (< 0.2 μg/g oxygen in the atmosphere), sorption of technetium was observed on iron-oxide inclusions in ferrous-iron-containing minerals (biotite, olivine, pyroxene, hornblende) and on iron-oxide coatings on microfractures in granite, but not on the ferrous-iron minerals within the granite themselves. Subsequent static sorption tests with crushed magnetite showed that sorption is a function of the composition of the solution and of the radionuclide concentration, and again occurred only in the absence of oxygen. This behaviour is in contrast with that observed with metallic iron, which sorbs technetium strongly, even in the presence of air.

These results show that technetium can be contained by magnetite in the geosphere, provided reducing conditions can be maintained. This can be aided, for example, by the incorporation of iron or iron oxides in the buffer and backfill materials in the waste disposal vault.

0097-6156/84/0246-0025$06.00/0

The fissioning of ^{235}U and ^{239}Pu in a nuclear reactor produces a large number of radioactive fission products. Most of these decay to stable isotopes within a few minutes to a few years after the fuel has been discharged from the reactor and therefore pose no problem in the management of nuclear fuel wastes. There are, however, a number of longer lived radionuclides that must be considered in assessing the environmental impact of any nuclear fuel waste disposal vault in the geosphere.

For example, the fission products technetium and promethium are unique, in that they do not have any stable isotopes and do not occur in nature in measureable amounts. While promethium has a number of chemical analogues in the other rare-earth elements, this is not the case for technetium, and it is thus difficult to predict its behaviour in the geosphere.

The technetium isotope of interest for nuclear fuel waste disposal is ^{99}Tc. It is a pure β-emitter (E = 0.293 MeV) with a half-life of 2.13×10^5 years. Its high fission yield of 6% accounts for the relatively high concentration (∿ 0.02% by weight) (1) in fuel discharged from a CANDU (CANada Deuterium Uranium) reactor (burnup ∿ 650 GJ/kg U).

Technetium is a Group VII B element. Its chemical behaviour is not well-known, but is expected to fall between that of manganese and rhenium, and is summarized for an aqueous medium in Figure 1. Under oxidizing conditions, technetium exists in solution as the anionic species TcO_4^-, in the 7+ valence state, and shows little sorption by geologic materials (2-5). For this reason, in previous safety and environmental assessments of geologic disposal of nuclear fuel wastes, technetium has been assumed to travel at the same rate as moving groundwater.

Under reducing conditions, or in the absence of oxygen, sorption of technetium has been noted in some cases. Bondietti and Francis (6) reported the removal of technetium from a nitrogen-sparged solution by basalt and granite, and Allard et al. (3) reported k_d values (defined as the ratio of sorbed to nonsorbed concentrations) well above zero, indicating sorption on ferrous-iron-containing minerals in contact with deaerated solutions, and on granite in a deaerated solution containing Fe^{++} ions.

Figure 2 is a combined E_h-pH diagram for technetium and iron and shows that, under certain conditions, technetium can be reduced by ferrous iron. To study the role of ferrous iron in the removal of technetium from solution, experiments were carried out with crystalline rock and ferrous minerals and TcO_4^--containing solutions, under both anoxic and oxic conditions.

Experimental

Geological Materials. Granite was obtained from a quarry located on the Lac du Bonnet batholith near the Whiteshell Nuclear Research Establishment. An olivine gabbro sample was obtained

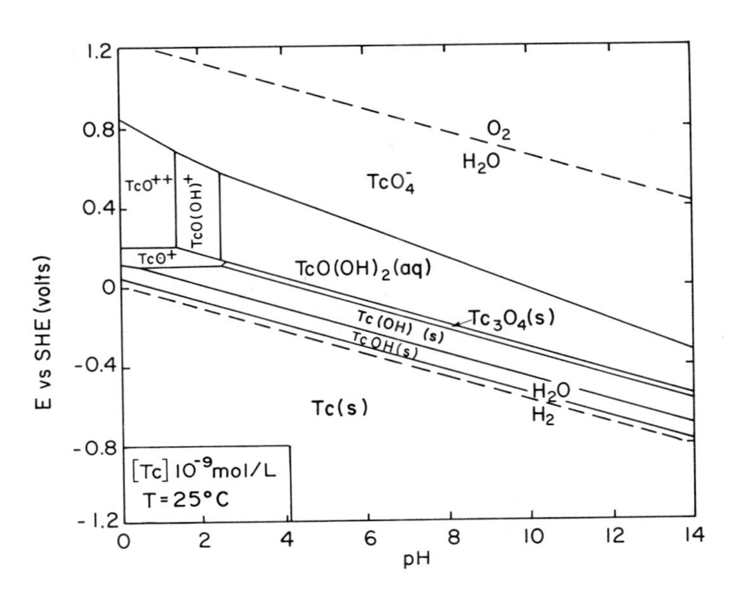

FIGURE 1. E_h-pH diagram for dissolved technetium species.

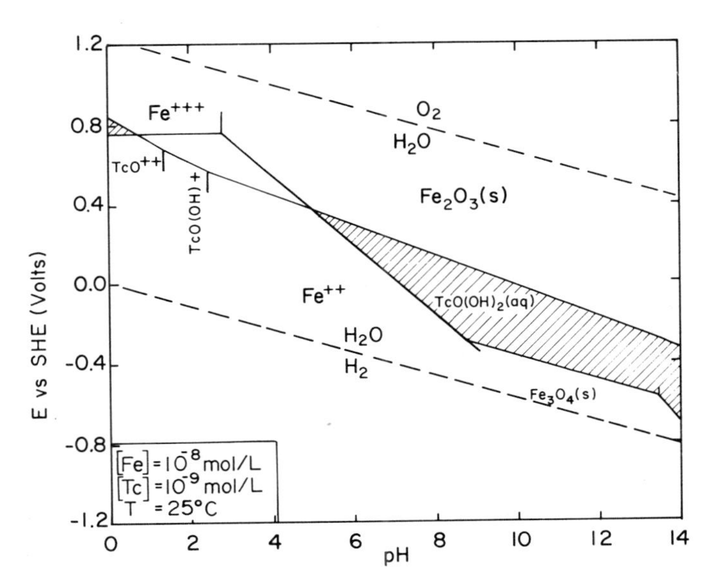

FIGURE 2. Combined E_h-pH diagram for technetium and iron. The shaded area indicates region where TcO_4^- can be reduced by Fe^{++}

through the Geological Survey of Canada (GSC) from the Rouyn-Noranda area of western Quebec. The GSC also supplied the biotite mica, hornblende, and pyroxene samples. The olivine, epidote, iron oxides and iron oxyhydroxides were obtained from Ward's Natural Science Est. Inc. In addition, some drill-core material from the Eye-Dashwa Lakes pluton near Atikokan in northwestern Ontario was used. The chemical and mineral compositions of the granite and gabbro specimens are given in Tables I(a) and I(b). The unconsolidated material was prepared by crushing the geologic material with a jawcrusher fitted with tungsten carbide teeth to avoid contamination with metallic iron, and was wet-sieved before using. Thin sections of some of the rocks and minerals were made for sorption/autoradiographic studies.

Table I(a). Chemical Composition of Lac du Bonnet Granite and of Olivine Gabbro used in Sorption Studies Concentration in wt%

	Granite		Gabbro	
Oxide	Bulk	80-150 mesh	Bulk	80-150 mesh
SiO_2	73.1	76.3	49.7	50.0
Al_2O_3	14.2	12.1	17.8	19.2
FeO	0.81	0.88	5.74	5.71
Fe_2O_3	0.76	1.06	2.34	1.90
CaO	1.43	1.05	11.7	11.7
MgO	0.46	0.49	8.80	8.46
Na_2O	4.23	3.66	2.29	2.62
K_2O	4.88	5.03	0.27	0.23
TiO_2	0.22	0.23	0.55	0.54
MnO	0.03	0.03	0.13	0.13
H_2O	n.d.*	n.d.	0.38	n.d.
CO_2	n.d.	n.d.	0.14	n.d.
Total	100.09	100.83	99.78	100.49

* n.d. = not determined.

Solutions. The following solutions were used in one or more of the experiments. With the exception of the distilled water, they reflect the composition of the solutions that may be expected to be present in and around a nuclear fuel waste vault. Their chemical compositions are given in Table II.

(1) Distilled deionized water (DDW)
(2) Standard granite groundwater (GGW), with a composition based on reported analyses of naturally occurring groundwaters associated with granitic formations (7).

Table I(b). Mineral Composition of Lac du Bonnet Granite and of Olivine Gabbro used in Sorption Studies (Modal wt%)

Mineral	Lac du Bonnet granite	Olivine gabbro
Quartz	25.8	-
K-feldspar	28.1	-
Plagioclase	33.9	46.2
Clinopyroxene	-	41.7
Olivine	-	6.5
Biotite	5.2	1.2
Muscovite	2.5	-
Opaques	0.7	3.6
Epidote	1.5	-
Chlorite	2.3	0.3
Scapolite (?)	-	0.5
Total	100.0	100.0

Table II. Chemical Composition of Solutions used in the Technetium Sorption Studies Concentration in mg/L

Ion	GGW	WN-1	SCSSS	HA	NAP
Na	8.3	1910	5 050	15	46
K	3.5	14	50	-	-
Mg	3.9	61	200	-	-
Ca	13.	2130	15 000	-	-
Sr	-	24	20	-	-
Fe	-	0.56	-	-	-
Si	-	-	15	-	-
HCO_3	58.	68	10	-	-
Cl	5.0	6460	34 260	-	-
SO_4	8.6	1040	790	-	-
NO_3	0.62	33	50	-	-
F	0.19	-	-	-	-
$HPO_4^{=}$*	-	-	-	-	95
Humic acid	-	-	-	85	-
pH	6.5±0.5	7.0±0.5	7.0±0.5	8.2	8.0

GGW : granite groundwater
WN-1 : saline groundwater based on WN-1 analyses
SCSSS : standard Canadian Shield saline solution
HA : humic acid (100 mg/L)
NAP : Na_2HPO_4 solution (140 mg/L)
* : in equilibrium with $PO_4^{=}$ and $H_2PO_4^{-}$

(3) Saline groundwater (WN1), based on groundwater obtained from the 455-m level of the WN-1 borehole in the Lac du Bonnet batholith (8).
(4) Standard Canadian Shield saline solution (SCSSS), with a composition approximately that of saline solutions obtained from various sources in the Canadian Shield.
(5) Humic acid (HA). A 100 mg/L solution of the sodium salt of humic acid, to simulate groundwater containing an organic complexing agent.
(6) Na_2HPO_4 solution (NAP) (140 mg/L), to provide a solution containing an inorganic complexing agent to form anionic species with reduced technetium (9). This anion was used instead of carbonate, as the purification system of the anaerobic chamber removes CO_2 from the atmosphere, and a bicarbonate-carbonate solution would not be stable.

Technetium Isotopes. $NH_4{}^{95m}TcO_4$ (half-life 61 days), obtained from New England Nuclear, and $NH_4{}^{99}TcO_4$, from Amersham-Searle, were used in this study. The presence of reduced technetium was determined by extracting the Tc(7+) into chloroform/tetraphenylarsonium chloride, and assaying the aqueous phase for technetium activity.

Experimental Details and Results. A series of experiments was carried out to study the behaviour of TcO_4^- in various solutions in contact with a number of rocks and minerals, under both oxic and anoxic conditions, to determine the conditions that lead to removal of technetium from solution and the role played by the various minerals in this process.

Experiments under anoxic conditions were carried out in a Vacuum Atmospheres Inc. anaerobic chamber containing a nitrogen atmosphere, with an oxygen concentration of $\sim$ 0.2 μL/L (as determined by a Teledyne Model 317-X trace oxygen analyzer).

Sorption on Crushed Whole Rock under Oxic Conditions

To determine the behaviour of TcO_4^- under oxic conditions (i.e. conditions similar to those expected in a nuclear waste disposal vault prior to removal of atmospheric oxygen by geologic and bacteriological processes), crushed and sintered granite and gabbro were contacted with granite groundwater containing $NH_4{}^{95m+99}TcO_4$ for 150 days. The particle size of the rock was in the range of 100-180 μm (80-150 mesh), and the solid-to-liquid ratio was 1 g of rock to 10 mL of solution. The initial technetium concentrations ranged from $3x10^{-12}$ mol/L to 10^{-4} mol/L. As a standard, crushed and sieved quartz was used. This was obtained from a single crystal, and washed with 6 mol/L HCl to remove any iron. Small amounts (< 20 mg) of iron-metal filings were added to one half of all samples to check the effect of inadvertently introducing this impurity during the crushing process. The solutions were sampled after 60 and 150 days. The sorption coefficients, expressed in mL/g and defined as the ratio between sorbed and solution concentrations are given in Table III. In the samples containing metallic iron, sorption was too great to give meaningful k_d values. Hence, the percent sorption is tabulated for those samples.

Table III. Sorption Coefficients for Technetium on Quartz, Granite, and Gabbro under Oxic Conditions

Material*	k_d** No iron metal present 35d (mL/g)	150d (mL/g)	Sorption† Iron metal present 35d (%)
Acid-washed quartz	0.2±0.2††	0.3±0.8††	99.6±0.3††
Granite	0.5±0.6	0.8±1.4	99.8±0.2
Gabbro	4.3±1.4	4.6±1.6	99.7±0.5

* particle size 100-180 μm (80-150 mesh).

** $k_d = \frac{[Tc]\ \text{sorbed (mol/g)}}{[Tc]\ \text{solution (mol/mL)}}$

† 99+% removal of the technetium from solution corresponds to a "calculated k_d" of > 2000 mL/g.

†† error at 2σ.

Sorption on Granite and Gabbro Coupons under Anoxic Conditions. Machined granite and gabbro coupons, 19x19x4 mm, were contacted with 10 mL of granite groundwater containing 3×10^{-12} mol/L ^{95m}Tc as TcO_4^- for 35 days in the anaerobic chamber. The sorption coefficients, this time expressed as k_a, where

$$k_a(\text{cm}) = \frac{\text{moles of technetium sorbed/cm}^2}{\text{moles of technetium remaining in solution/mL}},$$

are shown in Table IV. To determine the possible cause of the anomalous behaviour of one of the gabbro samples, all six coupons were autoradiographed using Kodak spectrum analysis glass plates #1 (Kodak catalogue number 156 7387). Representative autoradiographs are shown in Figure 3, and these will be discussed in Section 3.

Sorption on Granite and Ferrous-Iron-Containing Minerals under Anoxic Conditions

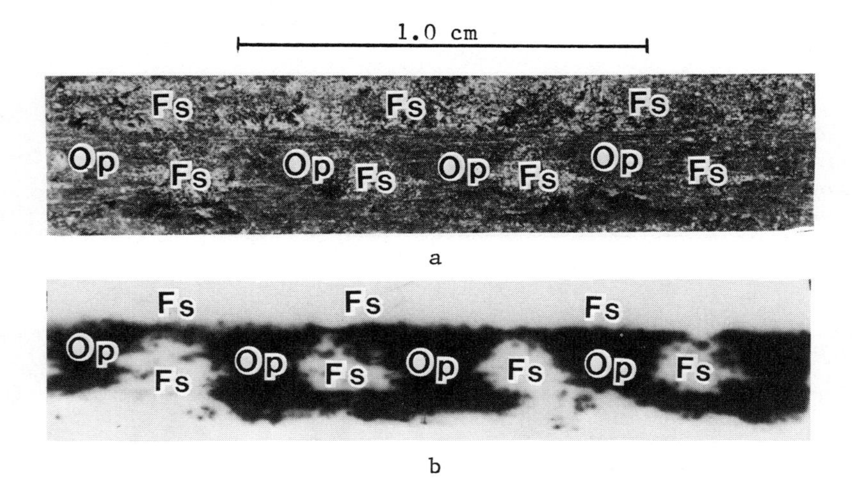

FIGURE 3. Photograph (a) and autoradiograph (b) of technetium sorbed on a gabbro coupon.

Table IV. Sorption Coefficients for Technetium on Granite and Gabbro Coupons under Anoxic Conditions

Material		k_a* (cm)	k_d** (mL/g)
Granite	1	0.0031	0.27
	2	0.0008	0.07
	3	0.022	1.9
Gabbro	1	0.018	0.77
	2	0.10	8.6
	2†	1.4	120
	3	- 0.022	- 1.9

* $k_a = \frac{[Tc]\ \text{sorbed}\ (mol/cm^2)}{[Tc]\ \text{solution}\ (mol/mL)}$

** k_d calculated assuming specific surface area of 43 cm^2/g.

† Considering only the edge of the coupon with sorbed technetium.

To study the effect of ferrous-iron-containing minerals on TcO_4^- in solution, granite, biotite, hornblende, epidote, olivine and pyroxene thin sections were contacted with granite groundwater containing ∼ 3×10^{-12} mol/L ^{95m}Tc, again as TcO_4^-, for one week in the anaerobic chamber, and subsequently autoradiographed using the procedure outlined elsewhere (10). These particular granite samples were obtained at depths of 72 and 1074 metres from the ATK-1 borehole in the Eye-Dashwa Lakes pluton near Atikokan, northwestern Ontario. This pluton has been used in geochemical and hydrogeological studies of large intrusive rock formations in the Canadian Shield. The samples were chosen because they contain iron-oxide infillings in minute fractures, due to hydrothermal alteration of the primary minerals of the rock matrix. The minerals were selected because they contain Fe(II) (see Table V). Some representative autoradiographs of thin sections contacted in the anaerobic chamber are shown in Figures 4 to 6.

Table V. Ferrous-Iron-Containing Minerals used in Technetium Sorption Studies

Mineral	General Chemical Formula
Biotite	$K(Mg,Fe^{++})_3Si_3AlO_{10}(OH)_2$
Olivine	$(Mg,Fe^{++})_2SiO_4$
Pyroxene	$(Ca,Fe^{++})SiO_3$
Hornblende	$(Na\ Ca_2)(Mg,Fe^{++})(Al,Fe^{+++})(Si_3AlO_{11})_2(OH)_2$
Ilmenite	$Fe\ TiO_3$

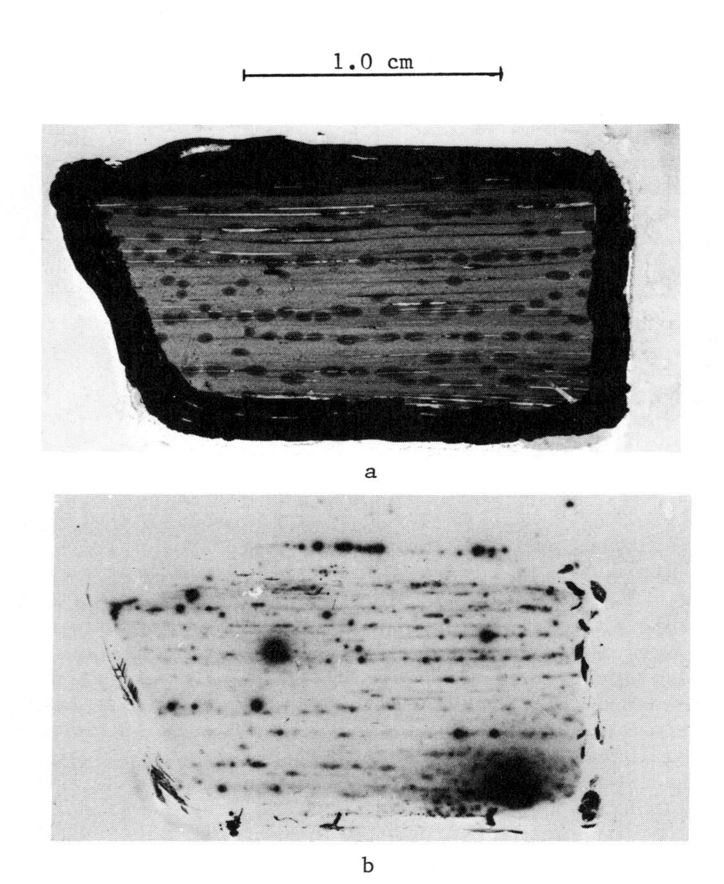

FIGURE 4. Photograph (a) and autoradiograph (b) of technetium sorbed on a biotite mica thin section.

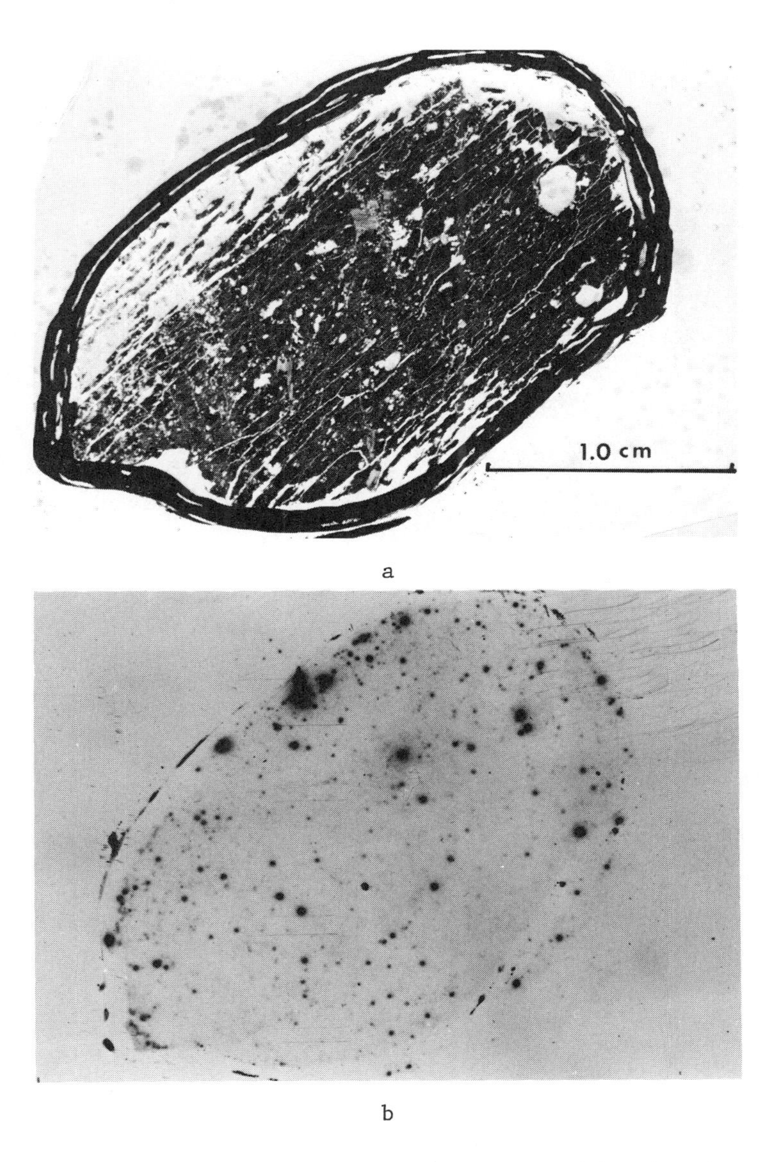

FIGURE 5. Photograph (a) and autoradiograph (b) of technetium sorbed on a hornblende thin section.

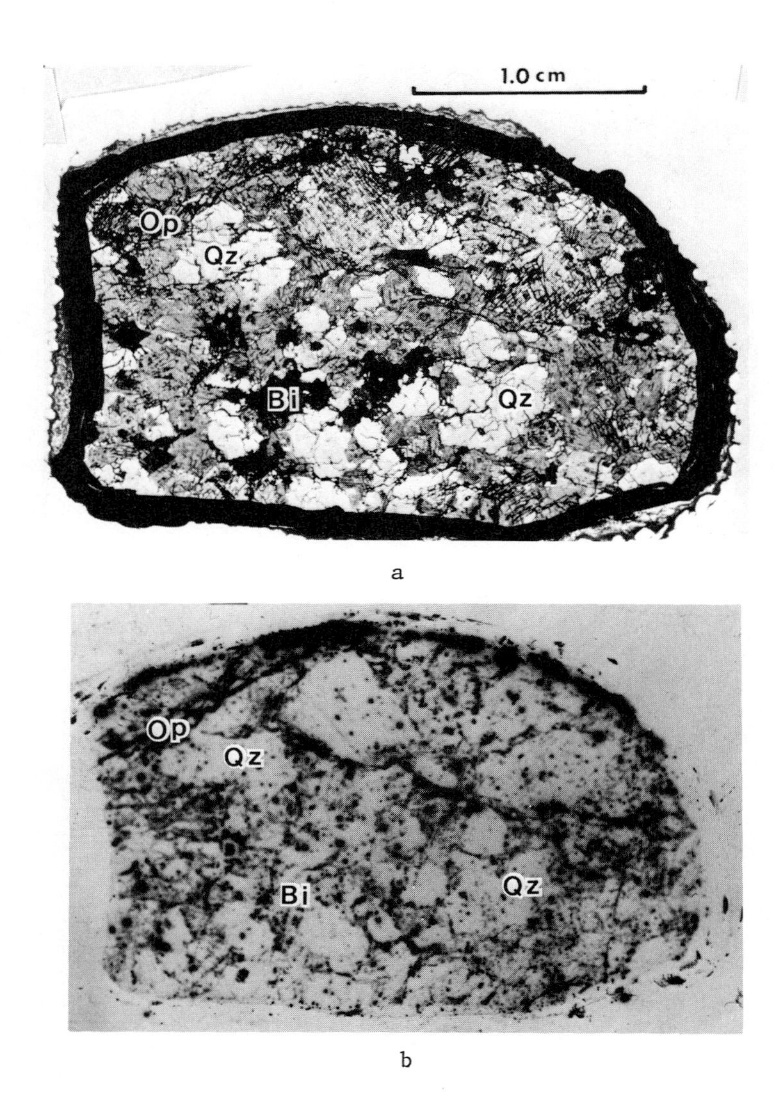

FIGURE 6. Photograph (a) and autoradiograph (b) of technetium sorbed on a granite thin section from the 1074-m level in the Eye-Dashwa Lakes pluton.

Sorption on Iron Oxides and Iron Oxyhydroxides under Oxic and Anoxic Conditions

Since iron oxides and iron oxyhydroxides are commonly found as fracture-infilling materials, samples with a particle size of 100 to 180 μm were contacted with granite groundwater containing ~ $5x10^{-12}$ mol/L ^{95m}Tc, as TcO_4^-, for 30 days in air, and in the anaerobic chamber, and the solutions sampled periodically. In addition, synthetic hematite was used, prepared by adding 2 mol/L NH_4OH to 1 mol/L $Fe(NO_3)_3$, washing and drying the precipitate at 105°C for one day, followed by heating at 400°C for two days. The minerals were analyzed for total iron, Fe^{++} and sulfur, and their concentrations are listed in Table VI. Figures 7(a) and (b) show the decrease in technetium concentration in solution as a function of time, under both oxic and anoxic conditions.

Table VI. Total Iron, Fe(II) and Sulfur. Concentrations in "Opaques" (Ilmenite, Iron Oxides, and Iron Oxyhydroxides used in Technetium Sorption Studies Concentrations in wt%

Mineral	Formula	Fe	Fe(II)	S
Ilmenite	$Fe TiO_3$	40.8±0.1	0.9±0.3*	0.55±0.01
Goethite	FeOOH	55.1±0.7	0.062±0.009	< 0.01
Hematite	Fe_2O_3	68.5±0.3	0.33±0.3	0.01±0.01
Specular Hematite	Fe_2O_3	49.1±0.5	0.45±0.28	< 0.01
Limonite	Fe OOH	21.8±0.1	0.027±0.004	0.02±0.01
Magnetite	$FeO{\cdot}Fe_2O_3$	65.3±0.4	20.3±0.1	< 0.01
Synthetic Hematite**	Fe_2O_3	72.6±2.2	< $3x10^{-3}$	-

* sample dissolution incomplete.
** prepared at WNRE.

Effect of Groundwater Composition on Sorption on Magnetite under Anoxic Conditions

One-gram samples of crushed, sieved magnetite (particle size 100 to 180 μm) were contacted with the six solutions described earlier in the anaerobic chamber for 50 days. The initial technetium concentrations ranged from $3x10^{-12}$ mol/L to 10^{-4} mol/L, and the solid-to-liquid ratio was 1 g/10 mL. The results for the two extreme starting concentrations are shown in Figures 8(a) and (b). At the end of the experiment, the Fe^{++}/Fe^{+++} ratios in solution were measured to estimate the E_h. In all cases where this ratio could be measured, $Fe^{++}/Fe^{+++} > 5$, indicating reducing conditions.

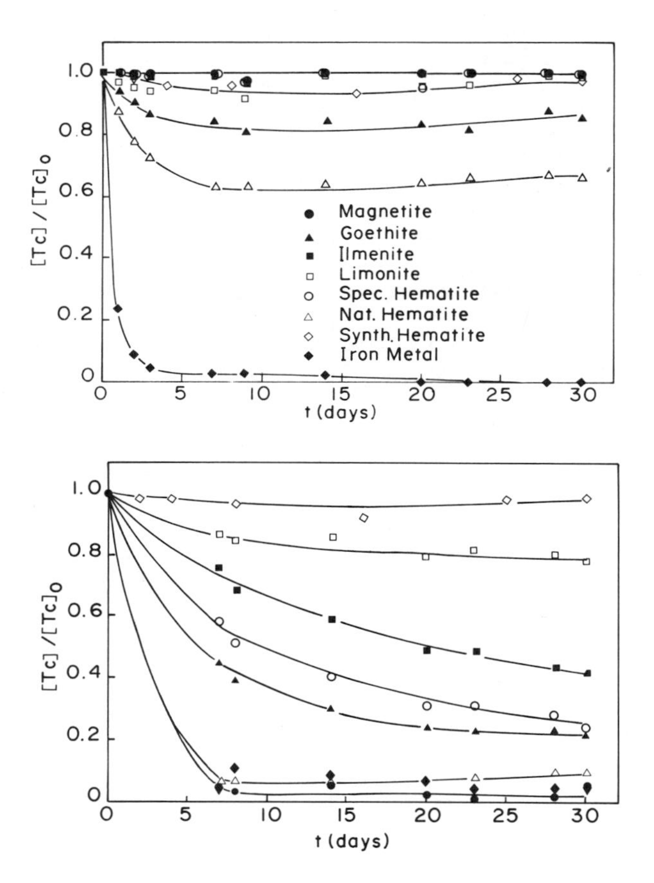

FIGURE 7. Decrease in technetium concentration as a function of time for solutions in contact with iron metal and iron minerals under (a) oxic and (b) anoxic conditions.

a

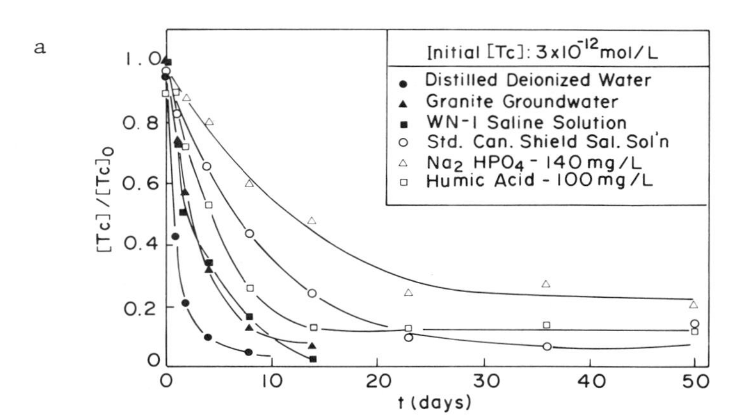

b

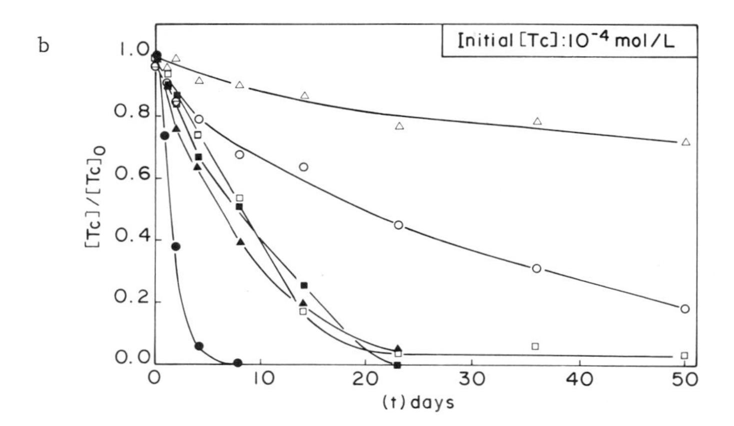

FIGURE 8. Decrease in technetium concentration as a function of time for six solutions in contact with magnetite ($FeO.Fe_2O_3$). Initial technetium concentration 3×10^{-12}mol/L (a) and 1×10^{-4} mol/L (b).

Discussion

The results obtained for crushed granite and gabbro under oxic conditions, compared to those obtained for the acid-washed quartz (see Table III), clearly show that no significant sorption takes place on granite, but that some technetium can be removed from solution by gabbro. Since small amounts (< 20 mg) of metallic iron are able to remove as much as 99+% of the technetium from solution, any sorption observed for granite over the time scale of the experiment may be attributed to the presence of metallic iron, inadvertently introduced in the crushing and grinding processes.

In the absence of oxygen, again no significant sorption was noted for granite. However, in one case, some sorption was observed for gabbro. Petrographic analysis of the area of the gabbro coupon that showed sorption, as revealed by an autoradiograph (Figure 3), indicated the presence of a thin iron-oxide band. Thus, even though gabbro contains ~ 42% pyroxene ($(Ca,Fe^{++})SiO_3$), the ferrous iron in this mineral is not able to remove the technetium from solution as effectively as the smaller amount of "opaques" (iron oxides and ilmenite). The iron oxide is not distributed homogeneously throughout the gabbro rock matrix, and this accounts for the wide variation in technetium removal from one coupon to the next. Crushing the rock frees these opaques and distributes them more uniformly, and it is most likely that this material is responsible for the significant k_d values obtained with crushed gabbro under oxic conditions.

The autoradiographs of the rock and mineral thin sections (Figures 4 to 6) also confirm the importance of iron oxides: although biotite ($K(Mg,Fe^{++})_3Si_3AlO_{10}(OH)_2$) and hornblende ($(Na,Ca_2)(Mg,Fe^{++})(Al,Fe^{+++})(Si_3AlO_{11})_2(OH)_2$) contain ferrous iron, sorption appears to take place solely on the small opaque (iron-oxide) inclusions. In the case of biotite, these oxides are located between the basal planes, and are randomly distributed in the hornblende. Similar distributions are observed for olivine, pyroxene, and epidote. The results for pyroxene further confirm the low sorption results obtained with gabbro, where it is one of the major minerals.

The autoradiographs also show some sorption on the granite obtained from the Eye-Dashwa Lakes pluton (Figure 6). Even though the iron concentration is low, hydrothermal alteration of the granite has resulted in the infilling of the minute fractures in the microcline feldspar with iron oxides, which show technetium sorption, while there is no sorption on the biotite crystals. Thus, while fresh, unaltered granite matrix rock has little or no iron in the form of iron oxides, alteration zones around fractures do, and technetium sorption may occur there. It should also be noted that sorption of technetium is limited to specific mineral surfaces. Thus, it appears that the reduction of TcO_4^- to a lower oxidation state occurs at or near the surface of the iron oxide and not in the bulk of the solution, by dissolved ferrous ions.

The experiment involving a suite of iron-oxide minerals showed that, with the exception of the synthetic hematite, some sorption took place on all minerals under anoxic conditions and also, in some cases, in the presence of air (Figures 7(a) and (b)). Subsequent chemical analyses indicated small amounts of Fe(II) in all natural minerals, although goethite, limonite, and hematite should not contain any. These minerals may have been formed by the oxidation and hydration of magnetite ($FeO{\cdot}Fe_2O_3$) and therefore contain residual amounts of unoxidized Fe(II). At any rate, this points out the importance of using purified and/or chemically analyzed material. It is again noteworthy that the rate of sorption on ilmenite is lower than that on natural hematite, even though the former contains more Fe(II). This can again be explained in terms of availability of the ferrous iron, as was shown to be the case with the thin sections.

Sorption on magnetite as a function of groundwater composition shows that, under anoxic conditions, technetium removal from solution is essentially complete after 50 days, with the exception of solutions containing phosphate ions. As pointed out earlier, phosphate was used instead of carbonate, as both are known to form anionic complexes with Tc(IV) (9). In these studies, the presence of humic acid did not affect its sorption. Strong saline solutions (up to 34 000 mg/L Cl) do not have a marked effect on the rate of technetium removal from solution either, as evident from Figure 8.

Conclusions

The results reported in this paper show that technetium is removed from solution by iron oxides, and not by minerals containing ferrous iron as an integral part of their crystal lattice, such as biotite, pyroxene, or hornblende. It was shown that there are cases where the small amount of iron in granite has a greater effect in removing technetium from solution than the larger amounts of iron in gabbro. Reduction of technetium occurs close to the mineral surface, and not in the bulk of the solution, by dissolved ferrous ions. Technetium is removed from groundwaters having widely different levels of total dissolved solids, and its removal is only affected by ligand-forming anions with a strong affinity for technetium, such as phosphate (and presumably also carbonate).

The significance of this study for nuclear fuel waste disposal is that iron-oxide-containing fractures in hydrothermally altered granite are capable of sorbing technetium. Technetium transport in the far-field region of a waste disposal vault can thus be impeded by iron-oxide coatings on hydrologically conducting fracture surfaces. If necessary, retention of technetium in the near-field region can be improved by incorporating rock containing large amounts of iron oxides in the backfill material.

The presence of saline solutions at depth in plutons in the Canadian Shield (11) should not be detrimental to retaining technetium in the valut, since the experiments showed that technetium removal occurs from highly saline solutions.

Throughout this paper, reference has beem made to reduced technetium species. Although it has been assumed that Tc(IV) is formed, there is no direct evidence for this. Experiments are now underway to determine the nature of the sorbed species using Fourier Transform Infrared Spectroscopy.

Acknowledgments

The authors would like to acknowledge the assistance of J. Paquette and N. Garisto in providing the E_h-pH diagrams for technetium and iron and of D.C. Kamineni, who performed the petrographic analyses. The minerals from the GSC were obtained through its curator, H.R. Steacy. R.F. Hamon, B.L. Sanipelli and K. Ross performed the various chemical analyses reported in this paper.

Literature Cited

1. Clegg, L. J; Coady, J. R., "Radioactive Decay Properties of CANDU fuel Volume 1. The Natural Uranium Fuel Cycle," AECL-4436/1, Chalk River, Ontario, Canada, 1977.
2. Palmer, D. A.; Meyer, R. E., "Adsorption of Technetium on Selected Inorganic Ion-Exchange Materials and on a Range of Naturally Occurring Minerals under Oxic Conditions," *J. Inorg. Nucl. Chem.* 1981, *43*, 2979.
3. Allard, B; Kipatsi, H.; Torstenfelt, B., *Radiochim. Radioanalyt. Letters*, 1979, *37*, 233.
4. Sheppard, M. I.; Vandergraaf, T. T.; Thibault, I. M.; Reid, J. A. K., Accepted for publication in *Health Physics*.
5. Landa, E. R.; Thorvig, L. M.; Gast, R. H., *J. Environ. Qual.* 1977, *6*, 181.
6. Bondietti, E. A.; Francis, C. W., *Science* 1979, *203*, 1337.
7. White, D. E.; Hem, J. D.; Waring, G. A., "Chemical Composition of Subsurface Waters," In: M. Fleischer (Editor), Data of Geochemistry, U.S. Geolog. Survey Prof. Paper, 440-F, 6th ed. 66 pp. 1963.
8. Analytical Science Branch, "Chemical Analysis of the Initial Groundwater Samples from the Lac du Bonnet Batholith," Whiteshell Nuclear Research Establishment, Pinawa, Manitoba, Canada, 1979.
9. Paquette, J.; Reid, J. A. K.; Rosinger, E. L. J., "Review of Technetium Behavior in Relation to Nuclear Waste Disposal," TR-25, Whiteshell Nuclear Research Establishment, Pinawa, Monitoba, Canada, 1980.

10. Vandergraaf, T. T.; Abry, D.R.M.; Davis, C. E. *Chem. Geol.* 1982., *36*, 139.
11. Fritz, P.; Frape, S. K. *Chem. Geol.*, 1982, *36*, 179.

RECEIVED October 20, 1983

3

Radionuclide Sorption Mechanisms and Rates on Granitic Rock

Determination by Selective Chemical Extraction Techniques

F. B. WALTON, T. W. MELNYK, J. P. M. ROSS, and A. M. M. SKEET

Atomic Energy of Canada Limited, Whiteshell Nuclear Research Establishment, Pinawa, Manitoba R0E 1L0 Canada

The gamma-emitting radionuclides ^{137}Cs, ^{144}Ce, ^{75}Se and ^{60}Co were simultaneously contacted with granite from the Lac du Bonnet batholith using both static and dynamic methods. Selective chemical extraction was then used to differentiate between the amounts of sorbed radionuclides that are (a) readily ion-exchangeable, (b) associated with amorphous oxyhydroxide deposits, and (c) "fixed" by other mineralogical or physical processes. Comparison of the experimental results from the dynamic tests with calculations from single sorption site kinetic models, using a variety of isotherms, showed that the models did not adequately describe the sorption reactions. Use of double sorption site models greatly improved the ability to describe solution concentrations and radionuclide surface inventories measured by extraction methods. Laboratory alteration of fresh granite surfaces was found to affect the sorption capacities and ratios of ^{60}Co, ^{75}Se and ^{144}Ce. Granite alteration on a laboratory time scale had no effect on ^{137}Cs sorption.

Rate constants determined for the various processes indicate that under oxidizing conditions ion-exchange processes are rapid, with equilibrium achieved within days. Reactions with oxyhydroxides or other mineralogical or physical processes take longer to achieve equilibrium, but, on time scales relevant to groundwater flow rates, will provide more significant retardation of the radionuclide migration than ion exchange.

0097-6156/84/0246-0045$06.25/0

The location and physical complexity of hard-rock fracture systems make it difficult to determine the mechanisms affecting radionuclide migration under field conditions. Techniques are needed that, under closely controlled conditions, provide data relevant to mass transport in the field. Development of two such techniques is described here.

Knowledge of sorption kinetics is essential for the design of laboratory or field radionuclide migration experiments. There are two requirements for the retardation of radionuclide migration: first, an interaction between the radionuclide and the geologic material, and second, sufficient time for the interaction to occur. Criteria for assessing the latter requirement have been developed (1) in terms of a dimensionless sorption rate parameter $\beta = kAt_w/V$, where k is the sorption (first order) rate constant, t_w is the water transit time (the time taken for groundwater to flow between two observation points), and A/V is the ratio of the granite surface area to the groundwater volume. The degree of retardation provided by a given chemical mechanism is related not only to the equilibrium distribution parameter, k_a, but also to the sorption rate parameter, β. For instance, a pulse injection of a radionuclide will travel with the groundwater and display only tailing for $0.1 < \beta < 1$. On the other hand, complete retardation, according to the value of k_a can be assumed for $\beta > 100$. For intermediate values of β, kinetic peak broadening will occur.

In these studies a dynamic testing method (the mixing-cell) has been used to measure the sorption kinetics of 4 different radionuclides on Lac du Bonnet granite from the Archean Superior Province of the Canadian Shield. This method has been used previously (2) for measurement of sorption kinetics without the interference of hydraulic dispersion common in many other dynamic techniques.

A variety of chemical extraction techniques has been developed by soil scientists to determine quantitatively the amount of trace metals bound to soil particles by various mechanisms. Multimechanism sorption has been suggested in order to explain partial irreversibility of radionuclide sorption and increased sorption with exposure time. The five main sorption mechanisms that have been reported are:

1. exchange with $CaCl_2$ or $MgCl_2$,
2. binding by carbonates,
3. binding by organic compounds,
4. sorption by iron or manganese oxyhydroxides, and
5. residual or "fixed" material.

Since the mixing-cell experiments were conducted with fresh or slightly weathered granite surfaces, techniques for determining sorption by carbonates and organic compounds were omitted. Using extraction techniques to determine radionuclide sorbed by mechanisms 1, 4 and 5 above on both mixing-cells and granite coupons, it was found that radionuclide sorption can be considered quantitatively to involve at least two mechanisms with varying rates of reaction. In addition, rock alteration kinetics - particularly when fresh rock is used - were found to strongly affect the sorption of some radionuclides.

Theory and Sorption Models

In order to assess the feasibility of any nuclear waste disposal concept, mathematical models of radionuclide sorption processes are required. In a later section kinetic descriptions of the three common sorption isotherms (3) are compared with experimental data from the mixing-cell tests. For a radionuclide of concentration C in the groundwater and concentration S on the surface of the granite, the net rate of sorption, by a first-order reversible reaction, is given by

$$dS/dt = k_1C - k_2S \tag{1}$$

where k_1 and k_2 are the sorption and desorption rate constants respectively, V is the fluid volume and A is the granite surface area. For a Freundlich isotherm the net rate of sorption can be expressed as

$$dS/dt = k_1C_o(C/C_o)^n - k_2S \tag{2}$$

where n is a constant and C_o is the initial tracer concentration in the groundwater.

For a Langmuir sorption isotherm the net sorption rate can be written

$$dS/dt = k_1C(S_o - S)/S_o - k_2S \tag{3}$$

where S_o is the maximum surface concentration of radionuclide.

As will be shown later, sorption of most radionuclides may be a function of two or more mechanisms. The combination of two first-order reactions has been successfully applied to ^{90}Sr migration over a twenty-year time period in a sandy-aquifer (1). The equations describing two parallel first-order reactions are

$$\left.\begin{aligned} S &= S_1 + S_2 \\ dS_1/dt &= k_1C - k_2S_1 \\ dS_2/dt &= k_3C - k_4S_2 \end{aligned}\right\} \tag{4}$$

The theory and verification of the mixing-cell mass balance equation has been reported previously (2). For a cell with initial concentration of tracer, C_o, flushed with tracer-free water at a volumetric rate, W, the mass balance is given by

$$VdC/dt = -WC - A(dS/dt) \quad (5)$$

For the special case of a non-reactive tracer, integration of Equation 5 gives

$$C(t)/C_o = \exp(-tW/V) \quad (6)$$

Experimental

Active Coupon Experiments. Coupons (19.8 x 19.8 x 4.00 mm) of Lac du Bonnet granite, obtained from the Cold Spring Quarry near the Whiteshell Nuclear Research Establishment (WNRE), Pinawa, Manitoba, were exposed to granite groundwater (GGW) and brine containing the gamma-emitting nuclides ^{137}Cs, ^{144}Ce, ^{75}Se, ^{125}Sb, ^{113}Sn and ^{60}Co. Petrographic and chemical analyses of the granite are given in Table I (4). Radionuclides were obtained carrier-free from New England Nuclear. Groundwater compositions and radionuclide starting concentrations are given in Tables II and III. A rock surface area to solution volume ratio of ~ 1 cm^2/mL, approximately the same as the ratio in the mixing-cell experiments, was used in these tests. The following procedures, in order of application, were used to determine the locations and quantities of radionuclides associated with various sorption mechanisms:

1. The coupons were contacted with granite groundwater or brine containing six radionuclides for 28 d. The coupons were removed and the solutions assayed to determine the amount of sorbed radionuclides.
2. The coupons were contacted with tracer-free groundwater solutions, which were assayed after 28 d to determine the amount of activity that is reversibly bound to the granite under normal groundwater conditions.
3. The coupons were contacted with 0.5 mol/L $CaCl_2$ solution for 72 h to displace all remaining exchangeable radionuclides.
4. The coupons were contacted for 24 h with a solution (termed KTOX) containing 0.1 mol/L potassium tetraoxalate and 0.1 mol/L hydroxylamine hydrochloride to remove radionuclides associated with iron and manganese oxyhydroxides.
5. The coupons were contacted with boiling Na_2CO_3 solution (5 wt%) for 15 min to remove radionuclides associated with hydrated silicates.
6. The coupons were contacted for a second 48 h period with KTOX solution to remove some of the radionuclides associated with lattice substitution into iron-bearing minerals.
7. The coupons were gamma counted to determine residual or fixed activity.

Inactive Coupon Experiments. A fresh cut surface of Lac du Bonnet granite contains both altered and unaltered ferromagnesian minerals (4). Exposing the unaltered or partially altered phases to groundwater solutions during an experiment will produce further phase alteration. Since KTOX solution selectively dissolves oxyhydroxide phases, it can be used to monitor their production during alteration. A second series of inactive experiments was performed to study alteration rates of fresh granite coupon surfaces. The coupons were ultrasonically cleaned and placed in a solution of GGW for 101 d. They were then leached with KTOX solution for incremental times of 1,2,3,6,12,24,24 and 48 h, with the solution being renewed at each time interval. The same extraction procedure was applied to freshly cut coupons. The KTOX solutions were analyzed by inductively coupled plasma (ICP) spectrophotometry for iron in order to calculate oxyhydroxide extraction rates as a function of cumulative exposure to KTOX solution. A granite surface area to extraction solution volume of $\sim 1\ cm^2/mL$ was used in all extraction procedures.

Table I. Mineral and Chemical Composition of Lac du Bonnet Granite (4)

Mineral	Modal Percent	Oxide	Weight Percent
Quartz	25.8	SiO_2	73.1
K-feldspar	28.1	Al_2O_3	14.2
Plagioclase	33.9	FeO	0.81
Biotite	5.2	Fe_2O_3	0.76
Muscovite	2.5	CaO	1.43
Opaques	0.7	MgO	0.46
Epidote	1.5	Na_2O	4.23
Chlorite	2.3	K_2O	4.88
		TiO_2	0.22
		MnO	0.03
		H_2O	n.d.
		CO_2	n.d.
Total	100.0	Total	100.09

n.d. = below level of detection

Table II. Groundwater Compositions (mg/L)

	GGW*1	Brine
Na^+	8.3	5,050
K^+	3.5	50
Mg^{2+}	3.9	200
Ca^{2+}	13	15,000
Sr^{2+}	-	20
Fe_t	-	-
Si	-	15
HCO_3^-	*2	10
Cl^-	5.0	34,260
SO_4^{2-}	8.6	790
NO_3^-	0.62	50
F^-	0.19	-
pH	6.5±0.5	7±0.5

*1 GGW = granite groundwater
*2 In equilibrium with CO_2 in atmosphere

Table III. Carrier-Free Radionuclide Starting Concentrations

Isotope	Starting concentration* (10^4 Bq/mL)
^{60}Co	26.9
^{144}Ce	53.3
^{137}Cs	27.7
^{125}Sb	2.05
^{75}Se	1.37
^{113}Sn	0.347

* Specific activities of isotopes in Bq/g are: ^{60}Co - $4.2 \cdot 10^{13}$; ^{144}Ce - $1.2 \cdot 10^{14}$; ^{137}Cs - $3.2 \cdot 10^{12}$; ^{125}Sb - $3.9 \cdot 10^{13}$; ^{75}Se - $5.2 \cdot 10^{14}$; ^{113}Sn - $3.7 \cdot 10^{14}$

Mixing-Cell Experiments. A schematic diagram of a typical cell used in these experiments is shown in Figure 1. The cells are circular in cross-section and are fabricated by a double diamond-drill coring operation on a slab of Lac du Bonnet granite.

Two sets of experiments were conducted using six radioactive tracers simultaneously. Starting concentrations are given in Table III. Radionuclides were eluted from the cells with groundwater solutions using a peristaltic pump. Effluent was collected at 90-min intervals by means of a fraction collector connected to the cell with small-bore Teflon tubing. Concentrations of the various isotopes in the effluent were measured by gamma spectrometry.

In the first set of experiments, six cells were ultrasonically cleaned to remove rock powder produced during coring, and then flushed before tracer injection. Three were flushed with GGW and three with brine. A second set of experiments was designed to determine whether mineral alteration rates and/or solution temperature affect the rate of radionuclide interaction with granite. (Test conditions are given in Table VI of the Results Section). The 60°C temperature was maintained by submerging the cells in a water bath. Two cells were flushed with GGW for several weeks and then allowed to sit, filled with GGW, for approximately five months before the experiment was started.

After both sets of experiments were completed, the groundwater was drained from the cells and selective chemical extractions of the granite cell walls were performed. The cells were filled with 0.5 mol/L $CaCl_2$ solution and stirred continuously for 72 h to displace exchangeable radionuclides. After a rinse with demineralized water to remove residual $CaCl_2$ solution, the cells were filled with KTOX solution and stirred for 24 h to remove radionuclides associated with oxyhydroxides. All solutions were analyzed by gamma spectrometry to determine the amounts of radionuclides extracted. Residual activity was measured by direct gamma counting of the cells.

Results

Extraction of Inactive Coupons with KTOX Solution. Average iron extraction rates are plotted as a function of cumulative extraction time in Figure 2. Where rates for the altered and unaltered granite differ, the confidence level at which the difference is significant is indicated in brackets.

During the first six hours of extraction, more iron is removed from the altered granite than from the fresh granite. The extra iron from the altered granite is believed to be that associated with the production of iron oxyhydroxides during the 101-d exposure of the granite to GGW.

The iron extraction rates for both fresh and altered granite decrease for the first 24 h and become the same constant rate for the remaining 24 to 120 h. Thus the iron associated with oxyhydroxides (formed during recent exposure to GGW or during hydrothermal alteration) is removed during the first 24-h exposure to

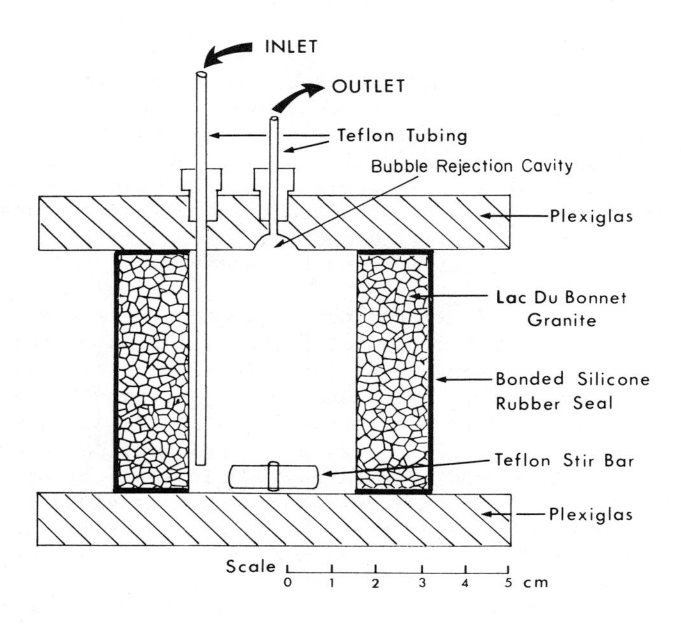

Figure 1. Cross-sectional view of cylindrical mixing-cell.

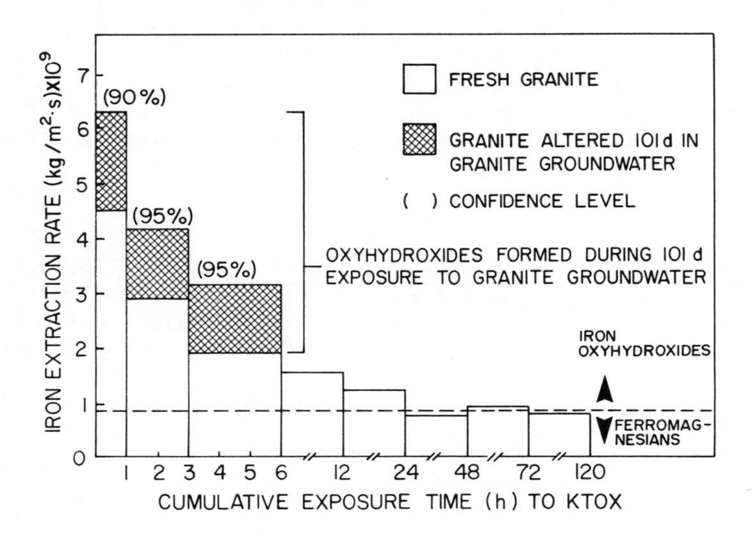

Figure 2. Average iron extraction rates for fresh and altered Lac Du Bonnet granite as a function of cumulative extraction time.

KTOX. Iron extracted during the subsequent 24 to 120 h is probably associated with the iron-bearing minerals (e.g. magnetite). This hypothesis is consistent with autoradiographic studies described in the following section.

<u>Selective Chemical Extraction of Active Coupons</u>. Using gamma spectrometric analysis of the groundwaters in the sorption and desorption steps, distribution coefficients (k_a) have been calculated for the isotopes ^{60}Co, ^{137}Cs, ^{144}Ce, ^{75}Se, ^{125}Sb and ^{113}Sn. Values of k_a for these isotopes in GGW and brine have been reported elsewhere (2). Figure 3(A) shows a photograph of a typical granite coupon that was exposed to the GGW solution. Autoradiographs of this coupon after a 28-d GGW desorption, a 72-h $CaCl_2$ extraction and a 24-h KTOX extraction (Figures 3(B), (C) and (D)) indicate that the areas of most highly concentrated activity (light areas in the autoradiographs) correspond to ferromagnesian and/or other dark, iron-bearing minerals in the photograph. However, Figure 3(B) shows that there is some sorption on all mineral phases. Figure 3(C) shows a darkening of all areas after $CaCl_2$ extraction, indicating the removal of exchangeable activity from all areas. Figure 3(D), taken after the 24-h KTOX extraction, shows a decrease in activity in the location of the iron-bearing minerals and virtual elimination of activity in all other areas. It is not possible to discern the behaviour of any individual nuclide from these autoradiographs; however, specific studies of ^{90}Sr, ^{137}Cs, ^{144}Ce, ^{237}Pu, ^{75}Se, ^{147}Pm and ^{241}Am, using a combination of autoradiographic and petrographic techniques, indicate a clear preference for sorption on ferromagnesians, opaques and their alteration products (4,5).

The amount of radionuclide removed from the granite in the groundwater desorption and selective extraction procedures was measured by gamma spectrometry. The data is summarized for ^{60}Co, ^{137}Cs, ^{144}Ce and ^{75}Se in Figure 4. In this figure the percentage of the initial radionuclide inventory sorbed during contact with the two groundwaters is indicated numerically. In the bar chart the total amount of <u>sorbed activity</u> has been normalized in order to compare the relative amounts of sorbed activity extracted by the different reagents. Activities for ^{125}Sb and ^{113}Sn were below background detection limits. The activity on the surface of the granite after the 28-d sorption has been divided into four categories: exchangeable, associated with oxyhydroxides, associated with hydrated silicates, and "fixed" to ferromagnesians, their alteration products and/or other opaque mineral phases.

There are three significant features in Figure 4. First, the various sorption mechanisms affect each nuclide differently. Second, the amount of each radionuclide that is exchangeable or reversibly sorbed is less than 50% for all nuclides tested. In 28 days, more than 50% of the sorbed radionuclides have undergone phase transformation or reacted with oxyhydroxides to form non-exchangeable phases. Third, for a given radionuclide, the

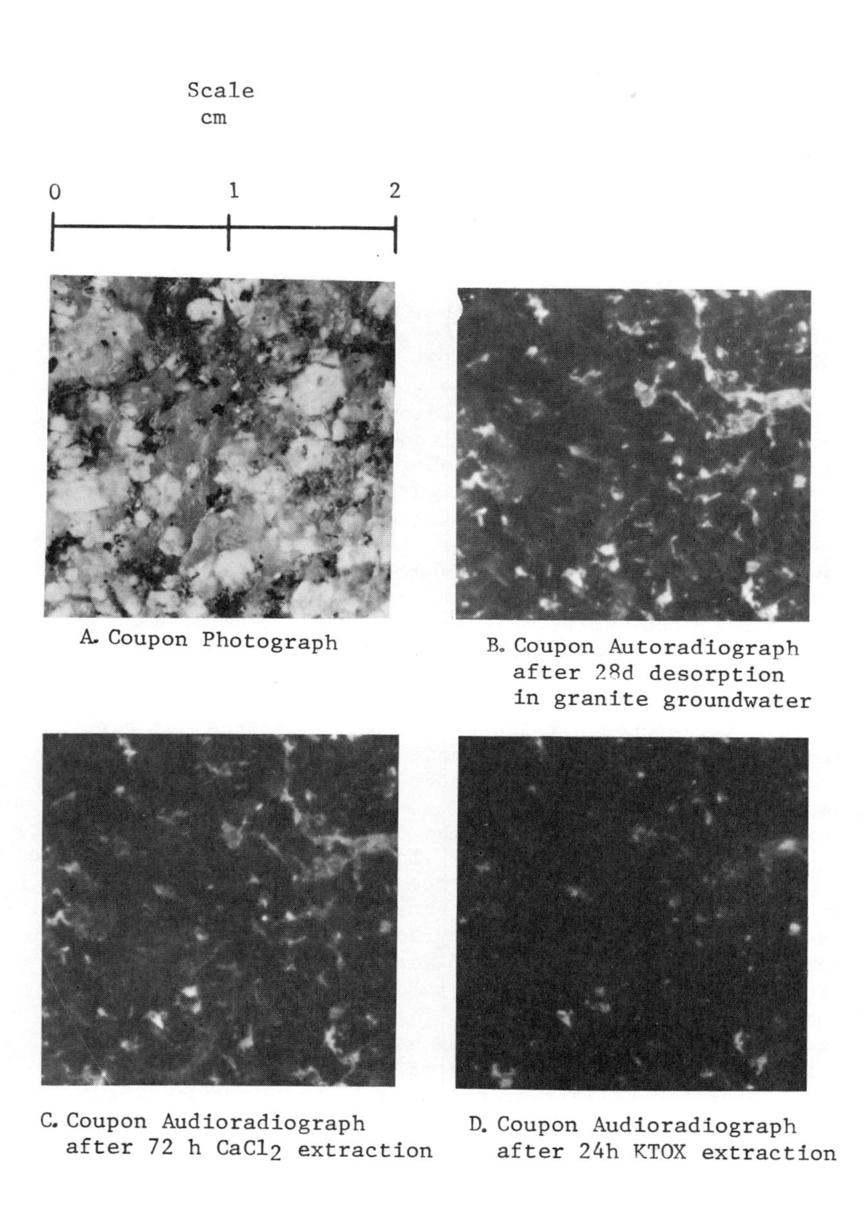

Figure 3. Photograph and autoradiographs of active granite coupon. Dark areas in photograph (A) indicate areas of high mafic mineral content. Light areas in autoradiographs (B), (C) and (D) indicate areas of radionuclide concentration.

relative amounts of exchangeable, bound to oxyhydroxides, and "fixed" is nearly the same for both GGW and saline groundwaters, although the absolute amounts sorbed in the two groundwaters differ. This observation indicates that the same mechanisms operate in both groundwaters, but are affected by ionic strength. The decrease in sorptive capacity in saline groundwater could be caused by a reactive-site competition effect or could be due to complexation in solution.

The iron extraction rates derived from the inactive coupon tests suggest that activity removed during the 24-h KTOX extraction is associated primarily with oxyhydroxides. A possible exception is cesium (see following section). Activity removed during the subsequent 48-h KTOX extraction is associated with the dissolution of iron-bearing minerals. This activity is considered part of the "fixed" activity inventory. Further attempts to remove this "fixed" activity were not very successful. Even partial dissolution of the coupon by 4-h contact with 43 vol.% HF removed only 50% of the "fixed" activity. Thus the "fixed" activity, localized in the areas of high ferromagnesian concentration, requires very aggressive chemical attack for removal. Gamma spectrometric measurements of the coupon indicate the "fixed" activity contains about equal quantities of ^{60}Co, ^{137}Cs and ^{144}Ce, and about ten times less ^{75}Se.

<u>Mixing-Cell Results</u>. Models representing kinetic versions of the three common sorption isotherms (first order, Equation 1; Freundlich, Equation 2 and Langmuir, Equation 3) were fitted to mixing-cell data for one set each of room temperature GGW and brine solutions. Model fit variances for these calculations are given in Table IV for ^{137}Cs, ^{144}Ce, ^{60}Co and ^{75}Se. Residual plots (not given) showed systematic deviations for all three models, indicating that all are inappropriate. Since selective extraction of the active granite coupons indicated the presence of at least two sorption mechanisms, a model utilizing two parallel first-order reactions, Equation 4, was fit to the data. The model fit variances are given in Table IV and calculated rate parameters for the two reactions in Table V. In nearly all cases an improvement in model fit was observed. This is typified by the comparison between the fit of a single first-order and double first-order (DFO) model to the ^{60}Co data, as shown in Figure 5. Although the DFO model shows considerable improvement over single-site models, residual plots indicate a small systematic deviation at higher concentrations. This is discussed further in the following section.

The residual activity in the cells after the dynamic experiment was selectively extracted using techniques previously described. The extracted activity, expressed as a percentage of starting inventory, is compared in Table VI to inventories predicted by the DFO model, which was fitted to tracer concentrations in the groundwater. Also shown in Table VI are the static

Table IV. Comparison of the Fits[*1] of Different Models to the Experimental Mixing-Cell Data

Model	Fit Variance[*2]							
	^{137}Cs		^{144}Ce		^{60}Co		^{75}Se	
	GGW(3)	Brine (4)	GGW(3)	Brine(4)	GGW(3)	Brine(4)	GGW(3)	Brine(4)
First Order	0.066	0.075	0.165	0.116	0.077	0.094	0.042	0.059
Freundlich	0.050	0.067	0.047	0.050	0.023	0.074	0.039	0.054
Langmuir	-	0.073	0.060	0.102	0.062	0.050	0.039	0.056
DFO	0.015	0.006	0.037	0.034	0.017	0.020	0.039	0.019

*1 Cell reference number is shown in brackets.

*2 $(\text{Fit Varience}) = n^{-1} \Sigma [(C_{model} - C_{experimental})/C_{model}]^2$

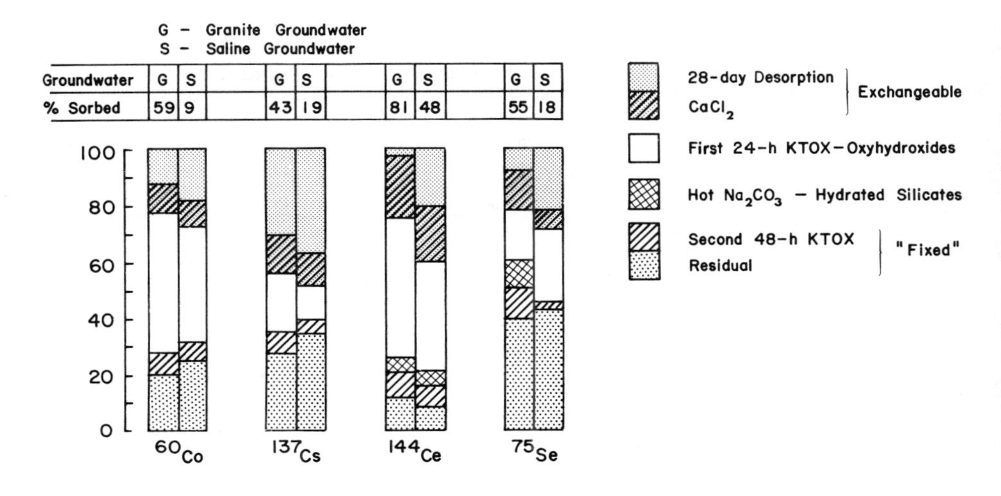

Figure 4. Percentage of sorbed radionuclides as a function of various sorption mechanisms for 28-d contact with ^{60}Co-, ^{137}Cs-, ^{144}Ce- and ^{75}Se-doped granite and saline groundwaters.

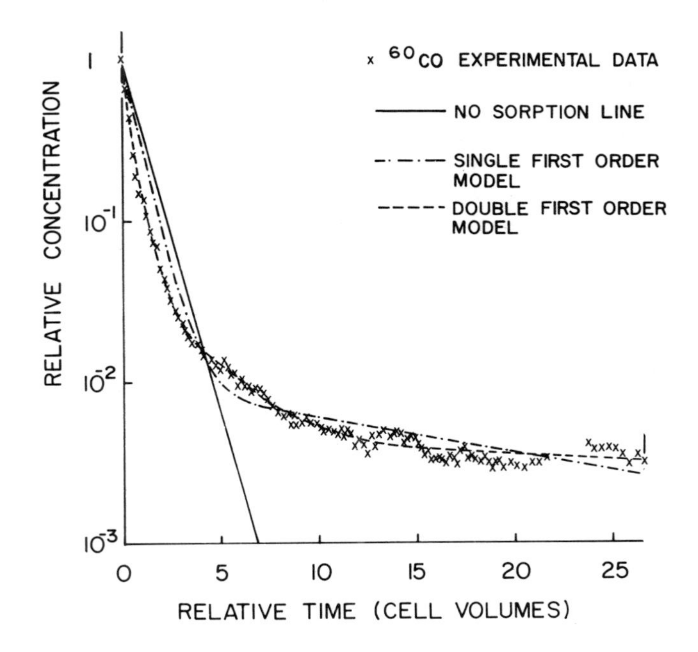

Figure 5. Comparison of single first-order and double first-order models to ^{60}Co mixing-cell data.

Table V. Double First-Order Model Parameters and Fit Variance

Isotope	Reference	Variance	Rate Constants x 10^3			
			$k_1(\frac{mL}{h \cdot cm^2})$	$k_2(h^{-1})$	$k_3(\frac{mL}{h \cdot cm^2})$	$k_4(h^{-1})$
^{60}Co	3	0.010	21.0	6.96	17.0	0.374
	37	0.010	1.87	11.6	1.61	1.67
	34	0.025	3.59	5.52	17.2	3.11
	33	0.056	265.	33.2	67.6	0.189
	4	0.020	3.27	9.87	2.19	0.496
	35	0.007	0.500	7.17	0.763	0.629
^{137}Cs	3	0.016	3.35	9.97	2.34	0.687
	37	0.020	17.2	33.2	2.08	2.48
	34	0.004	6.86	17.0	1.27	2.07
	33	0.007	2.90	16.6	0.952	2.09
	4	0.006	0.622	9.69	0.213	0.300
	35	0.013	0.482	10.0	0.102	1.75
^{144}Ce	3	0.037	113.	9.28	83.1	0.475
	37	0.050	1130.	58.8	46.9	0.211
	34	0.069	6.19	3.09	12.8	0.054
	33	0.218	227.	142.	30.1	0.671
	4	0.034	25.1	9.20	17.0	0.225
	35	0.017	2.31	9.89	1.18	1.36
^{75}Se	3	0.036	1.72	13.8	0.447	0.383
	37	0.034	1.90	15.6	0.703	1.19
	34	0.046	1.20	10.1	0.887	0.487
	33	0.069	0.602	9.73	1.66	0.225
	4	0.019	0.518	8.59	0.789	0.158
	35	0.030	0.856	13.2	0.994	0.309

equilibrium distribution coefficients (k_a) which can be calculated by assuming dS_1/dt and dS_2/dt are zero in Equation 4 as time becomes large.

$$\left.\begin{aligned} k_{a1} &= (S_1/C) = (k_1/k_2) \qquad t \rightarrow \infty \\ k_{a2} &= (S_2/C) = (k_3/k_4) \qquad t \rightarrow \infty \end{aligned}\right\} \tag{7}$$

Discussion

Selective Chemical Extraction. Since granite contains many minerals, the interpretation of selective extraction data requires knowledge of the sorptive capacity of those minerals and their susceptibility to alteration during the course of experiments.

Monitoring of Lac du Bonnet granite alteration rates at 20°C using a mixing-cell technique (2) has shown a period of high alteration rate during the first three weeks followed by a period of decreasing rate extending for more than five months. Continuation of these experiments showed that the alteration rate decreases for up to ~ 13 months and then becomes constant. A repeat of these tests at 60°C showed the same trends, with the alteration rates two to three times higher (6). During alteration of iron-bearing minerals, iron, because of its very low solubility, will precipitate as iron oxyhydroxides. Iron extraction rates for altered and unaltered granite (Figure 2) indicate a significant increase in iron oxyhydroxide inventory over a 101-d exposure to groundwater. The production at low temperatures on a laboratory time scale of other alteration products, such as clay minerals, is unlikely (7).

Correlation of total surface inventories, obtained by selective extraction of the mixing-cell, to the degree of granite alteration and to temperature, indicates that ^{60}Co and ^{75}Se show a positive correlation with both alteration and temperature. Cobalt-60 and ^{75}Se inventories for the experiments conducted with fresh granite at 20°C and at 60°C (Reference No. 37 and 34, Table III) indicate about a factor of two increase at 60°C, as would be expected from the production of oxyhydroxides based on the alteration rate data. Other residual inventories for ^{60}Co and ^{75}Se (Table III) follow a pattern consistent with increased oxyhydroxide inventories due to both alteration and elevated temperature.

The sorption on newly formed iron hydroxides is known to be reversible on short time scales (8). The differences in autoradiographs (B) and (C) of Figure 3 indicate that a large amount of exchangeable activity is associated with the ferromagnesian minerals; however, areas other than those of high iron content also show a decrease in activity after the $CaCl_2$ extraction. These same areas show virtually no activity after extraction with KTOX (Figure 3(D)). Since minerals like quartz have been shown

TABLE VI. Comparison of Model Predictions and Selective Chemical Extraction Data

Isotope	Reference No.	Ground-water	Temp. (°C)	Granite Precondition (wk)	Surface Inventories as Percentage of Starting Inventory — Selective Extraction: $CaCl_2$	KTOX	Fixed[*4]	Model Prediction: S_1	S_2	k_{a1} (mL/cm^2)	k_{a2} (mL/cm^2)
^{60}Co	3[*1]	GGW	20	3	1.80	39.	6.6	1.35	35	3.0	46.
	37	"	20	0	0.79	13.	3.6	0.02	9.1	0.16	0.97
	34	"	60	0	0.85	26.	15.	0.07	47.	0.65	5.5
	33	"	60	20	0.80	54.	7.3	1.3	73.	8.0	360.
	4[*2]	Brine	20	3	1.0	6.3	1.6	0.04	5.5	0.33	4.4
	35	"	60	0	<0.01	1.7	0.69	<.01	0.62	0.07	0.45
^{137}Cs	3[*1]	GGW	20	3	0.53	1.1	2.5	0.06	5.1	0.34	3.4
	37	"	20	0	0.32	1.1	2.4	0.04	0.58	0.52	0.84
	34	"	60	0	0.12	0.91	3.2	0.03	0.59	0.40	0.61
	33	"	60	20	0.22	0.53	2.4	0.01	0.36	0.20	0.48
	4[*2]	Brine	20	3	0.08	0.10	0.32	<.01	0.68	0.06	0.71
	35	"	60	0	0.09	0.38	0.55	<.01	0.05	0.05	0.06
^{144}Ce	3[*1]	GGW	20	3	21.	28.	19.	4.2	69.	12.	175.
	37	"	20	0	13.	8.5	20.	5.0	63.	19.	220.
	34	"	60	0	6.0	43.	23.	0.79	37.	2.0	240.
	33	"	60	20	1.7	12.	2.3	0.93	43.	1.6	45.
	4[*2]	Brine	20	3	6.4	5.3	3.4	0.71	37.	2.7	76.
	35	"	60	0	0.11	1.8	2.7	0.02	1.0	0.23	0.87

^{75}Se	3*1	GGW	20	3	n.d.*3	n.d	n.d	<.01	1.3	0.11	1.1
	37	"	20	0	0.33	5.6	n.d	<.01	0.80	0.12	0.59
	34	"	60	0	0.19	9.3	n.d	<.01	2.4	0.12	1.8
	33	"	60	20	0.22	18.	n.d	<.01	6.0	0.06	7.4
	4*2	Brine	20	3	n.d	n.d	n.d	<.01	2.9	0.06	5.0
	35	"	60	0	0.08	0.86	n.d	<.01	3.2	0.07	3.2

Notes: *1 Selective extraction data for cell 3 is not available. Average for replicate cells 1 and 2 is given.

*2 Selective extraction data for cell 4 is not available. Data for replicate cell 5 is given.

*3 n.d - not detectable due to isotopic decay.

*4 "Fixed" activity was determined by direct count of cell; all other inventories are based on solution assays.

to have negligible sorption (4), iron oxyhydroxides formed by iron-bearing mineral alteration during contact with groundwater were likely the predominant sites of exchangeable sorption for ^{60}Co and ^{75}Se in both coupon and mixing-cell studies.

Sorption of ^{60}Co onto oxyhydroxides precipitated in stream beds has been shown to be virtually irreversible (9). Diffusion into the crystal lattice of iron-bearing minerals has also been suggested (10). The slow removal of ^{60}Co during the second 48-h KTOX extraction is consistent with this mechanism.

Studies of selenium have shown that its geochemistry is largely controlled by that of iron under both oxidizing and reducing conditions at acid to neutral pH's. Both $HSeO_3^-$ and SeO_4^{2-} are strongly adsorbed by hydrated surfaces of ferric oxides (11). This is consistent with the data of Figure 4, which shows that only about 20 to 30% of the adsorbed selenium is exchangeable. Selenium has been found as ferroselite in goethite-hematite deposits at the migrating oxidizing/reducing interface in a roll-type uranium ore body (11). From the present studies it is not possible to tell the chemical form of the "fixed" selenium. However, its resistance to chemical extraction and concentration on iron-bearing minerals suggest the formation of a stable iron-selenium mineral phase.

Cesium inventories in Table III show no effect of granite alteration or temperature. This is consistent with the observation that cesium prefers micaceous and clay minerals which are not expected to form under the conditions of these experiments. It has been shown that cesium diffuses into the lattice of micaceous minerals, resulting in its entrapment (12). The $\sim$ 35 to 40% "fixed" cesium (Figure 4) is, therefore, likely bound internally at lattice sites, which would have to be destroyed in order to release the cesium. The extraction of cesium by KTOX in these experiments is likely due to dissolution of iron-bearing minerals, such as biotite and horneblende rather than oxyhydroxides. If sorption by oxyhydroxides had been involved, a trend of increasing sorption with granite alteration, similar to that exhibited by cobalt and selenium, should have been observed. These observations support the hypothesis that the similarity of the cesium elemental distribution to that of iron in natural granite weathering profiles is a result of coincidental clay mineral production rather than a direct association with oxyhydroxides (14).

In the coupon experiment, 60 to 75% of the sorbed ^{144}Ce was found to be associated with oxyhydroxides and "fixed" by lattice substitution (Figure 4). However, selective extraction data from the mixing-cell tests is ambiguous with respect to the relative importance of exchangeable inventory and that potentially associated with oxyhydroxides. Total surface inventories show a decreasing trend with increasing alteration. No clear trend with temperature is evident.

Cerium is associated with iron in some weathering profiles

(14). Static sorption studies with drill core material from two rock formations in the Canadian Shield have shown a positive correlation between ^{144}Ce sorption and mafic mineral concentration and transition metal oxide concentration in the rock (5).

The cerium solution concentration profiles from the mixing-cell experiments often show considerable data scatter (2). The largest scatter was observed in GGW at 60°C with the 20-week preconditioned surface (DFO fit variance of 0.218). Filtration of the solution gave ~ 98% recovery (> 0.45-μm particle size) of ^{144}Ce, indicating that the predominant cerium phase was colloidal. The least scatter occurred in the brine solution at 60°C with a fresh granite surface (DFO fit variance of 0.017). The relatively smoothly varying solution data is generally indicative of species in true solution. The dramatic difference in data scatter in these two experiments may indicate a strong effect of the brine on the colloidal behaviour of cerium at near neutral pH. Sorption of cerium (colloidal) onto granite is likely controlled by surface-charge effects, which vary with groundwater ionic strength and alteration of the granite surface. Further experiments are required to clarify the association of cesium with iron-bearing minerals.

The hot Na_2CO_3 extraction removed about 5% of the sorbed ^{144}Ce in coupon tests with both GGW and brine; about 10% of ^{75}Se was removed for GGW only, and no ^{60}Co or ^{137}Cs was removed. In the case of cerium, colloid coprecipitation with amorphous silica may explain these extraction results. The association of selenium with possible hydrated silicates is unknown. Further investigation of these associations will be required before any significance can be attached to these Na_2CO_3 extraction results.

Model Fitting to Mixing-Cell Data. Multiple-site kinetic models have been used to describe pesticide and herbicide movement in soils (15,16,17), cesium migration in columns (18), and strontium migration in a sandy aquifer over a twenty-year time period (1). The results of the selective extraction procedures in all experiments discussed here suggest that a multi-site model should provide a better fit of the data than a single-site model. This hypothesis is supported by the variances in Table I, with the possible exception of selenium.

Assuming, in Table III, that S_1 represents exchangeable activity and S_2 represents the net inventory of all other mechanisms, a comparison of these model-predicted inventories with those obtained by selective chemical extraction indicates that, while the overall mass balance trends are predicted, agreement in detail of individual extraction inventories with the model-predicted inventories is not always good. The overall mass balance is governed almost totally by the higher concentration (early time) data. In the case of ^{60}Co, inspection of residual plots indicates that, in all cases, the DFO model shows small systematic deviations at early times that account for the mass balance disagreements. Where the best fit at high concentrations was produced, the best overall mass balance agreement was obtained.

A comparison of ^{60}Co inventories (Table III) as a function of alteration history indicates that total inventories change by more than a factor of three, and k_{a1} and k_{a2} change by more than a factor of 100 for conditions ranging from fresh surfaces at 20°C to 20-week altered surfaces at 60°C. Hence, the alteration kinetics of the granite will need to be explicitly included in models to predict ^{60}Co sorption on fresh granite surfaces. Alteration kinetics may not be important for well-weathered fracture surfaces where oxyhydroxide inventories may be relatively stable over long time periods. Provided the rate constants for ^{60}Co sorption on natural fracture surfaces, including the effect of the inventory of oxyhydroxides, can be obtained, the DFO model may be useful for describing ^{60}Co migration.

Although selenium is observed in the geologic setting in association with iron, and the DFO model provides a good fit to the groundwater solution data from the dynamic experiments, the selective extraction inventories do not agree well with the model predictions, and the chemistry of selenium is complicated by redox effects (11). It is likely that the DFO model is still too simple to adequately describe selenium migration.

DFO model fit variances for cesium show a considerable improvement over the single-site model variances given in Table I. A two-site model has been used (18) to model cesium transport in soils. In this model a Langmuir-type model, Equation 3, was used to represent surface sorption and a first-order model, Equation 1, was used to approximate interparticle diffusion. Extraction with $CaCl_2$ was used to verify the exchangeable site inventory. The success of the two-site sorption models for dissimilar testing conditions clearly indicates its superiority over single-site sorption models for modelling cesium transport.

Using the criteria referred to in the introductory section, the deposition rate constants given in Table II can be used to estimate transit times necessary to achieve equilibrium in laboratory or field fracture flow studies. For example cesium sorption from GGW in a 100-μm aperture fissure in unweathered Lac du Bonnet granite requires a minimum water transit time of ~ 3 d for site 1, and 12 d for site 2, in order to be able to assume equilibrium sorption. Water transit times of the order of hours will produce only tailing. Transit times required in brine groundwaters are an order of magnitude higher than those in GGW.

Comparison of model fit variances for cerium indicates a definite superiority of the DFO model over single-site sorption models. The large data scatter in groundwater cerium concentrations, and the ambiguity in extraction inventory trends, indicate that systems producing significant colloidal material may require more specialized models and experimental methods to fully understand their transport mechanisms.

Conclusions

Using a combination of selective chemical extraction and autoradiographic techniques, sorption has been shown to be multimechanism. Dynamic experiments using mixing-cells have shown that significant changes in sorption kinetics and capacity can occur due to mineral alteration on a laboratory time scale. Iron oxyhydroxides, resulting from the alteration of iron-bearing minerals, have been shown to play an important role in the sorption of a variety of radionuclides.

The comparison of kinetic sorption models presented here was made possible by the use of the mixing-cell dynamic technique, which eliminates the masking effects of hydraulic dispersion. With the possible exception of selenium, a two-site, double first-order (DFO) model shows an improvement over single site-models for describing sorption of the radionuclides studied. The dependence of sorption on alteration history in the majority of cases indicates that experiments with systems representative of well-weathered fracture systems are necessary to obtain data applicable to actual disposal vault conditions.

Acknowledgements

The authors gratefully acknowledge the assistance of Dr. R. Hamon in providing the KTOX leaching data for the inactive coupon experiments. We are also indebted to the radiochemical analysis section of the Analytical Science Branch at WNRE for the gamma spectrometric analyses of solutions.
Issued as Atomic Energy of Canada Limited
Contribution No. AECL-8049.

Literature Cited

1. Melnyk, T.W.; Walton, F.B.; Johnson, L.H. "High-Level Waste Glass Field Burial Tests at CRNL. The Effect of Geochemical Kinetics on the Release and Migration of Fission Products in a Sandy Aquifer", AECL-6836, Chalk River, Ontario, Canada, 1983.
2. Walton, F.B.; Melnyk, T.W.; Abry, R.G. Fleury, K. *Chem. Geol.* 1982, *36*, 155.
3. Travis, C.C. "Mathematical Description of Adsorption and Transport of Reactive Solutes in Soil: A Review of Selected Literature", ORNL-5403, Oak Ridge National Laboratory, Oak Ridge, Tennessee, 1978.
4. Vandergraaf, T.T.; Abry, D.R.M.; Davis, C. E. *Chem. Geol.* 1982, *36*, 139.
5. Vandergraaf, T.T.; Abry, D.R.M. *Nucl. Tech.* 1982, *57*, 399.
6. Walton, F.B., unpublished data.
7. Krauskopf, K.B. "Introduction to Geochemistry", McGraw-Hill: 1979; p.150.

8. Leckie, J. O.; Benjamin, M. M.; Hayes, K.; Kaufman, G.; Altmann, S. "Adsorption/Coprecipitation of Trace Elements from Water with Iron Oxyhydroxides," Electric Power Research Institute (EPRI) report, Palo Alto, Cal. EPRI CS-1513, 1980.
9. Cerling, T. E.; Turner, R. R. *Geochim. Cosmochim. Acta* 1982, *46*, 1333.
10. Chao, T. T.; Theobald, P. K. Jr. *Econ. Geol.* 1976, *71*, 1560.
11. Howard, J. H. III *Geochem. Cosmochim. Acta* 1977, *41*, 1665.
12. Levi, H. W.; Mickeley, N. paper SM-93/12 in *Proc. Symp. Disposal of Radioactive Wastes into the Ground, Inter. At. Ener. Agency.* IAEA STI/PUB/156. 1967.
13. Helmke, P. A. *Nucl. Tech.* 1980, *51*, 182.
14. Koons, R. D.; Helmke, P. A.; Jackson, M. L. *Soil Sci. Soc. Amer. J.* 1980, *44*, 155.
15. van Genuchten, M. T.; Davidson, J. M.; Wierenga, P. J. *Soil Sci. Soc. Amer., Proc.* 1974, *38*, 29.
16. Rao, P.S.C.; Davidson, J. M.; Jessup, R. E.; Selim, H. M. *Soil Sci. Soc. Amer. J.* 1979, *43*, 22.
17. Cameron, D. R.; Klute, A. *Water Resour. Res.* 1977, *15*, No. 1, 183.
18. Fukui, M. *Health Phys.* 1978, *35*, 555.

RECEIVED October 20, 1983

4

Actinide and Technetium Sorption on Iron-Silicate and Dispersed Clay Colloids

J. W. SHADE, L. L. AMES, and J. E. MCGARRAH

Pacific Northwest Laboratory, Richland, WA 99352

Two different colloidal suspensions, representative of those found in waste package interaction tests, were prepared from iron metal and silica powders or sodium-bentonite at 90°C. Aliquots were spiked with ^{233}U, ^{235}Np, ^{237}Pu, or ^{95m}Tc at pH ranges from 2 to 12, then shaken for 24 hours followed by a 15Å filtration. Zeta potential measurements were made on unspiked samples. Similar sorptive properties were observed for both colloids. At 25°C both ^{233}U and ^{237}Pu exhibit maximum sorption (50-90%) near pH 6. Sorption drops by about a factor of 5 at pH >8. Slight sorption of ^{235}Np occurs at pH 11 and decreases to zero at lower pH values. ^{95m}Tc does not sorb on Fe-silicates and is only slightly sorbed (10%) on smectites.

As part of an effort to evaluate the effects of waste package components (canister, waste form, backfill) on waste form leaching behavior, interactive tests have been conducted that include monolithic specimens of iron in the same container with monolithic glass specimens. These experiments indicate that the leach rates of elements into groundwater from the glass are enhanced relative to rates observed in comparable glass-only tests (1). The enhanced leach rates are suggested to be the result of decreased soluble silica activity caused by reactions between iron and silica to form hydrated iron silicate reaction products. Similar experiments (2-3) conducted with sodium-bentonite backfill materials in place of iron also yield enhanced leach rates from glass. This might be attributable to the alkali sorption properties of the smectite component of bentonite or, because glass leach rates appear to be related to dissolved silica concentrations (4), possible silica-bentonite polymerization or condensation reactions similar to that suggested for silicic

0097-6156/84/0246-0067$06.00/0

acid (5) may be an influence. Although the interaction mechanisms are not understood, ICP analysis of sequentially filtered leachates (0.45 μm to 15Å) and micro-electrophoretic examination during zeta potential measurements have indicated the existance of silica-rich particles of colloidal size in both iron and bentonite systems (1), but no additional characterization was attempted.

If colloids generated by waste package component interactions readily flocculate or are otherwise removed from solution soon after formation, they may not represent a waste management problem because colloidal transport of radionuclides would be limited. In the previous interaction experiments (1-3), no evidence for flocculation or precipitation was reported over the pH range 6 to 9.5, which implies that the colloids remain as sols and could potentially be transported. If transport is possible, then it is desirable to know the extent to which sorbed radionuclides could also be transported.

Such a mechanism would require modifications to some concepts of solubility controlled release. This report concerns a preliminary effort to determine some of the sorption properties of colloidal species representative of those formed during waste form/waste package interaction tests. Sorption of actinides and Tc on those colloids as a function of pH at 25°C was studied.

The glass-iron and glass-bentonite interaction experiments (1-2) that resulted in colloid formation were conducted at 90°C in deionized water and in low ionic strength groundwater with final pH values generally in the range 8 to 9.5. On the basis of ICP analyses of filtrates (1), it is thought that colloids in the glass-iron system are silica-rich, iron-bearing materials (possibly similar to those described in Reference 6) rather than an iron hydroxide such as geothite. In the absence of further characterization, these colloids are simply considered as iron silicates. Thus, colloids generated with powdered iron and silica under conditions similar to the experiments in Reference 1, but at a slightly higher pH, are thought to be representative of iron silicate colloids from waste glass interaction systems but exclude possible complications from leachable waste glass elements. Moreover, colloids from the simple iron-silica systems provide a basis for comparison with glass-bearing systems in future work.

Experimental Procedure

Iron silicate colloids were prepared by placing fifty grams of iron powder and silicic acid or Na silicate in a liter of 0.01 M NaOH solution, then heating the solution at 90°C for four to five days. After this digestion period, the solution was centrifuged at 3500 rpm for one hour and the supernate decanted. Aliquots of the supernate were examined for the presence of colloids using a micro-electrophoresis unit, and the amount of colloids in suspension was estimated by weight loss after evaporation. This same procedure was used with Na-bentonite to generate colloidal size particles except that deionized water was used instead of a hydroxide solution. The Na bentonite was the same material used in Reference 7.

Separate 20 ml aliquots of supernate were then adjusted to a desired pH within the range 2 to 12 with HCl or NaOH and spiked with ^{237}Pu, ^{233}U, ^{235}Np, ^{95m}Tc, ^{137}Cs, or ^{85}Sr. Only one isotope was used for a given aliquot, and the amount of spike was on the order of 10^{-12} or 10^{-13} M for the actinides and 10^{-7} M for uranium--well below expected solubility limits. The spiked samples were shaken for 24 hours at 25°C, then filtered through 15Å filters. Counting the initial solution and the 15Å filtrate allowed a determination of the percentage of nuclide sorbed on the colloid. Corrections were made for sorption on container walls.

In addition, the surface charge (or zeta potential) of suspended colloids was measured on separate pH-adjusted aliquots using a laser micro-electrophoretic instrument. These measurements were made between the pH range of 2 to 12.

Results and Discussion

The percentages of radionuclides sorbed by iron silicate and sodium bentonite are listed in Tables I and II along with initial and final pH values. The percentages of ^{233}U and ^{237}Pu sorbed on iron silicate and bentonite colloids are shown graphically in Figure 1. It is apparent that similar sorption behavior occurs for both colloids with these isotopes. Both exhibit maximum sorption near pH 6 with less sorption at high pH and, to some extent, also at low pH. ^{237}Pu seems to sorb more than ^{233}U, while ^{233}U exhibits a somewhat greater pH dependence--at least under conditions of relatively dilute solutions.

In an attempt to offer a partial explanation for these observations, the zeta potential of the colloids measured as a function of pH is shown in comparison with possible speciation of U and Pu (Figure 2). The sorption experiments were conducted under oxidizing conditions, so only the +200 to +400 mV range of Eh is considered. The speciation of U and Pu shown was taken from References 8 and 9, respectively. It is recognized that dominance fields for specific species vary as functions of parameters such as ionic strength, total dissolved carbonate, etc., and also that actinides themselves may form polymers (10-11). For purposes of this discussion, however, the intent is to illustrate that dissolved species of U and Pu tend to be negatively charged at high pH values and positively charged at low pH. When viewed in these terms, simple electrostatic considerations suggest that low sorption would be expected at high pH because both colloid and dominant actinide species are negatively charged. If, at intermediate pH, the dominant actinide species change from negative to positive while the colloid zeta potential is still negative (i.e., before the isoelectric point), then maximum sorption would be

TABLE I. Sorption of Radionuclides by Iron Silicate at 25°C.

Solids-to-solution ratio was 108 mg/ℓ. Contact time was 24 hours. All samples were 15A filtered.

Radionuclides	Initial Concentration moles/	24-hour Sorption,%	Initial pH	Final pH
^{233}U	4.591×10^{-7}	0	11.10	11.20
		0	10.00	10.10
		2.92	8.00	8.05
		35.12	5.99	6.09
		42.72	5.00	5.30
		15.10	4.00	4.00
		6.96 0.0	2.99 2.00	3.00 2.0
^{235}Np	5.135×10^{-12}	5.82	11.20	11.25
		0	10.00	10.02
		0	8.00	8.01
		0	6.00	6.10
^{237}Pu	1.230×10^{-13}	44.37	11.10	11.25
		15.12	10.00	10.10
		23.27	8.00	8.00
		55.40	6.01	6.20
		61.19	5.00	5.40
		57.73	4.05	4.10
		56.75 50.15	2.98 2.00	2.90 2.60
^{95m}Tc	4.187×10^{-14}	0	11.03	11.15
		0	10.00	10.00
		0	7.99	7.99
		0	6.00	6.15
^{137}Cs	2.593×10^{-11}	19.77	10.00	9.92
		10.74	8.00	8.18
		11.09	6.00	7.45
		8.19	4.00	3.97
		0	2.00	1.99
^{85}Sr	2.370×10^{-13}	64.52	10.00	10.28
		48.08	8.00	8.24
		44.72	6.00	7.09
		20.74	4.00	3.82
		7.68	2.00	1.90

TABLE II. Sorption of Radionuclides by Na Bentonite at 25°C

Solids-to-solution ratio was 68.2 mg/ℓ. Contact time was 24 hours. All samples were 15Å filtered.

Radionuclides	Initial Concentration moles/ℓ	24-hour Sorption,%	Initial pH	Final pH
^{233}U	4.591×10^{-7}	21.33	11.20	11.00
		14.95	10.00	9.90
		41.18	8.00	8.00
		71.55	6.00	6.90
		86.94	5.02	6.26
		90.21	4.00	5.66
		76.51	4.00	4.05
^{235}Np	5.135×10^{-12}	14.23	11.20	11.15
		4.02	10.00	9.80
		6.94	8.05	8.05
		8.58	6.05	7.50
		3.65	5.00	6.04
		7.43	3.95	5.37
^{237}Pu	1.230×10^{-13}	56.79	11.20	11.00
		58.06	10.00	9.80
		82.71	8.00	8.10
		80.62	6.00	6.85
		75.30	5.00	6.06
		77.94	4.02	5.65
^{95m}Tc	4.187×10^{-14}	0.89	11.20	10.10
		1.50	10.00	9.85
		3.89	8.00	8.00
		13.40	6.00	7.00
		9.51	5.00	6.85
		9.41	4.02	5.68
^{137}Cs	2.593×10^{-11}	77.05	10.00	9.90
		73.57	8.05	8.09
		68.71	6.00	6.69
		62.69	4.00	5.53
		18.57	2.00	2.09
^{85}Sr	2.370×10^{-13}	98.04	10.00	9.83
		96.35	8.00	7.99
		95.02	6.00	6.48
		94.43	4.00	4.42
		35.48	2.00	1.94

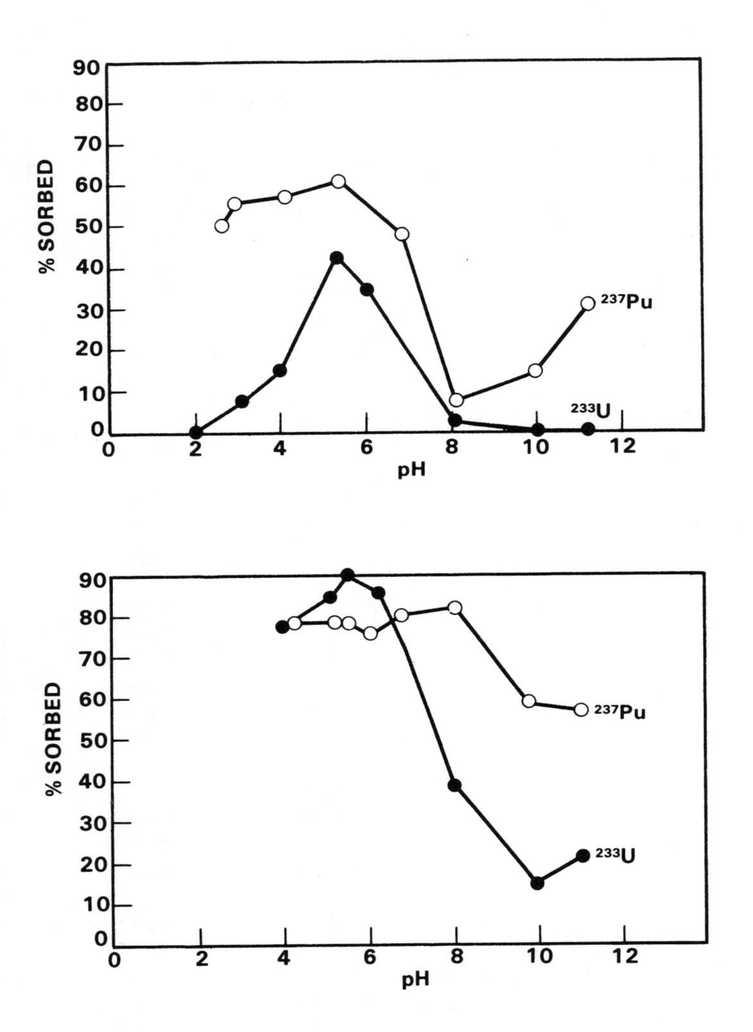

Figure 1. Sorption of ^{237}Pu and ^{233}U by Iron Silicate (top) and Na Bentonite (bottom) Colloids at 25°C.

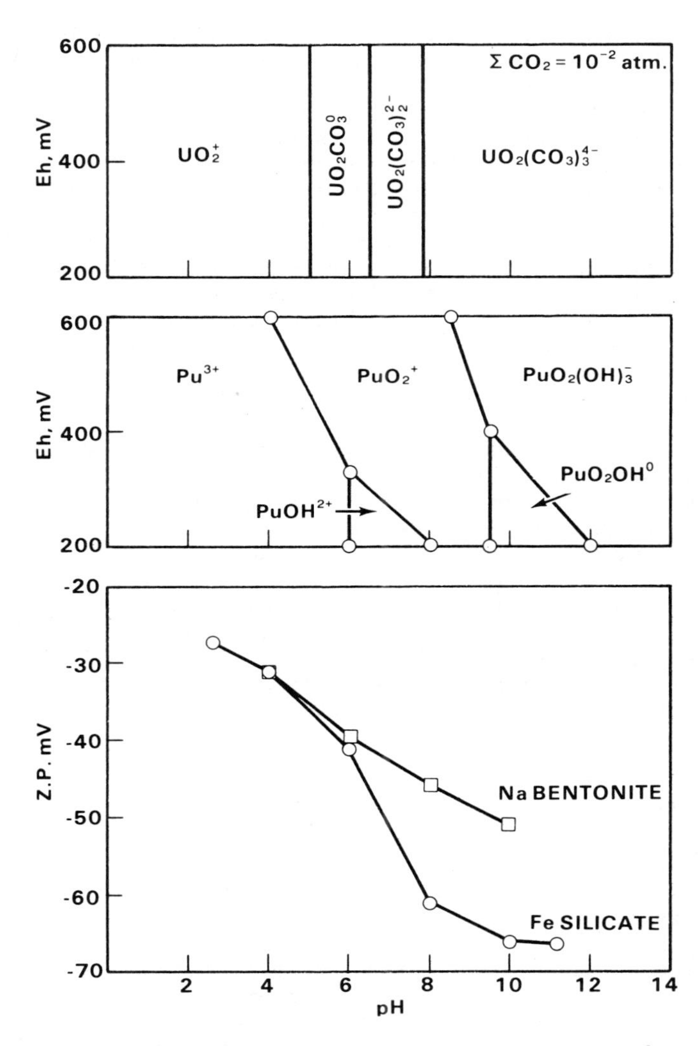

Figure 2. Comparison of Colloid Zeta Potential (bottom) with U and Pu Speciation in Oxidizing Conditions at 25°C.

expected. At even lower pH values, both colloid and actinide are likely to be positively charged so that sorption is reduced. A more detailed characterization of colloid-actinide systems would permit a quantitative interpretation, but this was not within the scope of this effort.

The same type of experiments were conducted using ^{95m}Tc and ^{235}Np spikes, and these results are shown in Figure 3. Very little sorption was observed on either colloid. The larger amounts of ^{235}Np sorption at high pH for both colloids suggest the presence of a positively charged species, but it apparently is not present in large amounts. The dominance of the pertechnetate ion at oxidizing conditions readily accounts for the low sorption on negatively charged colloids. The slight amount of ^{95m}Tc and ^{235}Np sorption on bentonite might be attributed to localized sorption at sites of ferrous iron in bentonite or to reduction by small amounts of organics. The iron associated with iron silicate colloids is likely ferric because they were formed under oxidizing conditions.

The results of additional experiments conducted with ^{85}Sr and ^{137}Cs spikes are shown in Figure 4. The well known sorption characteristics of bentonite for Sr and Cs ions is apparent (7). The sorption properties of bentonite are reduced at low pH, which is consistent with an electrostatic concept. ^{137}Cs sorption on the iron silicate colloids is considerably less than that observed with bentonite, even though the colloid zeta potentials are similar, which suggests that mechanisms other than simple electrostatic concepts may be involved. Also, the linear trend of data for Sr in iron silicate systems is considered to represent precipitation rather than sorption.

Conclusions

This preliminary work suggests that sorption properties of iron silicate colloidal size particles generated in waste package interaction experiments are similar to those of bentonite for U and Pu as well as for Tc and Np. Although no specific effort was made in this investigation to determine colloid stability, flocculation was not observed in the short 24-hour experiments within the pH range 2 to 12 for low ionic strength solutions. A significant amount of sorption was observed which justifies a more detailed characterization effort to evaluate colloid growth rates, composition and flocculation characteristics over a range of repository-relevant temperatures and radiation fields. This type of effort will be required before a full evaluation of the significance of colloids in waste package systems can be made.

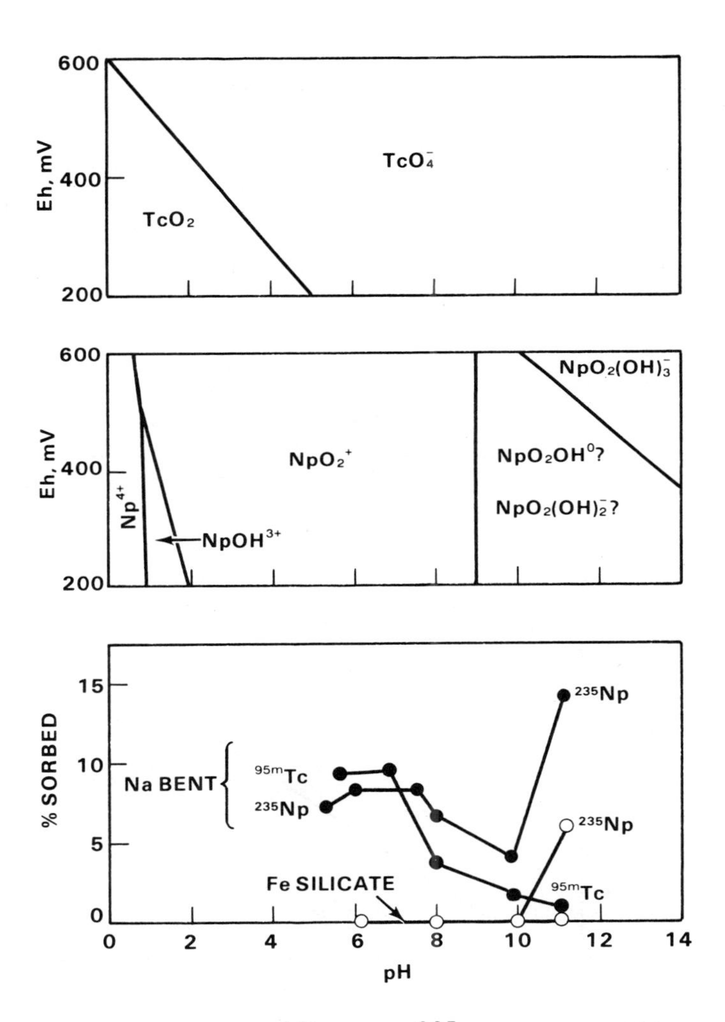

Figure 3. Sorption of ^{95m}Tc and ^{235}Np on Iron Silicate and Na-Bentonite Colloids at 25°C and Comparison with Possible Tc and Np Speciation.

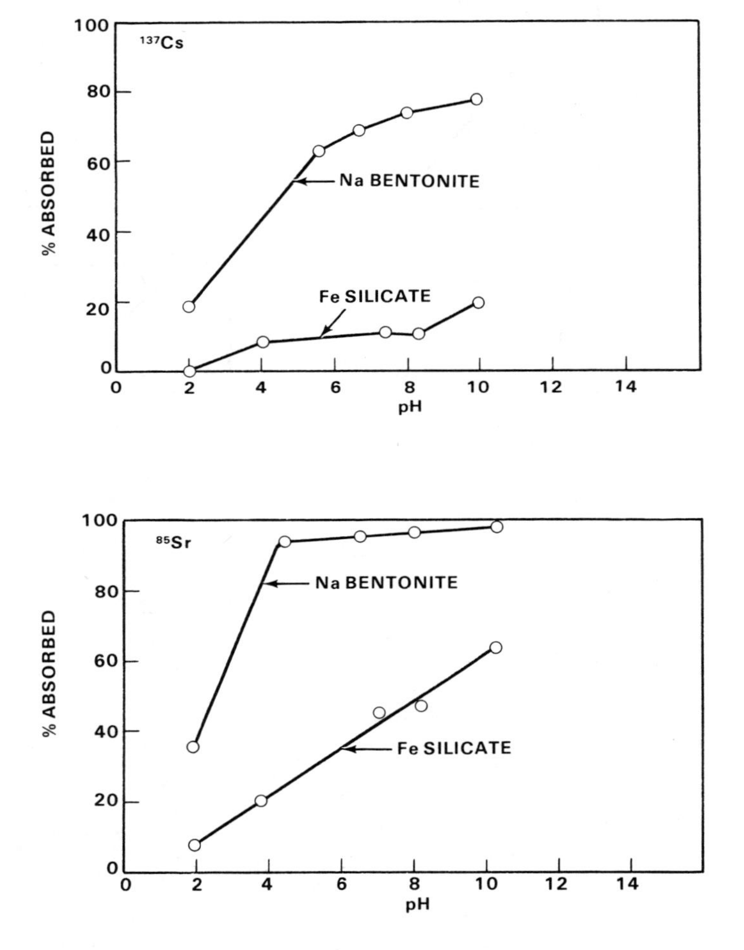

Figure 4. Sorption of ^{137}Cs and ^{85}Sr by Colloidal Iron Silicates and Na Bentonite at 25°C.

Literature Cited

1. McVay, G. L.; Buckwalter, C. Q. J. Am. Ceramic Soc. 1983, 66:3, 170-4.
2. Shade, J. W.; Chick, L. A. Abs. for American Ceramic Society Meeting: Cincinnati, Ohio, 1982; PNL-SA-10012, Pacific Northwest Laboratory, Richland, Washington
3. Lanza, F.; Ronsecco, C. in "Scientific Basis for Nuclear Waste Management"; Lutze, W., Ed.; Elsevier Publishing Company: New York, 1982; Vol. 5, pp. 125-133.
4. Pederson, L. R.; Buckwalter, C. Q.; McVay, G. L.; Riddle, B. L. in "Scientific Basis for Nuclear Waste Management"; Brookins, D. G., Ed.; Elsevier Publishing Company: New York, 1983; Vol. VI, pp. 47-54.
5. Stober, W. Adv. in Chem. Series 1967, 67, 161-182.
6. Yokoyama, T.; Nakazato, T.; Tarutani, T. Bull. Chem. Soc. Japan 1980, 53, 850-8.
7. Hodges, F. N.; Westsik, J. H.; Bray, L. A. in "Scientific Basis for Nuclear Waste Management"; Lutze, W., Ed.; Elsevier Publishing Company: New York, 1982; Vol. 5, pp. 641-8.
8. Langmuir, D. Geochim et Cosmechim Acta. 1978, 42, 547.
9. Rai, D.; Serne, R. J.; Swanson, J. L. J. Environ. Quality 1980, 9, 417-20.
10. Rai, D.; Swanson, J. L. Nucl. Tech. 1981, 54, 107.
11. Olofsson, U.; Allard, B.; Anderson, K.; Torstenfelt, B. in "The Scientific Basis for Nuclear Waste Management"; Topp, S. V., Ed.; Elsevier Scientific Publishing Company: New York, 1982; Vol. 6, p. 191.

RECEIVED December 6, 1983

5

Adsorption of Nuclides on Hydrous Oxides

Sorption Isotherms on Natural Materials

R. E. MEYER, D. A. PALMER, W. D. ARNOLD, and F. I. CASE

Oak Ridge National Laboratory, Oak Ridge, TN 37830

Hydrous oxides and minerals which have adsorbing groups that behave like hydrous oxides are ubiquitous components of geological formations and may dominate the adsorptive properties of the formations at conditions of natural groundwaters. An understanding of the adsorptive behavior of hydrous oxides is therefore necessary for reliable prediction of migration of nuclides through the formations. Various isotherms are derived from equilibrium ion-exchange theory for the sorption of non-hydrolyzed ions on hydrous oxides. These isotherms are compared with experimental isotherms for sorption of Cs^{+}, Sr^{2+}, Eu^{3+}, and TcO_4^{-} on several hydrous oxides. The experimental isotherms for cations show little dependency of sorption on the ionic strength at intermediate pH values but considerable dependence at higher pH values. In the pH range of negligible hydrolysis, sorption increases with pH for cation sorption, and the slope of this dependency on pH increases with the charge of the cation. General features of these isotherms are predictable from ion-exchange equilibrium theory as applied to hydrous oxides. By combining isotherms for hydrous oxides with those for layer-type clay minerals, many unusual features of isotherms found on geological materials can be explained.

Safety analysis of nuclear waste repositories requires realistic prediction of the rates of migration of nuclides from the repository through the host geological medium to the accessible environment. These predictions require sorption isotherms rather than single values of distribution coefficients, and there must be substantial confidence that the sorption isotherms

0097-6156/84/0246-0079$06.00/0

are realistic. Theoretical justification of the isotherms can provide this confidence, particularly if the justification is obtained from long-established principles such as chemical equilibrium theory.

Hydrous oxides and minerals with adsorbing groups which behave like hydrous oxides are ubiquitous components of geological formations. These formations may also contain other highly adsorbing materials, such as the clay minerals, and, in practice, samples of the formations will probably be mixtures of oxides, clays, and other materials. One approach to modelling these formations is to consider them to be mixtures of the individual minerals of the formations and to calculate sorption isotherms from sorption isotherms of the components. If chemical equilibrium theory, for example as applied to ion exchange, could be used to predict the form of these isotherms, the task of prediction of the overall isotherm of the formations would be considerably simplified. In this paper, we show that many of the features of sorption isotherms found with natural materials can be predicted with relatively simple relations derived from ordinary equilibrium ion exchange equations. Further, in the pH range 6-9 of most natural groundwaters, the sorption properties of many natural materials are dominated by the sorption properties of hydrous oxides. Here, we will refer primarily to three types of isotherms: plots of equilibrium distribution coefficient, D, vs concentration of the nuclide on the adsorbent; D vs ionic strength; and D vs pH. The first of these types may also be plotted as an essentially equivalent graph of concentration of the nuclide on the adsorbent vs concentration in the solution.

In deriving the shapes of these isotherms, we first must define isotherms for ideal clays and ideal oxides. We then combine these ideal functions into overall isotherms and compare the derived functions to experimental isotherms determined for various adsorbents. We will show that observations which have been said to preclude ion exchange are, in fact, quite consistent with ion-exchange behavior. We will not attempt to derive actual values of equilibrium distribution coefficients, but rather we seek only to define the shapes of the isotherms.

Isotherms for Ideal Clay

We will define an ideal clay as one with a fixed negative charge in the lattice and in which the charge is completely independent of pH. Further, we will stipulate that this charge is the only source of the sorption capacity of the exchanger. This implies that adsorption is entirely independent of pH except for the inclusion of the hydrogen ion as one of the exchangeable cations. Using this assumption, we can calculate isotherms from normal ion exchange equilibrium equations. For exchange of a cation, A^{n+}, with a univalent cation, B^{+}, we can write the following equations:

$$A_s^{n+} + nB_{ads}^+ = A_{ads}^{n+} + nB_s^+ \quad (1)$$

$$K = [A_{ads}][B_s]^n / \{[A_s](C-n[A_{ads}])^n\} \quad (2)$$

$$D = [A_{ads}]/[A_s] = K\,(C-n[A_{ads}])^n/[B_s]^n \quad (3)$$

where K is the equilibrium constant for the ion exchange reaction, $[A_{ads}]$ and $[A_s]$ are the concentrations of the nuclide on the clay and in the solution, $[B_{ads}]$ and $[B_s]$ are the concentrations of the exchanging cations, n is the charge of the nuclide, C is the capacity of the clay, and D is the equilibrium distribution coefficient. Isotherms derived from these equations are quite familiar. Values of D are constant as the concentration of the cation is increased until the amount adsorbed approaches the capacity. Equivalently, the slope of a plot of $\log[A_{ads}]$ vs $\log[A_s]$ is one at low capacity and approaches zero as the capacity of the exchanger is approached. These equations show the sensitivity of this type of sorption to competing ions, and this sensitivity is often used as a test for the presence of ion exchange. If the solution contains ions of only one charge, m, e.g., all monovalent or all divalent, and if the nuclide, A^{n+}, is present at trace concentrations, then plots of log D vs $\log[B_s{}^{m+}]$ are equal to $-n/m$. This relation is illustrated in Figure 1 for exchange with monovalent cations, calculated for a hypothetical clay, from Equation 3. Unfortunately, it is sometimes assumed that the converse is true, i.e. that if these relations are not found, ion exchange is not the sorption process involved. [In Figure 1 corrections are applied for the known activity coefficients in mixtures of the chlorides of sodium and the adsorbing nuclides, Cs(I), Sr(II), and Eu(III). This correction produces some curvature on these plots.]

Isotherms for Ideal Hydrous Oxides

Sorption on oxides and hydrous oxides has been extensively studied both experimentally and theoretically. In order to define an ideal oxide or hydrous oxide exchanger, we will rely on experiments with well-defined sorbents such as chromatographic alumina (1-4). Briefly, the adsorption characteristics of oxides and hydrous oxides are: (1) At high pH, they act as cation exchangers but with less sensitivity to the total salt concentration than clay minerals. This behavior is illustrated in Figure 2, where sorption of strontium on alumina is shown. At high pH, the sensitivity of the distribution coefficient to the concentration of the competing cation is somewhat less than

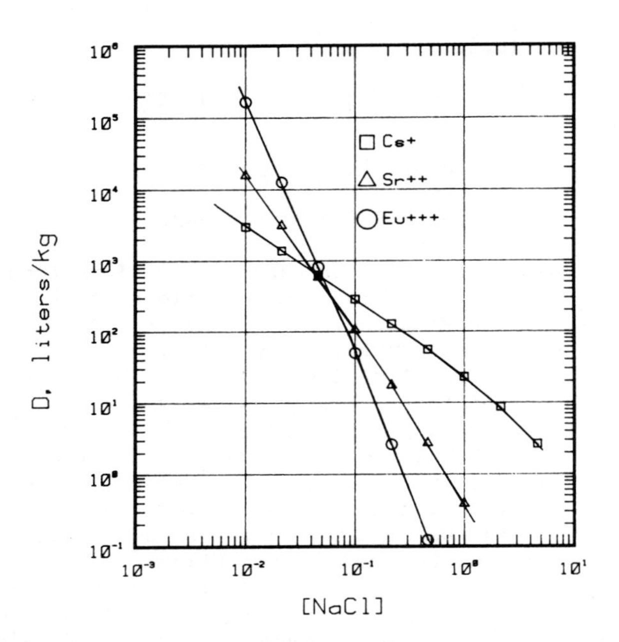

Figure 1. Theoretical plots of distribution coefficient vs concentration of NaCl for sorption of Cs(I), Sr(II), and Eu(III) on a hypothetical clay with a capacity of one equivalent per kilogram.

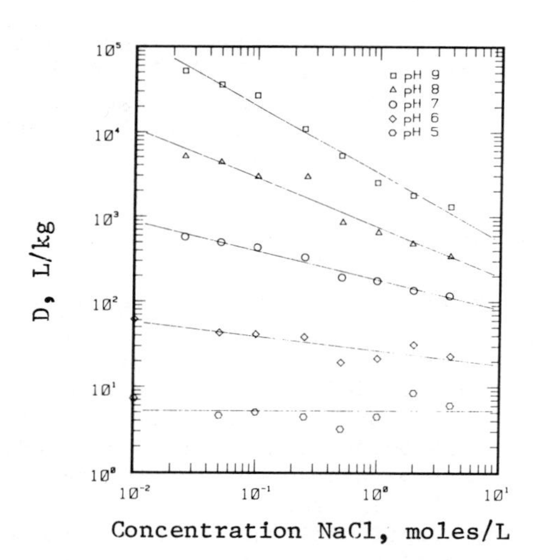

Figure 2. Adsorption of Sr(II) on the sodium form of alumina. (Trace Sr(II), equilibration for over 62hr.) Slopes: pH 9, -0.80; pH 8, -0.58; pH 7, -0.34; pH 6, -0.17; pH 5, +0.02.

half of what one would expect for an ideal clay. (2) At low pH, they become anion exchangers. This characteristic is illustrated by the plots of TcO_4^- sorption shown in Figure 3, where it is shown that distribution coefficients increase at lower pH values (3). (3) For cation exchange, the sensitivity to salt concentration decreases as the pH is lowered (Figure 2). For anion exchange, the opposite is true (Figure 3). (4) As shown in Figure 4, the slopes of log D vs pH increase with the charge of the adsorbing nuclide. In Reference 2, we have shown from ion exchange theory that this slope should, under certain ideal conditions, be equal to n/2 where n is the charge of the adsorbing nuclide. This is approximately true for the data shown in Figure 4. Slopes of log D vs pH are from 0.4 to 0.5 for Cs(I), 0.8 to 1.0 for Sr(II), and 1.5 to 1.65 for Eu(III) (1). (5) The capacities tend to increase with the charge of the ion as shown in Figure 5, where plots of the distribution coefficient vs loading (concentration of the nuclide on the sorbent) are shown for adsorption of Cs(I), Sr(II), and Eu(III) on alumina. In these experiments, pure chromatographic alumina was used. The alumina was equilibrated at least three times to convert all exchangeable cations to the sodium ion and sorption measurements were conducted at trace loading for all experiments except when an isotherm relating equilibrium distribution coefficient to concentration of the nuclide in the solution was desired. If pure adsorbents are not used and if the adsorbent is not properly pretreated, then it is quite difficult to obtain meaningful comparisons for various experiments.

The results of all of these measurements can be approximated by an empirical expression that explains most of the features of our experiments on oxides. In this paper, we will not attempt to give theoretical justification, but we and others (2,4) have shown that most of the five characteristics given follow directly from equilibrium ion exchange considerations if certain assumptions are made relative to the origin of the lattice charge on the oxide. Other approaches have been used to explain these observations on hydrous oxides. Among these are approaches which associate enhanced sorption with hydrolysis of the nuclide (5-7) and with formation of surface complexes to specific sites (8-10). Some of these approaches are quite elaborate making extensive use of computer calculations and include double layer theory. The approach that we have used (2) is relatively simple, and explains many of the characteristics of sorption on hydrous oxide with equilibrium theory.

For the data that we have collected so far for sorption on oxides, Equation 4 is a good empirical representation of the results:

$$[A_{ads}] = [A_s]K_{ox}(C-n[A_{ads}])[H^+]^{(-n/2)}[B_s]^p \qquad (4)$$

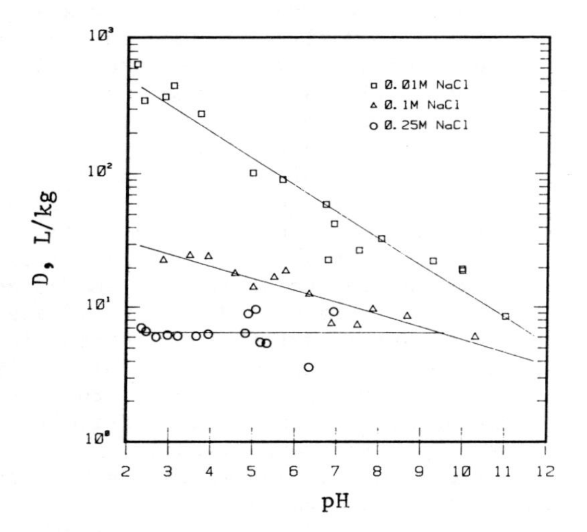

Figure 3. Plot of adsorption vs pH for TcO_4^- on ZrO_2 in NaCl solutions.

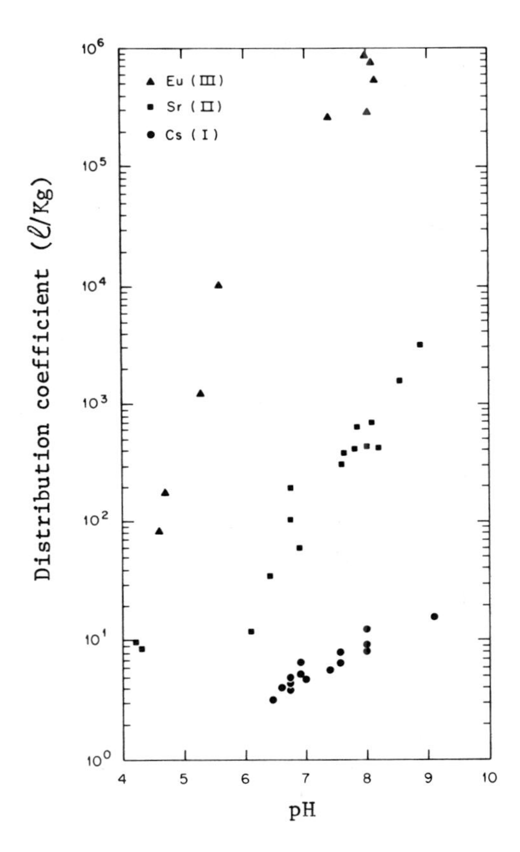

Figure 4. Adsorption of Cs(I), Sr(II), and Eu(III) on the sodium form of gamma-alumina. (Trace loading, 60-240 hr equilibration, 0.5 M NaCl).

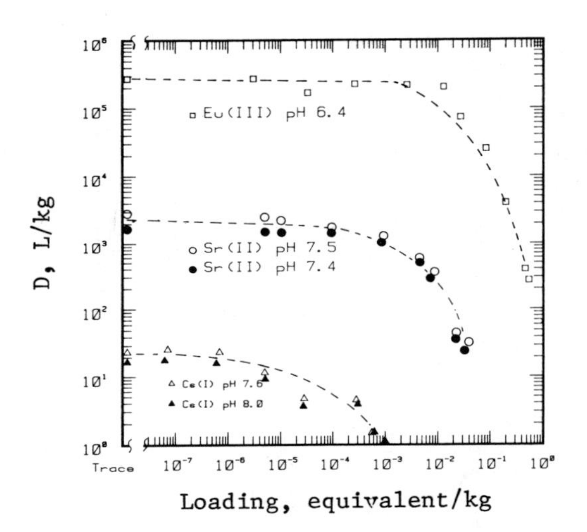

Figure 5. Effect of loading on distribution coefficients of Cs(I), Sr(II), and Eu(III) on the sodium form of chromatographic alumina, (0.05 M NaCl, equilibration for more than 165 hr).

where K_{ox} is a constant, $[H^+]$ is the hydrogen concentration, C is the capacity of the oxide for the particular nuclide that is adsorbed, and p is a factor that expresses the dependence of sorption of the nuclide on the salt concentration. For alkaline conditions, the parameter, p, seems to be about -0.4 to -0.6 (2). At low pH, it appears to be very small or in some cases even negative. Usually at low pH, equilibrium distribution coefficients are small, and it is difficult to obtain data accurate enough for meaningful comparisons. This parameter is therefore dependent on pH. The capacity, C, is given as being independent of pH; however, we have not investigated this point enough to say for sure. The concentration adsorbed, $[A_{ads}]$, will normally depend strongly on the pH as expressed by Equation 4. It is also possible that the expression $(C-n[A_{ads}])$ will have an exponent as it does in Equation 2 for simple ion exchange, but our data are not accurate enough to determine whether this is true. For our purposes here, this equation is sufficient, and we will consider C and K_{ox} as empirical constants.

Combined Isotherms

There are several ways to form combined isotherms. In this paper we will simply assume that we can combine them in a linear fashion such that each isotherm is completely independent of the other. The independence of the isotherms on the different adsorbents implies the following equations:

$$[A_{ads}](total) = f_1[A_{ads}](1) + f_2[A_{ads}](2) \qquad (5)$$

or

$$D(total) = f_1D(1) + f_2D(2) \qquad (6)$$

where f_1 and f_2 are weight fractions of the various adsorbers, $[A_{ads}](1)$ and $[A_{ads}](2)$ are the concentrations on the individual minerals. This equation can readily be extended to any number of adsorbers by adding terms of the form of the first two terms. In calculating the overall isotherms, we first calculate $[A_{ads}]$ for the various components of a mixture and then combine them with Equation 5 or 6.

Equation 6 was used to predict experimental measurements of binary mixtures of adsorbents for sorption of Cs(I) and Sr(II) from measurements on the individual minerals. (11) For some of the cases investigated, the distribution coefficients could be calculated with considerable precision from measurements on the individual minerals. In other cases, there appeared to be interaction between the minerals. However, these experimental results appear to confirm Equations 5 and 6 for cases where the sorbents do not interact.

In Figure 6, we show a theoretical calculation using Equations 3-6 of the effect of pH on the distribution coefficient for a case where the principal adsorbent is a hypothetical clay which contains a total of 0.8% oxides. In this case, we chose a mixture of three oxides as follows: oxide #1, 0.1%, $K_{ox} = .01$, $C = .5$; oxide #2, 0.2%, $K_{ox} = .005$, $C = .3$, oxide #3, 0.5%, $K_{ox} = 0.01$, $C = 1$. These constants were chosen arbitrarily to approximate roughly the data that we observe in Figure 7. These constants are not unique; other sets of constants could also represent the data. The calculations shown are for sorption of a trivalent ion from a solution of a monovalent salt like NaCl. Unless otherwise stated in these Figures, the concentration of the salt is 0.01 M, the capacity of the clay is assumed to be one equivalent per kilogram, $p = 0$ and K (for the clay) is assumed to be 1.0. These numbers give calculated values of D which are similar to those that we observe in our experiments. In our calculations, D is expressed in L/kg, C in equivalents/kg, concentrations in solution as moles/L, and concentrations on the solid (loading) as moles/kg, and units of K or K_{ox} which are consistent with these units.

The calculations show a strong dependence on concentration of the monovalent ion at low pH but a very small dependence at the upper pH values shown in Figure 6. The reason for this is that, because of the strong pH dependence of sorption on oxides, at low pH the clay-like behavior dominates and at the upper pH region, the oxide-like behavior dominates.

In Figure 7, we show experimental data for adsorption of Eu(III) on montmorillonite. As a number of investigators have suggested (12-14), layer-type clay minerals generally contain both a fixed-layer charge which corresponds to the constant capacity, and groups at the edges of the layers which behave like oxide groups. A natural clay can therefore be thought of as a mixture of a clay and an oxide. It is also quite possible that natural clay minerals contain small quantities of oxides. The curves shown for Eu(III) sorption on montmorillonite show the same features as the calculated curves of Figure 6. These distribution coefficients are equilibrium distribution coefficients, and although we show only one of the experiments, essentially the same behavior was obtained when the experiment was started at low pH, high pH, or intermediate pH and cycled through the range shown in Figure 7. (15) In addition, experiments were also done which showed that the distribution coefficients did not vary substantially with time. Because of the unusual shape of the curves which show the transition from a pH range dependent to a region independent of the competing ion, it is difficult to account for their shape. The assumption that the sample behaves like a mixture of oxide and clay adsorbing sites does provide a simple explanation of this behavior. The similarity between Figure 7 and Figure 6 does not prove that

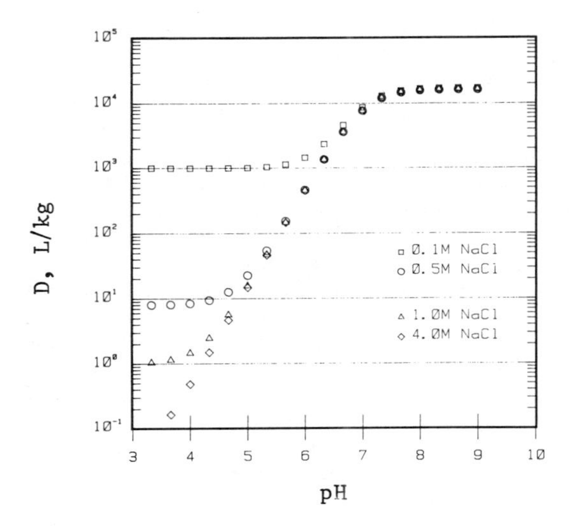

Figure 6. Theoretical distribution coefficients for trivalent ion sorption on sorbent containing 99.2% clay and 0.8% oxide. Arbitrary constants (see text).

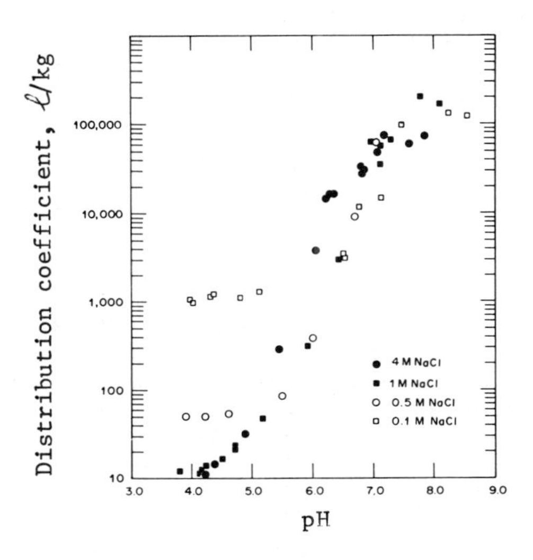

Figure 7. Adsorption of Eu(III) on the sodium form of purified Wyoming montmorillonite (trace loading, equilibration for 96 hr).

this is the correct explanation, but it does say that one cannot eliminate ion exchange as the mechanism of adsorption, because the equations are derived from ion exchange principles.

These equations also predict that plots of the concentration of the nuclide on the adsorbent vs concentration of the nuclide in the solution will not be linear in many cases. The slope of such plots will depend on the particular mixture present, the pH, and the range of the solution concentration of the nuclide. To illustrate this point, we show in Figure 8 several plots of calculations for various assumptions. The same constants given above were used to calculate these curves. When plotted as log D vs log $[A_{ads}]$, there is a clear indication at pH 7 of two regions, at low values of $[A_{ads}]$ up to about 10^{-4} moles/kg and the upper region above 10^{-4} moles/kg. The lower region corresponds to domination of sorption by the oxide component and the upper by the clay component. These calculations were replotted as log $[A_{ads}]$ vs log $[A_s]$ and the results are shown in Figure 9. The large hump observed in Figure 8 is observed to be a line above the pH 4 line (slope = 1). Normally, when such an experiment is carried out, it may not be possible to cover as wide a range of concentration of the nuclide shown in this hypothetical case. Solubility or other constraints may limit the accessible range. In the case of trivalent ions like Eu(III), it is not possible to increase the concentration of the ion much above tracer levels at higher pH because of the possibility of precipitation and the formation of hydrolyzed and polynuclear species which would change the nature of the sorption process. We have so far not been able to determine a reproducible sorption isotherm for Eu(III) on our samples of montmorillonite above about pH 5-6.

Although it is not possible to increase the concentration of Eu(III) enough at higher pH levels to determine experimentally an isotherm over a large concentration range, it is possible to do so with divalent ions like Ni^{2+} and Co^{2+}. Egozy (12) has reported curves similar to Figure 5 for Co^{2+} sorption on montmorillonite and Triolo and Meyer (16) have observed similar curves for sorption of Ni^{2+} and Zn^{2+} on montmorillonite.

Finally, we show in Figure 10 some recent data on adsorption of Eu(III), at trace concentrations, on a natural specimen of corundum. Here, we see a pattern somewhat similar to that observed for montmorillonite. At low pH levels, there is a strong dependence on salt concentration, and at pH 6-7, there is a rapid increase in D. As with our results with montmorillonite, this sorbent may be modelled as a mixture of clay-like and oxide-like sorbents. We do not know why the two points at low pH in 2.5 M NaCl are the highest. There are only two such points in Figure 9; perhaps these are anomalous. The results at high pH probably represent a mixture of precipitation and sorption.

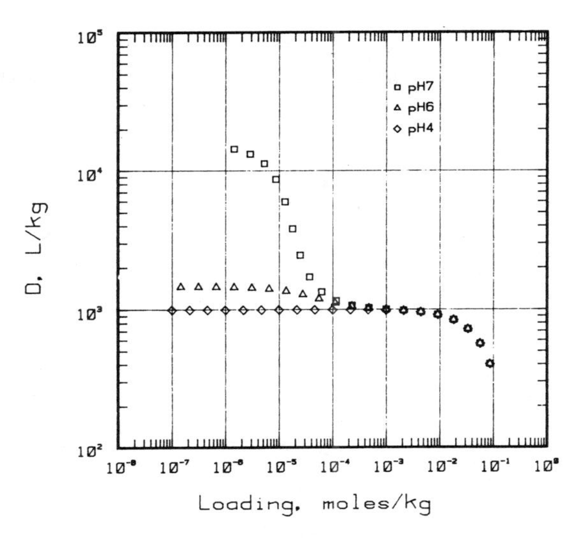

Figure 8. Theoretical distribution coefficients for trivalent ion sorption on sorbent containing 99.2% clay and 0.8% oxide. Arbitrary constants (see text).

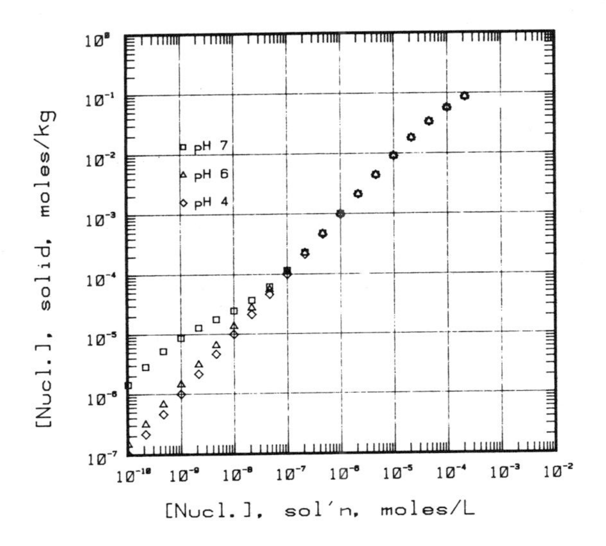

Figure 9. Theoretical isotherms for trivalent ion sorption on sorbent containing 99.2% clay and 0.8% oxides. Arbitrary constants (see text).

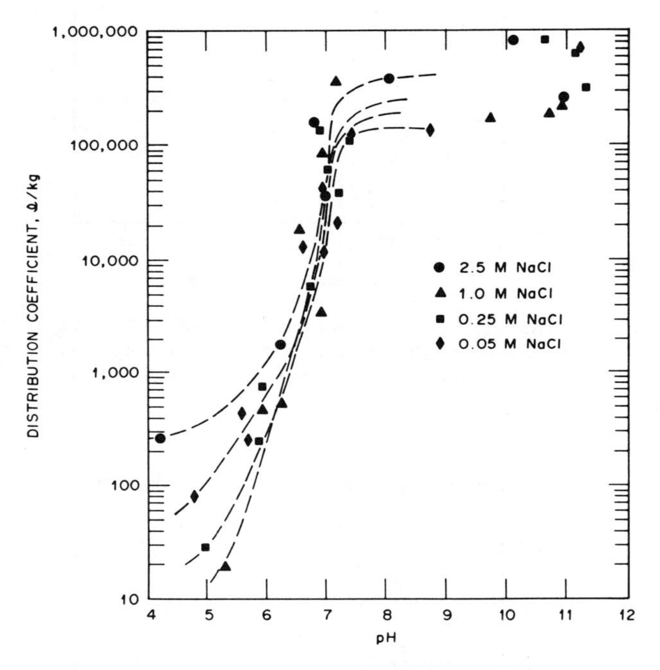

Figure 10. Adsorption of Eu(III) on the sodium form of corundum (loading: 1.96E-6 to 8.1E-5 mole Eu(III)/kg, equilibration for more than 192 hours).

Summary and Conclusions

Sorption of polyvalent ions increases very rapidly with pH on oxides and hydrous oxides. Because natural materials generally contain oxides or at least adsorbing groups that behave like oxides, sorption of polyvalent nuclides at intermediate pH will tend to follow behavior typical of oxides. Thus, a strong dependence of sorption on salt concentration may not be observed. This lack of dependence on salt concentration does not preclude ion exchange as a mechanism however, since this type of behavior follows directly from equilibrium equations.

Natural minerals and rocks can be modelled as mixtures of clay-like adsorbents and oxide-like adsorbents. Various isotherms calculated from these assumptions are similar to experimentally observed isotherms for adsorption of Eu(III) on montmorillonite and corundum. Montmorillonite is of course a clay mineral, but it does have oxide-like groups.

It is possible that these results could be generalized to natural geological formations. Rather than describing a formation as a mixture of specific minerals, it might be possible to describe the formation as a mixture of types of adsorbents. These adsorbents could be characterized by experiments at low pH where oxide groups have little influence on sorption behavior and experiments at higher pH where the pH has a significant effect on sorption.

Acknowledgments

Research sponsored by the Division of Chemical Sciences, Office of Basic Energy Sciences, U. S. Department of Energy, under contract W-7405-eng-26 with Union Carbide Corporation.

Literature Cited

1. Shiao, S. Y.; Egozy, Y; Meyer, R. E. J. Inorg. Nucl. Chem. 1981, 43, 3309.
2. Shiao, S. Y.; Meyer, R. E. ibid, 1981, 43, 3301.
3. Palmer, D. A.; Meyer, R. E. ibid, 1981, 43, 2979.
4. Kraus, K. A.; Phillips, H. O.; Carlson, T. A.; Johnson, J. S. "Ion Exchange Properties of Hydrous Oxides"; Proc. 2nd Int. Conf. Peaceful Uses Atomic Energy; Geneva, 1958; Vol. 28, p. 3,
5. James, R. O.; Healy, T. W. J. Colloid Interface Sci. 1972, 40, 42.
6. James, R. O.; Healy, T. W. ibid. 1972, 40, 55.
7. James, R. O.; Healy, T. W. ibid 1972, 40, 65.
8. Stumm, W.; Hohl, H.; Dalang, F. Croatica Chemica Acta 1976, 48, 419.

9. Yates, D. E.; Sevine, S.; Healy, T. W. Chem. Soc. Faraday Trans. 1974, 70, 1807.
10. Davies, J. A.; James, R. O.; Leckie, J. O. J. Colloid Interface Sci. 1978, 63, 480.
11. Palmer, D. A.; Shiao, S. Y.; Meyer, R. E.; Wethington, J. A. J. Inorg. Nucl. Chem. 1981, 43, 3317.
12. Egozy, Y. Clays and Clay Minerals 1980, 28, 311.
13. Maes, A. and Cremers, A., "Thermodynamics of Transition Metal Ion Exchange in Montmorillonite"; Proc. Int. Clay Conf., Mexico City; Bailey, S. W., Ed.; Applied Publishing: Wilmette, Illinois, 1975.
14. Peigneur, P.; Maes, A.; Cremers, A. Clays and Clay Minerals 1975, 23, 71.
15. Triolo, R.; Shiao, S. Y.; Meyer, R. E. "Adsorption of Eu(III) on Montmorillonite", in preparation.
16. Triolo, R.; Meyer, R. E. "Adsorption of Zn(II) and Ni(II) on Montmorillonite", in preparation.

RECEIVED October 20, 1983

6

Effects of Hanford High-Level Waste Components on the Solubility and Sorption of Cobalt, Strontium, Neptunium, Plutonium, and Americium

C. H. DELEGARD, G. S. BARNEY, and S. A. GALLAGHER

Rockwell International, Energy Systems Group, Richland, WA 99352

High-level radioactive defense waste solutions, originating from plutonium recovery and waste processing operations at the U.S. Department of Energy's Hanford Site, currently are stored in mild steel-lined concrete tanks located in thick sedimentary beds of sand and gravel. Statistically designed experiments were used to identify the effects of 12 major chemical components of Hanford waste solution on radionuclide solubility and sorption.

The chemical components with the most effect on radioelement solubility and sorption were NaOH, $NaAlO_2$, EDTA, and HEDTA. The EDTA and HEDTA increased Co, Sr, and Am solubility and decreased sorption for almost all radioelements studied. Sodium hydroxide and $NaAlO_2$ increased Pu solubility and decreased Np and Pu sorption. Sodium nitrite decreased Np solubility, while Na_2CO_3 and HEDTA increased it. These observations give evidence for the formation of radioelement complexes which are soluble and are not strongly sorbed by the sediments near the waste tanks.

High-level defense waste solutions resulting from plutonium recovery and waste processing activities currently are stored in mild steel-lined concrete tanks located underground at the U.S. Department of Energy's Hanford Site. Low radioelement solubility and extensive radioelement sorption on surrounding sediment help maintain isolation of hazardous radionuclides from the biosphere in the event of tank failure.

Chemical components in the waste solutions potentially could affect radioelement solubility and sorption reactions, and thus enhance or reduce radionuclide transport. The effects of 12 chemical components on the solubility and sorption of cobalt, strontium, neptunium, plutonium, and americium were studied to

0097-6156/84/0246-0095$06.00/0

aid in judging the feasibility of continued storage of Hanford high-level waste solutions in existing tanks, as well as in predicting the effects of possible future waste processing operations. The results of the solubility and sorption studies are presented in this report.

Experimental

Experimental Design. To identify which of the 12 high level waste (HLW) components significantly affected radioelement solubility and sorption, the Plackett-Burman design, a statistical 20-run screening test, was employed (1). The Plackett-Burman design is a two-level fractional factorial design having internal replication and is effective for screening variables (in this case, chemical components) for significance when the interactions between the variables are not important.

The effects of the significant HLW components on solubility and sorption were quantified using three-level statistical designs. Full factorial designs were run if only two variables (HLW components) were to be studied and fractional factorial Box-Behnken designs were run for three or four variables (2). The 2-variable design required 12 experiments while the 3- and 4-variable Box-Behnken designs required 15 and 27 experiments, respectively. The three-level designs yielded quadratic equations predicting solubility and sorption values in terms of the concentrations of the significant components.

Materials. Twelve chemical components of Hanford alkaline HLW solutions were studied. These were $NaNO_3$, $NaNO_2$, NaOH, $NaAlO_2$, Na_2CO_3, Na_2SO_4, Na_3PO_4, NaF, Na_3HEDTA (N-2-hydroxyethylethylenediaminetriacetic acid, trisodium salt), Na_4EDTA (ethylenediaminetetraacetic acid, tetrasodium salt), NaO_2CCH_2OH (sodium hydroxyacetate), and $Na_3C_6H_5O_7$ (trisodium citrate). Reagent chemicals and distilled and deionized water were used to prepare all test solutions.

The 12 waste components were selected for study based on their quantities in the Hanford HLW solutions and their abilities to complex, or influence the complexation of metallic radioelements. The range of chemical component concentrations studied, given in Table I, was broadly representative of concentrations found in HLW.

Three Hanford sediments were used in the sorption studies. Each of these sediments contained significant quartz, feldspar, vermiculite, mica, and montmorillonite and were typical of the Hanford sediment in which the HLW tanks are located. Properties of the sediments used are given in Table II.

Five radioelements were selected for study: cobalt, strontium, neptunium, plutonium, and americium. These radioelements were selected because of their presence in the HLW, their potentials to form complexed species, and their radiological hazards. Table III summarizes the radioelement concentrations estimated and measured in some HLW solutions and the concentrations used in these experiments.

Procedures. Each solubility experiment was conducted by adding dried tracer solids to 5.00 mL portions of test solution, and by capping and equilibrating the mixture for 2 weeks. The tracer solids were obtained by drying precise volumes of stock nitric acid solutions of the radioelement under a heat lamp. Thus, the radioelements were present initially as Co(II), Sr(II), Np(VI), Pu(IV), and Am(III). Following equilibration, the mixtures were filtered through 0.003 µm pore size ultrafilters to remove undissolved tracer solids. Spectrometry of gamma and low energy X-ray emissions was used to determine ^{60}Co, ^{85}Sr, ^{237}Np, ^{238}Pu, and ^{241}Am concentrations.

Each sorption experiment was conducted by adding 5.0 mL of the appropriate traced solution, prepared as described above, to a weighed (~1 g) portion of Hanford sediment. To simulate advancement of a radioelement plume from a failed tank through previously waste-wetted sediment, each sediment sample was pre-equilibrated twice with the relevant untraced solution prior to introduction of the traced solution. Each pre-equilibration lasted at least 2 hr. Following a 7-day equilibration with the traced solution, each sediment-solution mixture was centrifuged, the solution was filtered through an ultrafilter, and the radionuclide solution concentration was determined. Distribution coefficients and fractions of radionuclides sorbed were determined for each sorption experiment. The distribution coefficient, K_d, is the activity per gram of sediment divided by the activity per mL of solution at equilibrium.

Results

Solubility Screening Tests. Table IV summarizes the results of the solubility screening tests performed using the Plackett-Burman statistical design. Cobalt was found to be significantly more soluble in the presence of EDTA and HEDTA, and, to a lesser extent, Na_2SO_4. Cobalt probably is present as Co(II) in strongly basic solution (3). As expected, high formation constant values are known for Co(II) complexes of EDTA and HEDTA (4). These stable complexes enhanced the solubility of cobalt. The effect of Na_2SO_4 on increasing cobalt solubility was unexpected since the complex $CoSO_4{}^{o}$ has a very low formation constant (5).

Strontium, present as Sr(II), also showed increased solubility in the presence of EDTA and HEDTA due to the formation of stable soluble complex species (4).

Table I. Component Concentration Values

Component	Concentration Range Studied (M)
$NaNO_3$	0-2
$NaNO_2$	0-2
NaOH	1-4
$NaAlO_2$	0-0.5
Na_2CO_3	0-0.05
Na_2SO_4	0-0.01
Na_3PO_4	0-0.01
NaF	0-0.01
Na_3HEDTA	0-0.1
Na_4EDTA	0-0.05
Na hydroxyacetate	0-0.1
Na_3 citrate	0-0.03

Table II. Properties of Hanford Sediments Studied

Sediment	Texture (wt%)			CEC (meq/100g)	$CaCO_3$ (mg/g)
	Clay	Silt	Sand		
L	1.9	8.3	89.8	3.6	34
P	1.3	6.9	91.8	3.5	19
S	3.0	10.3	86.7	6.8	0

Table III. Radioelement Concentrations

Radionuclide	Radionuclide concentration (μCi/L)	Total element concentration (μMol/L)
Concentration in Hanford HLW[a]		
^{60}Co	153	0.002[b]
^{90}Sr	3,190 - 4,800	0.75 - 1.1[b]
^{237}Np	2.8	17
$^{238,240}Pu$	0.46 - 6.8	0.023 - 0.38[b]
^{241}Am	0.12 - 21	0.00016 - 0.025[b]
Maximum Possible Concentration in Solubility Experiments (no precipitation)		
^{60}Co	-	3.8
^{85}Sr	-	3,000
^{237}Np	-	6,600
$^{238,239,240}Pu$	-	90
^{241}Am	-	8.6
Concentration in Sorption Experiments		
^{60}Co	-	0.08
^{85}Sr	-	0.8
^{237}Np	-	30
^{238}Pu	-	0.002
^{241}Am	-	0.0006

[a]Filtered (0.45μm) solutions. No neptunium analyses available. It was assumed that all Np inventory in HLW was dissolved uniformly in waste liquor.

[b]Assume total Co/^{60}Co = 1. Ratios of total/active radioelement are 3, 1, and 1 for strontium, plutonium, and americium, respectively.

Table IV. Significant Components in Radioelement Solubility *

Radioelement	$NaNO_3$	$NaNO_2$	NaOH	$NaAlO_2$	Na_2CO_3	Na_2SO_4	Na_3PO_4	NaF	HEDTA	EDTA	Sodium hydroxy -acetate	Sodium citrate
Cobalt						+3			+2	+1		
Strontium									+2	+1		
Neptunium	-9	-1		+7	+3	+8	+5		+4	-2	+6	
Plutonium	+3		+1	+2								
Americium							+2		+1	+5	+4	+3

*Significant at >80% confidence interval. Numbers indicate rank in importance; + indicates component increased solubility, - indicates component decreased solubility.

The oxidation-reduction behaviors of neptunium, plutonium and americium in basic solution have been determined via polarographic and coulometric studies (6-9). These studies, which showed that the more soluble (V), (VI), and (VII) oxidation states of these actinides are stable in alkaline solution under certain redox conditions, helped identify possible actinide species and oxidation states in our experiments. Actual identification of radioelement oxidation states was not done in the present experiments.

As shown in Table IV, neptunium exhibited a complicated solubility behavior in HLW. Nine of the 12 waste components studied were significant in affecting neptunium solubility. By far, $NaNO_2$ had the greatest influence and significantly decreased the solubility of neptunium. According to the neptunium redox potentials determined in alkaline solution (6-7) and the potential established by the NO_2^--NO_3^- couple (3), nitrite apparently reduced Np(VI) to Np(V). The (V) state of neptunium has been shown to be less soluble than the (VI) state in NaOH solution (6). The probable existence of both Np(V) and (VI) in the screening tests complicated the interpretation of the data. The increase in neptunium solubility caused by Na_2CO_3, Na_2SO_4, Na_3PO_4, HEDTA, and hydroxyacetate probably was due to complexation. The decreased solubility of neptunium in the presence of EDTA was not anticipated since EDTA generally forms stable soluble complexes with metal ions. Sodium nitrate also decreased neptunium solubility while $NaAlO_2$ increased it. Electromigration studies have shown that neptunium (V), (VI), and (VII) are anionic species when in alkaline solutions (10). Sodium salts of Np(V) have precipitated from alkaline solution (10). Thus, $NaNO_3$ may have precipitated Np(V) as a sodium neptunate salt in the screening tests. However, an increase in sodium hydroxide concentration should have led to increased neptunium concentrations in the screening tests due to increased formation of the more soluble anionic hydroxide complexes of neptunium. Detailed analysis of the test data showed that in reducing conditions (i.e., with $NaNO_2$), NaOH did increase neptunium solubility; without $NaNO_2$, NaOH decreased solubility. Further studies must be done to deduce specific neptunium solubility dependencies with respect to the oxidation state.

The plutonium solubility increased in the presence of increased $NaNO_3$, NaOH, and $NaAlO_2$ concentrations. According to the literature, Pu(V) should be the stable oxidation state in alkaline NO_2^--NO_3^- solutions (8). It has been observed that the solubility of Pu(V) increases as the NaOH concentration increases (8,11); probably this occured due to formation of the more soluble anionic hydroxide complexes of Pu(V) such as $PuO_2(OH)_2^-$ (11). Sodium nitrate and $NaAlO_2$ may have increased Pu(V) solubility through complexation. Sodium nitrate also may have increased plutonium solubility by oxidizing the less soluble Pu(IV), initially present in the tracer solids, to Pu(V).

The screening tests showed that americium solubility increased in the presence of the four organic components tested: HEDTA, EDTA, sodium citrate, and sodium hydroxyacetate; as well as with Na_3PO_4. According to electrochemical studies, Am(III) should be the oxidation state present in the screening tests (9). In light of the high formation constants found for the respective americium complexes (12), the strong effect of the organic complexants increasing Am(III) solubility was reasonable. Literature data were not sufficient to confirm the effect of phosphate on americium solubility.

Solubility Prediction Equations. Using results obtained in the solubility screening tests, three-level statistically designed solubility prediction equation tests were run for strontium, plutonium, and americium. As shown above, strontium solubility depended principally on the presence of HEDTA and EDTA. Plutonium solubility depended on $NaNO_3$, NaOH, and $NaAlO_2$ concentrations. Americium solubility depended on concentrations of Na_3PO_4, HEDTA, EDTA, sodium hydroxyacetate, and sodium citrate. Tests were designed using concentrations of these HLW components as variables while the remaining components were maintained at constant intermediate concentrations representative of HLW. The component EDTA was not included in the americium tests because the number of experiments (46) required in the 5-variable Box-Behnken design would have been too large and because EDTA probably would have behaved similarly to HEDTA.

The solubility prediction equations derived from the three-level tests are given in Table V with their respective correlation coefficients (R^2) to assess goodness of fit.

Judging from the R^2 values, plutonium solubility was described adequately by its prediction equation while the strontium and americium solubility equations were less satisfactory. Analysis of the individual test data was found to be more useful than the prediction equations in assessing strontium and americium behavior.

Inspection of individual test runs showed that the strontium spike, corresponding to $3x10^{-3}$M total strontium, dissolved entirely when HEDTA or EDTA concentrations were 0.1M or 0.05M, respectively. However, in the presence of 10^{-5}M or lower HEDTA or EDTA concentrations, strontium was at saturation with concentrations of about $4x10^{-5}$M (~0.17 Ci ^{90}Sr/L in Hanford HLW). Strontium concentrations in Hanford HLW, as shown in Table III, are about 0.004 Ci ^{90}Sr/L and probably not solubility limited.

Inspection of the americium test data showed that the entire spike (about $8x10^{-6}$M americium) dissolved in the presence of 0.1M HEDTA. Subsequent synthetic HLW test solutions having 0.1M HEDTA completely dissolved an $8x10^{-4}$M americium spike (0.6 Ci ^{241}Am/L), thus confirming the strong effect of HEDTA on americium solubility. Citrate and hydroxyacetate also increased americium solubility. At 0.03M citrate, with $\leq 10^{-5}$M HEDTA,

Table V. Radioelement Solubility Prediction Equations

Radioelement	Solubility prediction equation (log $\underline{M}$)[a]	R^2
Sr	log (Sr) = -1.12 + 0.50 log (HEDTA) + 0.483 log (EDTA) + 0.050 [log (HEDTA)]^2 + 0.045 [log (EDTA)]^2 - 0.023 [log (HEDTA)] [log (EDTA)]	0.92
Pu	log (Pu) = -5.67 + 0.14 log ($NaNO_3$) - 0.18 log (NaOH) + 0.12 log ($NaAlO_2$) + 0.012 [log ($NaNO_3$)]^2 + 2.10 [log NaOH)]^2 + 0.0090 [log $NaAlO_2$)]^2	0.98
Am	log (Am) = -3.22 + 0.31 log (HEDTA) + 0.15 log (OAc)[b] + 0.51 log (citrate) + 0.39 [log (HEDTA)]^2 + 0.21 [log (OAc)]^2 + 0.048 [log (citrate)]^2 - 0.030 [log (HEDTA)] [log (OAc)] - 0.043 [log (HEDTA)] [log (citrate)]	0.96

[a]To convert mol/L total strontium to Ci^{90}Sr/L, change the constant in equation from -1.12 to 2.52.

To convert mol/L total plutonium to Ci^{239,240}Pu/L, change the constant in equation from -5.67 to -4.42.

To convert mol/L total americium to Ci^{241}Am/L, change the constant in equation from -3.22 to -0.30.

[b]OAc represents hydroxyacetate.

americium concentrations were solubility limited to about $3x10^{-6}$M. For test solutions containing 0.1M hydroxyacetate, with $\leq 10^{-5}$M HEDTA and $\leq 3x10^{-6}$M citrate, americium concentrations were solubility limited to about $3x10^{-7}$M. Solution containing $\leq 10^{-5}$M organic complexant had americium solubility limited to about $3x10^{-8}$M (or ~25μCi ^{241}Am/L), roughly equivalent to the maximum concentration found in Hanford HLW (see Table III). Sodium phosphate had no significant effect on americium solubility as determined by the prediction equation tests.

Interpretation of the plutonium solubility prediction equations (and the data used to generate those equations) showed that increase in $NaNO_3$ and $NaAlO_2$ concentration from ~10^{-4}M to ~1M increased plutonium concentrations about threefold. However, increase in NaOH concentration from 1 to 4M increased plutonium solubility by a factor of five. Plutonium concentration in these test solutions ranged from about 2 to 10μM, equivalent to about 40 to 200μCi ^{239,240}Pu/L in Hanford HLW, as NaOH concentration increased from 1 to 4M. Plutonium concentrations in actual HLW (see Table III), are as high as 7μCi/L and may be limited by the solubility of plutonium.

Sorption Screening Tests. The results of the Plackett-Burman sorption screening tests are summarized in Table VI. Comparison of the results for the three sediments show nearly identical rankings of components that are significant in each element's sorption. This similarity in rankings indicated that sorption mechanisms for the three sediments also were similar.

As shown in Table VI, many waste components affected cobalt sorption. The components HEDTA, $NaNO_2$, $NaNO_3$, and Na_3PO_4 decreased sorption while NaOH, $NaAlO_2$, and EDTA increased it. Most likely, HEDTA decreased cobalt sorption through complexation, forming a poorly sorbed anionic species. The salts $NaNO_3$ and $NaNO_2$ contributed 2M of sodium ion to each of the screening tests and may have decreased cobalt sorption by competing with cobalt for sorption sites. Nitrite also has been shown to decrease cobalt sorption on Hanford sediment at pH 8 (13), possibly due to formation of an anionic complex. The effect of phosphate on cobalt sorption was not strong and may have been anomalous.

Since NaOH and $NaAlO_2$ had no apparent effect on cobalt solubility, it seemed likely that cobalt sorption was increased by these components due to effects they had on the sediment minerals. Studies of the effects of NaOH and $NaAlO_2$ on the sediment minerals are required to identify possible new mineral phases which might cause increased cobalt sorption.

The increased cobalt sorption caused by EDTA was unexpected in light of cobalt's increased solubility with EDTA present and was opposite the sorption behavior found with HEDTA present. Further studies are required to explain the effect of EDTA on cobalt sorption.

Table VI. Significant Components in Radioelement Sorption *

Radio-element	Sedi-ment	Component											
		$NaNO_3$	$NaNO_2$	NaOH	$NaAlO_2$	Na_2CO_3	Na_2SO_4	Na_3PO_4	NaF	HEDTA	EDTA	Sodium hydroxy-acetate	Sodium citrate
Cobalt	L	-6	-4	+2	+3		+8	-7		-1	+5		
	P	-6	-4	+2	+3			-7		-1	+5		
	S	-5	-4	+3	+2		+7	-8	-9	-1	+6		
Strontium	L									-2	-1		
	P									-2	-1		
	S									-1	-2		
Neptunium	L			-2	-3		-5			-1	-4	-6	
	P			-2	-3		-5			-1	-4	-6	
	S			-2	-3					-1	-4	-5	
Plutonium	L	-5		-1	-2					-3	-4		
	P	-5		-1	-2					-3	-4		
	S			-2	-1	+5		+6		-3	-4		
Americium	L			+3						-1	-2		-4
	P			+2						-1	-3		
	S			+2				+4		-1	-3		

*At $\geq$80% confidence interval. Numbers indicate rank in importance; + indicates component increased sorption, - indicates component decreased sorption.

Strontium sorption behavior was consistent with the strontium solubility behavior described earlier. Both HEDTA and EDTA significantly decreased strontium sorption through formation of poorly sorbed anionic complexes (4). Close inspection of the test data suggested that increased sodium ion concentration might have led to decreased strontium sorption. The competition of strontium with sodium for sorption sites had been observed previously for Hanford sediments (13).

Neptunium and plutonium sorption behaviors were remarkably similar, implying that they had similar sorption reactions and solution species. Both NaOH and $NaAlO_2$ decreased neptunium and plutonium sorption. Several explanations can be offered to rationalize this behavior. First, NaOH and $NaAlO_2$ may have reacted with the sediment minerals to yield solids of lower sorptive capacity. Aluminate ion, as an anionic species, also may have competed with the similar neptunate and plutonate anions for sorption sites. Finally, sodium hydroxide may have stabilized the hydrolyzed $NpO_2(OH)_2^-$ and $PuO_2(OH)_2^-$ species in solution, as was shown in the solubility tests, and prevented sorption. Explanation of the effect of NaOH and $NaAlO_2$ on neptunium and plutonium sorption will require further investigation.

The presence of either HEDTA or EDTA resulted in significantly lower neptunium and plutonium sorption. Complexation of the neptunium and plutonium by HEDTA and EDTA may have caused the reduced sorption. However, this evidence for complex formation was not consistent with the observations made in the solubility studies (HEDTA increased and EDTA decreased neptunium solubility; neither affected plutonium solubility). Thus, HEDTA and EDTA may have decreased neptunium and plutonium sorption through some undetermined effect on the sediment minerals.

Americium sorption was decreased significantly when HEDTA or EDTA were present. Complexation of americium by these strong chelating agents was responsible for this behavior. Sodium hydroxide increased americium sorption but, again, its effect on americium was probably a manifestation of its effect on the sediment minerals.

Sorption Prediction Equations. Equations predicting radioelement distribution coefficients, K_d's, as arithmetic functions of component concentrations were obtained for sorption of strontium, neptunium, plutonium, and americium on two Hanford sediments. These equations, presented in Table VII and derived from statistical fits of Box-Behnken experimental designs, were generated for strontium in terms of sodium ion, HEDTA, and EDTA concentrations. Prediction equations for neptunium and plutonium sorption were derived from NaOH, $NaAlO_2$, HEDTA, and EDTA concentrations. Americium sorption prediction equations were based on NaOH, HEDTA, and EDTA concentrations.

Table VII. Radionuclide Sorption Prediction Equations

Radioelement/ sediment	Sorption prediction equation	K_d relative error estimate, 1σ*
Sr/P	$\log(K_d) = 2.55 + 90[EDTA] + 0.62[Na^+] - 25[Na^+][EDTA]$	5.3
Sr/S	$\log(K_d) = 1.33 - 17[HEDTA] - 34[EDTA] + 444[HEDTA][EDTA]$	2.2
Am/P	$\log(K_d) = 1.88 - 29[HEDTA] + 175[HEDTA]^2$	2.6
Am/S	$\log(K_d) = 2.00 + 0.12[NaOH] - 27[HEDTA] + 158[HEDTA]^2$	1.9
Np/P	$\log(K_d) = 1.30 - 0.12[NaOH] - 0.37[NaAlO_2] - 2.22[HEDTA] - 1.76[EDTA] + 0.60[NaOH][HEDTA]$	1.14
Np/S	$\log(K_d) = 1.80 - 0.088[NaOH] - 0.77[NaAlO_2] - 6.78[HEDTA] - 8.04[EDTA] + 7.78[NaAlO_2][HEDTA] + 106[HEDTA][EDTA]$	1.14
Pu/P	$\log(K_d) = 1.97 - 0.28[NaOH] - 1.76[NaAlO_2] - 1.81[HEDTA] - 4.02[EDTA] + 1.45[NaAlO_2]^2$	1.27
Pu/S	$\log(K_d) = 2.08 - 0.24[NaOH] - 0.77[NaAlO_2] - 4.98[HEDTA] - 8.40[EDTA] + 1.52[NaAlO_2][HEDTA] + 98[HEDTA][EDTA]$	1.25

*To obtain $\pm 1\sigma$ error bounds of K_d values, multiply and divide K_d values by the figure in this column. For example, the $\pm 1\sigma$ error bounds for a K_d of 50 for americium on sediment P would be 50 x 2.6 = 130 and 50/2.6 = 19.

As shown in Table VII, the equations modeled the data better for neptunium and plutonium than for strontium and americium. The lack of fit for strontium and americium reflected the strong effect that the increased HEDTA and EDTA concentrations had on decreasing the K_d. The ranges of K_d's observed for strontium and americium were broader, and more difficult to model, than the ranges observed for neptunium and plutonium.

The retardation factors of the four radioelements for four hypothetical HLW compositions were derived using the prediction equations. (The retardation factor is the ratio of the solution velocity to the radioelement velocity in a system of solution flow through a porous medium and increases linearly with K_d.) The four hypothetical HLW solutions broadly represented dilute/non-complexed, dilute/complexed, concentrated/noncomplexed, and concentrated/complexed HLW. Dilute waste had low concentrations while concentrated waste had high concentrations of Na^+, NaOH, and $NaAlO_2$. Non-complexed waste had no HEDTA or EDTA while complexed waste had 0.1M HEDTA/0.05M EDTA.

As shown in Table VIII, strontium and americium retardation factors were more dependent on whether the waste was complexed (had high concentrations of HEDTA and EDTA) than on whether the waste was concentrated (had high Na^+, NaOH, and $NaAlO_2$ concentrations). Neptunium and plutonium retardation factors were more dependent on waste concentration. Comparison of the retardation factors showed strontium migration rates could increase by up to a factor of 45 in going from noncomplexed to complexed waste compositions. Americium migration rates could increase by nearly 30-fold by changing waste composition to high complexant concentrations. Neptunium migration rates ranged over 6-fold while plutonium rates changed by a factor of 40 for the 4 HLW compositions.

Conclusions

In this study, the effects of 12 Hanford waste solution components on the solubility of cobalt, strontium, neptunium, plutonium, and americium were determined. Also determined were the effects of the 12 components on the sorption of 5 radioelements on 3 representative Hanford sediments. A number of general observations were drawn from these studies.

First, in accord with literature studies, the radioelements' behavior is consistent with their existence as Co(II), Sr(II), Np(V), Pu(V), and Am(III) in HLW. The existence of these oxidation states strongly influenced the effects the waste components exerted on the solubility and sorption reactions.

Table VIII. Predicted Radionuclide Distribution Coefficients and Retardation Factors for the Hanford Waste Solution Types

Radioelement/ sediment	Dilute noncomplexed		Dilute complexed		Concentrated noncomplexed		Concentrated complexed	
	K_d (mL/g)	R*	K_d (mL/g)	R	K_d (mL/g)	R	K_d (mL/g)	R
Sr/P	0.64	3.8	0.39	2.7	11	49	0.022	1.1
Sr/S	21	92	1.5	7.3	21	92	1.5	7.3
Am/P	76	330	5.6	25	76	330	5.6	25
Am/S	130	560	10	46	300	1,300	24	100
Np/P	16	68	8.7	38	4.6	21	3.9	18
Np/S	52	220	14	64	12	51	8.0	36
Pu/P	49	210	20	89	2.2	10	0.89	4.9
Pu/S	71	300	26	120	5.8	26	2.6	12

*R = retardation factor. The retardation factor is the ratio of the solution velocity to the radioelement velocity in a system of solution flow through a porous medium. The retardation factor $R = 1 + K_d (\rho/\phi)$ where ρ is bulk density of Hanford sediment (≈ 1.65 g/cm^3) and ϕ is the fraction of void volume in the sediment (≈ 0.38).

The chelating organic complexants, EDTA and HEDTA, significantly increased the solubilities of cobalt, strontium, and americium in synthetic HLW solutions. Americium solubility also was increased by hydroxyacetate and citrate. Cobalt solubility in HLW was increased to at least $4x10^{-6}$M in the presence of 0.1M HEDTA or 0.05M EDTA. The organic complexants, HEDTA and EDTA, increased strontium solubility to at least $3x10^{-3}$M, corresponding to 13 Ci ^{90}Sr/L in HLW. With HEDTA/EDTA concentrations below 10^{-5}M, strontium concentrations were limited by solubility to about $4x10^{-5}$M (0.2 Ci ^{90}Sr/L in HLW). Test HLW solutions having 0.1M HEDTA could dissolve at least $8x10^{-4}$M americium, corresponding to 0.6 Ci ^{241}Am/L. In solutions containing 10^{-5}M or lower organic complexants, americium concentrations in synthetic HLW were about $3x10^{-8}$M (25μCi ^{241}Am/L).

In contrast, the solubilities of the neptunyl species were increased by HEDTA and unexpectedly decreased by EDTA. Plutonyl(V) solubility was not affected by HEDTA or EDTA. The component most significant to neptunium solubility was $NaNO_2$. Sodium nitrite apparently decreased neptunium solubility by reducing the more soluble neptunyl(VI) to the less soluble (V) state. The existence of two oxidation states complicated interpretation of the neptunium data. Sodium carbonate, phosphate, aluminate, and sulfate all increased neptunium solubility, probably through complexation. Separation of the data according to reduced (with $NaNO_2$) and oxidized (without $NaNO_2$) conditions showed that sodium hydroxide increased neptunium(V) solubility but decreased the solubility of neptunium(VI). Plutonium(V) solubility increased when $NaNO_3$, $NaAlO_2$, and NaOH concentrations increased, probably due to complexation. From these data and other literature studies, it appeared that neptunium and plutonium exist as anionic hydroxide complexed actinyl (V) ions in HLW solutions having $NaNO_2$.

As a result of the existence of these actinyl anions, solubilities of neptunium and plutonium were less sensitive to variation in composition of HLW than cobalt, strontium, and americium. Neptunium solubility in synthetic HLW containing $NaNO_2$ ranged from about 2 to $40x10^{-5}$M (about 3 to $60x10^{-6}$ Ci ^{237}Np/L). Plutonium solubility in synthetic HLW ranged from about 2 to $10x10^{-6}$M (corresponding to 4 to $20x10^{-5}$ Ci ^{239,240}Pu/L).

The solubility results were reflected in the sorption behaviors observed for the five radioelements. Sorption of strontium and americium was decreased by HEDTA and EDTA. Cobalt sorption also decreased in the presence of HEDTA. Sorption of both cobalt and strontium was decreased by increasing sodium ion concentration in the HLW due to competition of the cations for sorption sites.

The sorption behaviors of neptunium and plutonium were similar, thus confirming their suspected similarities in solution chemistry. Both NaOH and $NaAlO_2$ decreased neptunium and plutonium sorption, possibly by favoring the formation of the poorly sorbed anionic complexes. Sorption of neptunium and plutonium also was decreased by HEDTA and EDTA.

The sorption results suggested that further studies of waste/sediment reactions were required. Sodium hydroxide and aluminate probably altered the sediment minerals and affected the sorption reactions. The complexants HEDTA and EDTA also could have changed the sediment minerals.

Predicted rates of the radioelement transport through sediment were found to depend strongly on HLW compositions. Compositions of HLW that are high in HEDTA/EDTA concentrations resulted in 30 to 40 times faster migration rates for americium and strontium. Neptunium and plutonium migration rates increased by factors of 6 to 40 by changing HLW from dilute/noncomplexed to concentrated/complexed compositions.

The three Hanford sediments had similar sorption behavior with respect to changes in HLW composition. Degree of sorption, however, did vary for different sediments.

In summary, the solubility and sorption reactions of cobalt, strontium, neptunium, plutonium, and americium were found to be dependent on HLW compositions. Evidence revealed the formation in HLW of organic complexes of cobalt, strontium, and americium, and of hydroxide complexes of neptunium(V) and plutonium(V). Sorption reactions were dependent on radioelement complex formation and suspected waste/sediment reactions. These data can aid in assessing effects of future HLW processing operations as well as in judging the feasibility of continued storage of HLW in existing tanks.

Acknowledgments

The authors thank J. R. Smith, K. J. Moss, and M. M. Sartain for their able laboratory assistance and C. P. McLaughlin and L. L. Weaver for radionuclide analyses.

Literature Cited

1. Plackett, R. L.; Burman, J. P., "The Design of Optimum Multi-Factorial Experiments," Biometrika, 1946, 33, 305-325.
2. Box, G. E.; Behnken, D. W., "Some New Three Level Designs for the Study of Quantitative Variables," Technometrics, 1960, 2, 455-475.
3. Pourbaix, M., "Atlas of Electrochemical Equilibria," National Association of Corrosion Engineers, Houston, USA and Cebelcor, Brussels, Belgium 1974.
4. Sillen, L. G.; Martell, A. E., "Stability Constants of Metal-Ion Complexes," Special Publication No. 17 and Special Publication No. 24 (Supplement to Special Publication No. 17), The Chemical Society, London 1964, 1971.

5. Smith, R. M.; Martell, A. E., "Critical Stability Constants," Vol. 4, Plenum Press, New York 1976.
6. Peretrukhin, V. F.; Alekseeva, D. P., "Polarographic Properties of Higher Oxidation States of Neptunium in Aqueous Alkaline Media," Sov. Radiochem., 1974, 16, 816-822.
7. Simakin, G. A.; Matyashchuk, I. V.; Vladimirova, N. A., "Potential of the Couple Np(VI)-Np(V) in Sodium Hydroxide Solutions," Sov. Radiochem., 1973, 15, 96-98.
8. Peretrukhin, V. F.; Alekseeva, D. P., "Polarographic Properties of Higher Oxidation States of Plutonium in Aqueous Alkali Solutions," Sov. Radiochem., 1974, 16 823-828.
9. Peretrukhin, V. F.; Nikolaevskii, V. B.; Shilov, V. P., "Electrochemical Properties of Americium Hydroxides in an Aqueous Alkaline Medium," Sov. Radiochem., 1974, 16, 813-815.
10. Cohen, D.; Fried, S., "Some Observations on the Chemistry of Neptunium in Basic Solution," Inorg. Nucl. Chem. Letters 1969, 5, 653-663.
11. Bourges, J., "Preparation et Identification du Plutonium a l'Etat d'Oxydation - V en Milieu Basique," Radiochem. Radioanal. Letters 1972, 12, 111-119.
12. Schulz, W. W., "The Chemistry of Americium," TID-26971, DOE Technical Information Center, Oak Ridge, Tennessee, 1976.
13. McCarthy, G. J., Ed., "Scientific Basis for Nuclear Waste Management," Vol. 1, Plenum Publishing Corporation, New York, 1979; p. 435.

RECEIVED December 2, 1983

LABORATORY STUDIES OF RADIONUCLIDE SOLUBILITY AND SPECIATION

7

Hydrolysis Reactions of Am(III) and Pu(VI) Ions in Near-Neutral Solutions

J. I. KIM, M. BERNKOPF, CH. LIERSE, and F. KOPPOLD

Institut für Radiochemie, Technische Universität München, 8046 Garching, Federal Republic of Germany

Hydrolysis reactions of Am(III) and Pu(VI) ions in CO_2-free solutions of 0.1 M $NaClO_4$ were studied by means of solubility experiments using the oxide or hydroxide of ^{241}Am and ^{238}Pu. The pH of solutions was varied from 3 to 13.5. All experiments were carried out under an argon atmosphere. The speciation of dissolved species was determined as far as possible by spectrophotometry. Various ultrafiltration membranes were applied to examine the proper phase separation. Stability constants of all possible hydrolysis products are presented and compared with literature data.

Hydrolysis reactions are common to all actinide ions in near-neutral solutions, and take place either in parallel with or predominantly over other complexation reactions. In connection with the migration studies of actinide ions in natural waters, attention recently has been focused on hydrolysis reactions of actinides since these reactions are important in determining the solubility of the actinide hydroxide or oxide. Although numerous studies have been made (1-4) to determine stability constants of various hydrolysis products, much of the necessary data are still lacking. The acquisition of these data and further improvement or verification of the existing data is desirable.

In a series of investigations to assess geochemical parameters of various complexation reactions, the present paper concentrates on hydrolysis reactions of the Am^{3+} and PuO_2^{2+} ions in the pH range from 3 to 13.5. For the Am^{3+} ion, a number of papers (5-10) report the formation constant of the monohydroxide species $Am(OH)^{2+}$. While the present experiment was in progress, Rai et al. (5) presented the data for various hydrolysis products. In contrast to the Am^{3+} ion, much more data are available for the hydrolysis reactions of the PuO_2^{2+} ions including its polymerization reactions (11-19). The present study therefore complements and verifies the hitherto existing data for hydrolysis reactions of the two actinide ions.

0097-6156/84/0246-0115$06.00/0

Experimental

The present experiment comprises solubility measurements as a function of the pH (3 ~ 13.5). In order to obviate the dissolution of CO_2 from air, especially in solutions with pH > 7, all experiments were carried out under an argon atmosphere. For the same reason, only CO_2-free chemicals, e.g., NaOH and $NaClO_4$, were used and handled in an inert gas (argon) box. Sodium hydroxide from Baker Co. was heated in an inert gas oven for several hours at 150°C and the $NaClO_4$ solution was prepared by mixing NaOH and $HClO_4$ solutions in the inert gas box. The doubly distilled water was pretreated by deaerating with argon and stored in the inert gas box prior to preparation of desired experimental solutions.

Am(III) Hydrolysis. For studying hydrolysis reactions of the Am^{3+} ion, two solid compounds of ^{241}Am, namely AmO_2 and $Am(OH)_3{\cdot}nH_2O$, were used. The AmO_2 was treated in a thermogravimetric oven at 1,000°C for one hour under oxygen gas to ensure formation of the dioxide. The $Am(OH)_3{\cdot}nH_2O$ was prepared by dissolving AmO_2 in 0.1 M $HClO_4$ and precipitating the hydroxide through addition of a slightly over stoichiometric amount of concentrated NH_4OH. The precipitate was filtered, washed with doubly distilled water until the wash solution reached a pH of 7, and dried in the inert atmosphere box to eliminate surface moisture. Experimental solutions of different pH containing 0.1 M $NaClO_4$ were prepared by addition of 0.1 M $HClO_4$ or 0.1 M NaOH. To each 25 mL of solution in a glass vial, about 3 mg AmO_2 or about 6 mg of $Am(OH)_3{\cdot}nH_2O$ (15 mg AmO_2 to solutions of pH < 4) was introduced and all solutions were stored under argon gas.

In a separate experiment, the hydroxide precipitate of americium was produced directly in each experimental solution: 10 mL solution of 0.1 M $HClO_4$ containing 1.3×10^{-3} M americium was mixed with 10 mL of 0.1 M $NaClO_4$ and made alkaline to produce the hydroxide precipitate by adjusting the pH with addition of NaOH. Solutions thus prepared were left under argon gas to attain the solubility equilibrium.

Pu(VI) Hydrolysis. For easy radiometric measurement, ^{238}Pu was chosen for this experiment. The PuO_2 was dissolved in boiling 1 M $HClO_4$ (8.6×10^{-3} M plutonium). The dissolved plutonium ion appeared to be hexavalent as verified by spectrophotometric analysis. One milliliter of this solution was diluted to 40 mL with 0.075 M $NaClO_4$ and the pH adjusted by addition of dilute NaOH or $HClO_4$. The PuO_2^{2+} ion originally introduced is immediately precipitated by increasing the pH of the solution. The solutions were stored under argon gas for several days before beginning solubility measurements.

Sampling and Measurements. The determination of dissolved actinide concentration was started a week after the preparation of solutions and continued periodically for several months until the solubility equilibrium in each solution was attained. Some solutions, in which the solubilities of americium or plutonium were relatively high, were spectrophotometrically analyzed to ascertain the chemical state of dissolved species. For each sample, 0.2 to 1.0 mL of solution was filtered with a Millex-22 syringe filter (0.22 µm pore size) and the actinide concentration determined in a liquid scintillation counter. After filtration with a Millex-22, randomly chosen sample solutions were further filtered with various ultrafilters of different pore sizes in order to determine if different types of filtration would affect the measured concentration. The chemical stability of dissolved species was examined with respect to sorption on surfaces of experimental vials and of filters. The experiment was performed as follows: the solution filtered by a Millex-22 was put into a polyethylene vial, stored one day, filtered with a new filter of the same pore size and put into another polyethylene vial. This procedure was repeated twice with two new polyethylene vials and the activities of filtrates were compared. The ultrafiltration was carried out by centrifugation with an appropriate filter holder. The results show that the dissolved species in solution after filtration with Millex-22 (0.22 µm) do not sorb on surfaces of experimental materials and that the actinide concentration is not appreciably changed with decreasing pore size of ultrafilters. The pore size of a filter is estimated from its given Dalton number on the basis of a hardsphere model used in the previous work (20).

Results and Discussion

Solubility of Americium Hydroxide. Solutions containing greater than 10^{-6} M americium were examined by spectrophotometry to verify whether or not polymers were present. The molar extinction coefficient measured for the 503 nm absorption band was found to remain constant at $\varepsilon = 390 \pm 5$ for all investigated solutions at pH = 3 ~ 6. Spectra recorded up to 850 nm in this pH range show the typical absorption bands of the Am^{3+} ion, which are similar to those observed in a dilute $HClO_4$ solution, without broadening or displacement. The concentration of americium ions in equilibrium solutions, [Am]s, is, therefore, considered a composite of only monomer species:

$$[Am]s = [Am^{3+}] + [Am(OH)^{2+}] + --- + [Am(OH)_i^{3-i}] \quad (1)$$

Since experiments are carried out at a constant ionic strength (I) and equilibrium concentrations of americium in solutions are found to be less than $10^{-4.8}$ M, all calculations are made on the concentration basis. Equation 1 can be transformed into a function of the free proton or hydroxide ion concentration which contains the solubility product Ksp and formation constant β_i:

$$Am(OH)_3 \rightleftharpoons Am^{3+} + 3\,OH^- \qquad Ksp = [Am^{3+}][OH^-]^3 \quad (2)$$

$$Am^{3+} + iOH^- \rightleftharpoons Am(OH)_i^{3-i} \qquad \beta_i = \frac{[Am(OH)_i^{3-i}]}{[Am^{3+}][OH^-]^i} \quad (3)$$

and hence gives

$$[Am]_s = \frac{Ksp}{[OH^-]^3} \sum_{i=o} \beta_i[OH^-]^i \quad (4)$$

The free hydroxide concentration is calculated from the pH measurement at the equilibrium state of each solution, using the ion product of water $Kw = 10^{-13.78}$ at $I = 0.1$ (1).

Calculation of the constants Ksp and β_i, using Equation 4, is made with the help of a least square fitting of the well known Simplex algorithm. As is apparent from the experimental points plotted as a function of the OH^- concentration, shown in Figure 1 (curve a), the i value does not exceed 3. The hydrolyzed species of americium therefore consist of $Am(OH)^{2+}$, $Am(OH)_2^+$ and $Am(OH)_3^0$. Calculated Ksp and β_i values are given in Table I, together with the literature data which are converted by the relation:

$$\beta_i = \frac{\beta_i^*}{Kw^i} \quad (5)$$

where β_i^* is the constant for the reaction: $Am^{3+} + iH_2O \rightleftharpoons Am(OH)_i^{3-i} + iH^+$. For the conversion, Kw is taken from the literature (1) acording to the ionic strength used by each author. The values of Ksp and β_i reported for this work were calculated from an average of two experimental runs (the second run is not given). Estimated values for Ksp and β_i reported in the literature (2) are not included in this table. Only experimental values (6-10) are given. Except the work of Rai et al. (5), four other authors report β_1 values, one β_2 and none β_3. The solubility product of $Am(OH)_3$ by Rai et al. appears somewhat higher than ours, by 4 logarithmic units. Given the difficulties involved in the experiment, the discrepancy observed in the results from the two laboratories is not surprising. There is no doubt, however, that a better agreement is desirable. As described above, the present experiment was conducted using two different methods, either starting with a solid americium hydroxide present in solution or

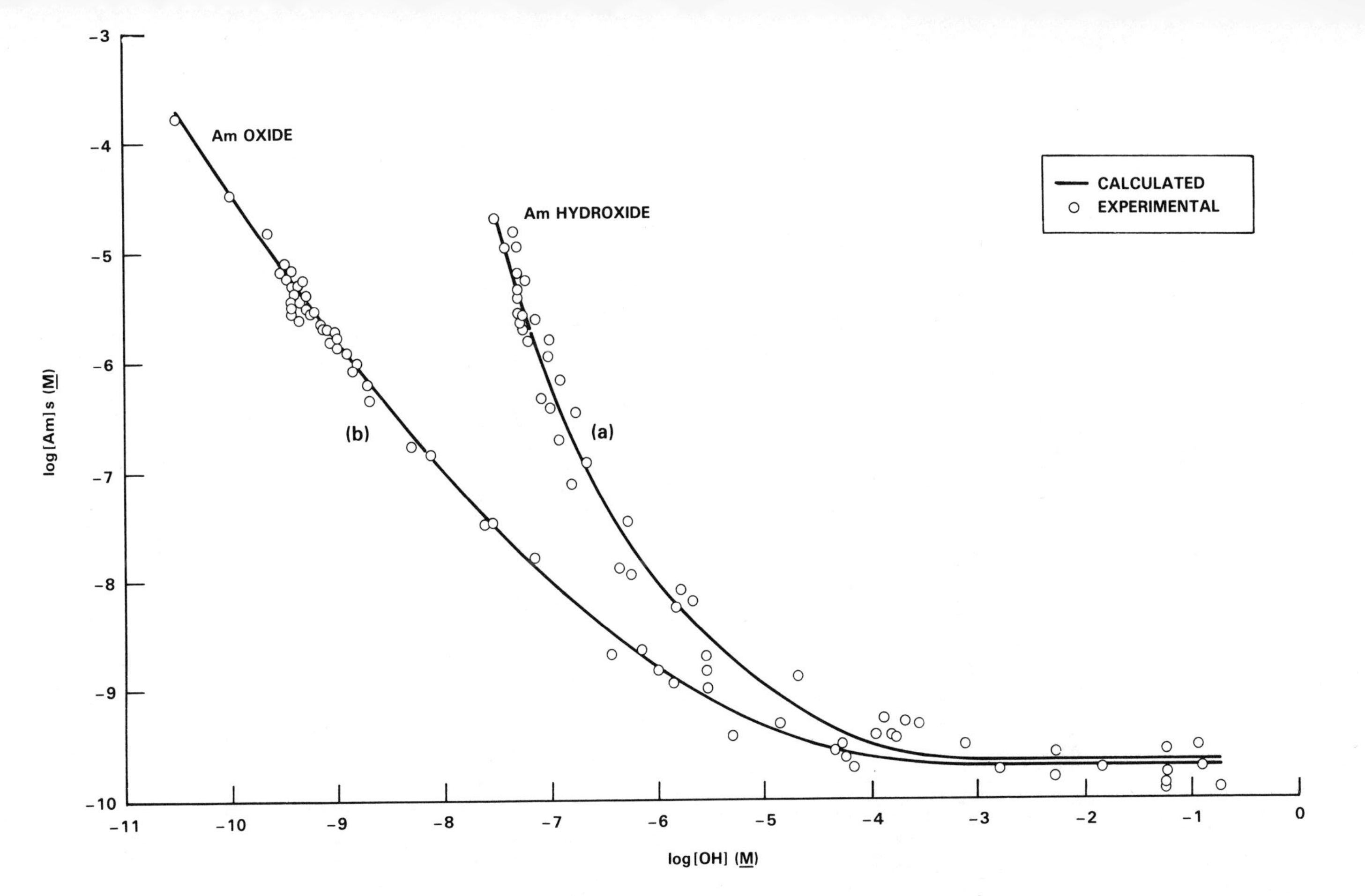

Figure 1. Solubilities of AmO_2 and $Am(OH)_3{\cdot}nH_2O$ as a Function of the Equilibrium OH^- Concentration. The Solid Lines represent Calculations by Equations 4 and 12.

Table I. The Stability Constants and Solubility Products of Am(III) Hydroxides (cf. Equations 2 and 3)[a]

log Ksp	log β_1	log β_2	log β_3	Medium	Reference
-27.16 ± 0.47 (-28.56)[c]	7.44 ± 0.83 (8.03)	13.92 ± 0.63 (14.93)	18.47 ± 0.52 (19.70)	0.1 M $NaClO_4$	This work
-24.34 (-24.62)	≥5.96 (≥5.96)	10.94 (10.94)	≥14.53 (≥14.53)	0.003 M $CaCl_2$	5
–	10.7 (10.9)	20.9[b] (21.2)	–	0.005 M NH_4ClO_4	6
–	11.3 (11.5)	–	–	0.001 M KCl	7
–	6.77 (7.90)	–	–	1 M $NaClO_4$	8
–	7.86 (8.47)	–	–	0.1 M $NaClO_4$	9
–	–	12.8 (14.6)	–	0.2 M $NaClO_4$	11

[a]The literature values are converted with the Kw values from Reference (1).
[b]Estimated value.
[c]The values in parentheses are thermodynamic data calculated by the method and input constants given in Reference (1).

producing a precipitate directly in each experimental solution. Both processes result in the same solubilities over a wide range of pH. Therefore, the present experiment involves reversible as well as reproducible reactions. Should polymers be produced in these solutions, the experiment in which the precipitation is made directly in solution by introducing an excess amount of the americium ion would result in higher solubility due to polymer formation than the experiment that is started with the solid americium hydroxide in solution. Once polymers are formed, they are not expected to depolymerize easily in near neutral solutions.

Spectrophotometric experiments, however, do not provide straightforward evidence for the existence of small amounts of americium hydroxypolymers. In order to compare the literature data with one another and with our data, they are transformed to thermodynamic data at $I = 0$ according to Baes and Mesmer (1), using the salt effect constants assessed by them. These data are given in parentheses in Table I. The first hydrolysis constants given in the literature differ from one another considerably. For lanthanide ions, log β_1 values are reported from 3.7 to 6.3 (1), which are distinctly lower than that of americium shown in Table I.

It has been suggested that the lattice parameter or ionic radii may be correlated with the hydrolysis constants of lanthanides and solubility products of $Ln(OH)_3$ (1). Whether such a correlation can be extended to trivalent actinides is not certain because supporting evidence is lacking. The similarity of Am^{3+} to Nd^{3+} assumed for hydrolysis behavior (5) still requires experimental foundation. The β_1 values known hitherto for americium (5-9) by no means corroborate this assumption. The β_1 value from this work is found to be close to the values from Desire et al. (9) and Nair et al.(8). Only two β_2 values are given in the literature, (5)(11) except for the estimated one (6). Our results are in fair agreement with the value of Bidoglio (10). The β_3 value of Rai et al. is given as a lower limit, so that the real value may be higher. In this respect our β_3 value seems to be in the correct direction. The log β_2 and log β_3 values for lanthanide ions are about 12 and 17 (1), respectively, which are somewhat lower than our values for the americium ion. In the present experiment there is no indication of formation of anionic hydrolysis products, not even the $Am(OH)_4^-$ species, although an analogous product is considered to form with lanthanides (1).

Solubility of Americium Oxide. The dissolution process of AmO_2 under normal or anoxic conditions in near neutral to alkaline solution includes the reduction of Am^{4+} to Am^{3+}, which turns the solution phase gradually into an oxidizing medium as AmO_2 dissolves. For this reason, the solubility of AmO_2 with respect to pH differs distinctively from that of $Am(OH)_3$ at pH < 9. The primary dissolution process may be expressed as follows:

$$AmO_2 + 2H_2O \rightleftharpoons Am^{4+} + 4OH^- \qquad K_1 \qquad (6)$$

$$Am^{4+} + e^- \rightleftharpoons Am^{3+} \qquad K_2 \qquad (7)$$

$$1/2\ H_2O \rightleftharpoons H^+ + e^- + 1/4\ O_2(g) \qquad K_3 \qquad (8)$$

$$AmO_2 + 5/2\ H_2O \rightleftharpoons Am^{3+} + H^+ + 4OH^- + 1/4\ O_2(g) \qquad Ks \qquad (9)$$

and the dissolution constant Ks is given by

$$Ks = K_1K_2K_3 = [Am^{3+}][OH^-]^3\ Kw[O_2(g)]^{0.25} \qquad (10)$$

The total concentration of trivalent americium species produced in the dissolution process is equivalent to $[O_2(g)]^{0.25}$. If all of these species are assumed to remain in solution, the following relation can be postulated:

$$[Am]_s^x = [O_2(g)]^{0.25} \qquad (11)$$

where $[Am]_s^x$ is the total concentration of americium in solution dissolved from AmO_2. In solution, all dissolved species are considered to be the same as those given by Equation 1, so that the solubility function of AmO_2 can be expressed on the basis of Equations 1, 3, 10, and 11 such that:

$$[Am]_s^x = \frac{Ks}{Kw[Am]s\,[OH^-]^3} \sum_{i=0} \beta_i[OH^-]^i \qquad (12)$$

The method used for solution of Equation 12 for Ks and β_i values was similar to that used for Equation 4.

As mentioned above, because of the dissolution of AmO_2, the redox potential of an equilibrium solution is changed as a function of pH as well as $[Am]_s^x$. A combination of Equations 8 and 11 leads to the relationship:

$$pe + pH = 20.78 + \log[Am]_s^x \qquad (13)$$

where 20.78 is calculated from the free energy of formation for a half mole water [$1/2\ \Delta G_f^\circ\ (H_2O) = 56.69$ kcal/mole and the redox potential, Eh(mV) = 59.2 pe]. In the present experiment pe + pH is constant only at the region where the $[Am]_s^x$ remains unchanged with pH (>9) and hence pe + pH = 11.08. This value increases gradually with increasing $[Am]_s^x$ and consequently the solution becomes more oxidizing with decreasing pH. Therefore, the dissolution of AmO_2 differs from that of $Am(OH)_3$ as observed in Figure 1 (curve b).

The situation may be comparable to deep groundwater, which is also an anoxic system. The change of pe + pH in the present experiment is shown in Figure 2.

The hydrolysis constants of the Am^{3+} ion from the AmO_2 solubility experiment are only in fair agreement with those from the $Am(OH)_3$ experiment. They are determined to be $\log \beta 1 = 8.82$, $\log \beta 2 = 14.68$, and $\log \beta 3 = 19.00$, which are slightly higher than the values given in Table I with the difference in order: $\Delta\beta_1 > \Delta\beta_2 > \Delta\beta_3$. To a certain extent, but not satisfactorily, these results justify the postulation of Equation 11. The reason for these results is not yet fully understood.

Solubility of PuO_2^{2+} Hydroxide. Solubilities of plutonium in equilibrium solutions are given in Figure 3 as a function of the pH. Dissolved hydrolysis products of the PuO_2^{2+} ion are examined up to the limit of spectrophotometric sensitivity, namely $[PuO_2^{2+}] > 10^{-5}$ M, to characterize their species. A few samples at pH = 5.5 to 6.5 show the concentration of dissolved plutonium to be high enough to be analyzed by spectrophotometry. A typical analytical result is shown in Figure 4. After the filtration with a Satorius 14529 membrane, the solution having $10^{-4.2}$ M plutonium at pH ≃ 5.8 manifests no typical features of the PuO_2^{2+} ion spectrum in the UV and NIR region. When the solution is brought to pH ≃ 1 by addition of a few drops of 1 M $HClO_4$, the peak at 830 nm appears after 25 minutes, growing slowly, and after 24 hours turns into a spectrum that can be ascribed to the PuO_2^{2+} ion. Upon acidification with HNO_3, the same result is obtained after 2 days. The observation at pH ~6 differs completely from studies reported in the literature (14)(18) which show the distinctive spectra of Pu(VI) solutions at the same pH. Their plutonium concentrations are $10^{-3.6}$ M (14) and $10^{-2.8}$ M (18), which are higher than can be maintained in the solution at pH ~6. The concentration we observed for this condition is about $10^{-4.2}$ M and does not change during several weeks of dissolution. At pH >6, the concentration of plutonium in solution decreases rapidly and at pH >9 reaches $[Pu]_s = 10^{-7.5}$ M. Spectra presented in the literature (14)(18) at pH = 6.5 to 11.0 cannot be reproduced in these experiments. The present experiment demonstrates that the solution at pH ~6 contains polymers, possibly devoid of the free PuO_2^{2+} ion which can be produced only by acidification. If the acidified solution is again made alkaline to pH >9, the plutonium concentration decreases to $< 10^{-7.5}$ M.

This result supports the conclusion that in the reasonably short time of these experiments no oxidation states are involved other than the Pu(VI) species, and that the hydrolysis reaction of the PuO_2^{2+} ion includes polymerization. Since there is no straightforward method to specify each polymer species involved, the most plausible species known in the literature (1)(17)(19) are considered (i.e., x,y = 2,2 and 3,5 of $M_x(OH)_y$).

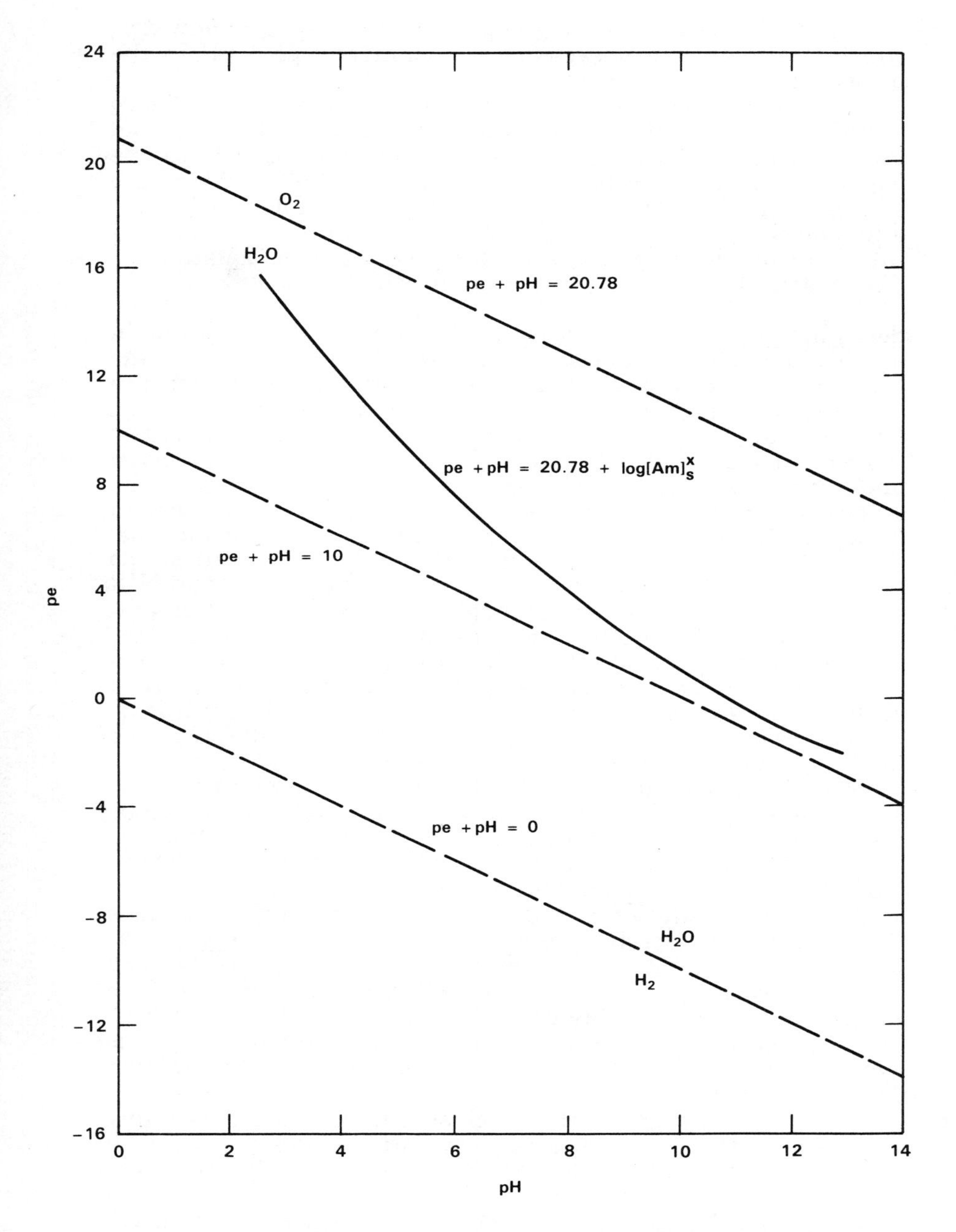

Figure 2. The Change of pe + pH in Dissolution Equilibria of AmO_2 (see Equation 13).

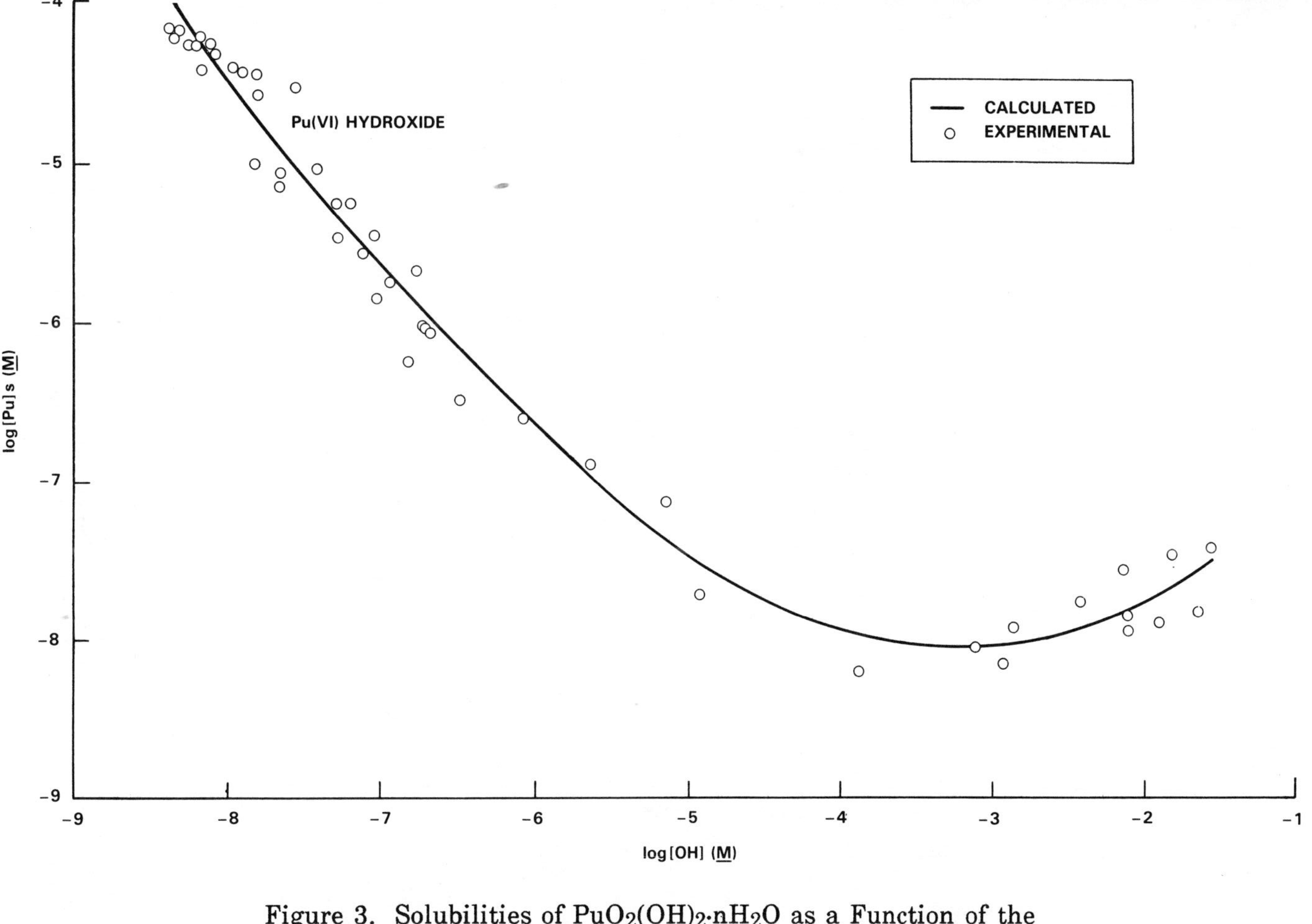

Figure 3. Solubilities of $PuO_2(OH)_2 \cdot nH_2O$ as a Function of the Equilibrium OH^- Concentration. The Solid Line Represents Calculation by Equation 15.

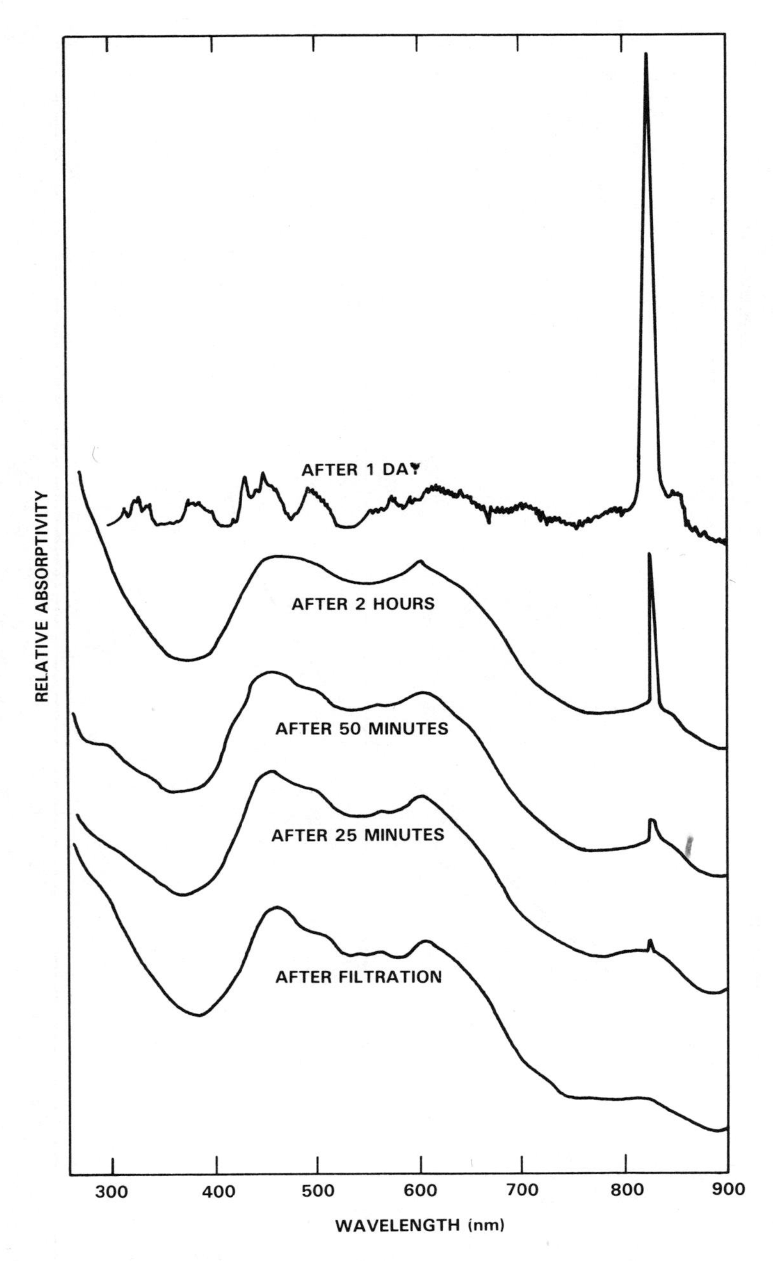

Figure 4. Absorption Spectra of the Pu(VI) Solution ([Pu]s = $10^{-4.2}$ M) Which Change Gradually After Acidification to pH = 1. The Initial Solution at pH = 5.8 is Filtered with a Satorious 14528 Filter (~0.6 nm pore size).

The dissolved Pu(VI) species are therefore considered to include the following hydrolysis products:

$$[Pu]s = [PuO_2^{2+}] + [PuO_2(OH)^{+}] + \cdots + [PuO_2(OH)_i^{2-i}] + [(PuO_2)_2(OH)_2^{2+}] + [(PuO_2)_3(OH)_5^{+}] \qquad (14)$$

which can be expressed by a general solubility function:

$$[Pu]s = \frac{Ksp}{[OH]^2} \sum_{i=o} \beta_i [OH]^i + \frac{Ksp^2}{[OH]^2} \beta_{2,2} + \frac{Ksp^3}{[OH]} \beta_{3,5} \qquad (15)$$

Calculation of experimental constants in Equation 15 was performed by the same least square fitting computer program as used for the americium experiments. The stability constants of hydrolysis products thus determined for the PuO_2^{2+} ion are given in Table II together with the data from the literature. The solubility product from the present work is fortuitously in good agreement with the literature values (14)(18) although somewhat better with the data of Moskvin and Zaitseva (14) than that of Musante and Porthault (18). These data, including our value, can be considered only as tentative ones since the determination of the solubility product in the presence of polymers is not absolutely reliable. The stability constant of the first hydrolysis product is found to be closer to the values of Cassol et al. (17) and Kraus and Dam (12) than those of other authors (13-14)(18), whereas for the second and third hydrolysis products our values are lower than the literature data. The present experiment demonstrates the higher stability of the polymers in question as shown by their formation constants which appear to be larger than those from the literature (17)(19). Obviously, the formation of polymers is enhanced at the expense of monomer hydrolysis products. Due to the arbitrary choice of the polymer species for the calculation with Equation 15, it is not ventured to evaluate errors for the constants given in Table II. For the same reason, assessment of thermodynamic values is not attempted.

Speciation of Dissolved Species. Since there is no straightforward method to identify the dissolved species in solubility experiments, indirect approaches were applied in this experiment.

To verify whether or not colloid species were present in solution, phase separation was thoroughly examined with different ultrafilters. Tables III and IV show results from various ultrafiltrations for the Am(III) and Pu(VI) solutions, respectively. The solution at solubility equilibrium was first filtered with the Millex-22 (0.22 μ) filter and further passed through different ultrafilters of nearly the same pore size (~2 nm). Table III demonstrates that the americium concentration in filtrates remains constant, although a small decrease of concentration is observed after filtration with Amicon CF 25 and Amicon YMT. This effect may be due to surface sorption. A similar experiment for the Pu(VI) solution with other ultrafilters of different pore sizes shown is presented in Table IV. The plutonium concentration was unchanged by filtration. From these filtration experiments it may be concluded that the dissolved species in filtrates are monomer and polymer hydrolysis products and do not include colloidal species.

The sorption of dissolved species onto the surface of experimental vials was also examined for Pu(VI) solutions. The results, shown in Table V for different pH values, indicate that after filtration with the Millex-22 filter (0.22 μ) the solutions remain stable, even if the solutions are transferred several times to new vessels.

The solutions from solubility experiments have shown a continuous decrease in pH with time due to radiolysis and/or hydrolysis. As shown in Figure 1, a large number of experimental points were obtained at the lower pH which were originally solutions of higher pH (up to 13.5). The same tendency was also observed in the Pu(VI) solutions (Figure 3). The solubility was changed as a result of this change in pH. When the americium solution was made alkaline to pH >9 by addition of NaOH, the solubility was decreased and it follows the curve shown in Figure 1. The reaction is therefore reversible with respect to pH. The results of this experiment imply that no colloid forms in these experimental solutions.

Direct speciation of dissolved americium and plutonium is possible only for solutions with appreciable concentration ([Am(III)] $\geq 10^{-6}$ M and [Pu(VI)] $\geq 10^{-5}$ M) using a spectrophotometer with cumulative data recording (21). Typical spectra measured for the Am^{3+} ion at pH = 6.5 are shown in Figure 5 for 1, 5, 10, and 40 times cumulation at 503 nm. With this method it is shown that only trivalent americium ions are present in both equilibrium solutions from AmO_2 and $Am(OH)_3{\cdot}nH_2O$ solids. For plutonium solutions, the spectrophotometric study indicates the presence of polymers as shown in Figure 4.

Calculated concentrations of possible hydrolysis species as a function of the equilibrium pH are given in Figures 6 and 7 for americium ions and plutonium ions, respectively. In the americium solution, the predominant species at pH <7 are the Am^{3+} and $Am(OH)^{2+}$ ions, whereas in the plutonium solution the PuO_2^{2+} ion is not the dominating species at near neutral media but polymers as postulated: $(PuO_2)_2(OH)_2^{2+}$ and $(PuO_2)_3(OH)_5^{+}$.

Table II. The Stability Constants and Solubility Products of Pu(VI) Hydroxides (cf. Equations 14 and 15) *

log Ksp	$\log \beta_1$	$\log \beta_2$	$\log \beta_3$	$\log \beta_{2,2}$	$\log \beta_{2,3}$	$\log \beta_{2,5}$	$\log \beta_{3,5}$	$\log \beta_{7,4}$	Medium	Reference
-23.00	8.26	14.91	16.90	21.98	-	-	56.28	-	0.1 M $NaClO_4$	This work
-	-	-	-	19.83	-	-	-	27.01	3 M $NaClO_4$	19
-24.05	9.93	16.31	20.27	-	30.74	-	-	-	0.1 M $NaClO_4$	18
-	7.80	-	-	19.01	-	-	46.64	-	1 M $NaClO_4$	17
-22.74	10.62	19.36	23.85	-	35.72	47.90	-	-	NH_4ClO_4	14
-	10.57	23.75	-	-	-	-	-	-	0.01 M $NaNO_3$	13
-	8.05	16.10	20.16	-	-	-	-	-	1 M $NaClO_4$	12

*The literature values are converted with the Kw values from Reference (7).

Table III. Comparison of Different Filtrations for the Determination of Solubilities of AmO_2: The Data are $\log [Am]_s$ (M) Values

Sample number	pH	Filters used			
		Millex-22[a]	Amicon CF 25[b]	Amicon YMT[c]	Satorius Sm 14549[d]
1	4.34	-4.68 ± 0.01[e]	-4.80 ± 0.02	-4.72 ± 0.01	-4.68 ± 0.01
2	5.33	-6.00 ± 0.01	-6.12 ± 0.02	-6.26 ± 0.02	-6.01 ± 0.01
3	7.97	-8.94 ± 0.05	-8.96 ± 0.13	-9.08 ± 0.08	-8.96 ± 0.10

[a] Average pore size 220 nm.
[b] Average pore size 1.9 nm.
[c] Average pore size 2.1 nm.
[d] Average pore size 1.8 nm.
[e] Average from four experiments.

Table IV. Comparison of Different Filtrations for the Determination of Solubilities of $PuO_2(OH)_2$: The Data are log $[Pu]s$ (M) Values

Sample number	pH	Filters used				
		Millex-22[a]	Millipore CX-30[b]	Millipore CX-10[c]	Satorius SM 14539[d]	Satorius 14529[e]
1	4.12	-4.19 ± 0.03	-4.19 ± 0.03	-4.36 ± 0.03	-4.20 ± 0.01	-4.21 ± 0.01
2	6.72	-4.22 ± 0.02	–	-4.36 ± 0.03	-4.22 ± 0.03	–
3	9.36	-5.47 ± 0.05	–	-5.47 ± 0.05	–	–
4	9.97	-6.20 ± 0.01	–	-6.24 ± 0.02	–	–

[a]Average pore size: 220 nm.
[b]Average pore size: 2.1 nm.
[c]Average pore size: 1.4 nm.
[d]Average pore size: 1.2 nm.
[e]Average pore size: 0.6 nm.

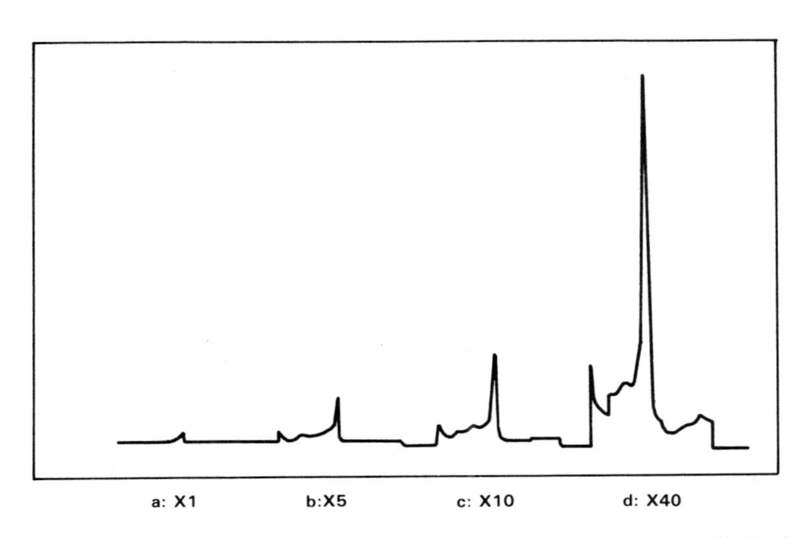

Figure 5. Absorption Spectra of the Am(III) Solution ([Am]s = $10^{-5.5}$ M) in the Region of 503 nm Recorded by a Spectrum Accumulator. The Spectrum a: x1, b: x5, c: x10, and d: x40 Accumulations.

Table V. Sorption Test of the Dissolved PuO_2^{2+} Ion on the Surface of Different Experimental Vials: The Data are log [Pu]s (M) Values

Sample number	pH	Different vials			
		Polyethylene I*	Polyethylene II	Polyethylene III	Polyethylene IV
1	4.12	– 4.14 ± 0.03	– 4.13 ± 0.03	– 4.19 ± 0.03	– 4.18 ± 0.03
2	6.72	– 4.18 ± 0.02	– 4.18 ± 0.02	– 4.23 ± 0.02	– 4.20 ± 0.02
3	9.36	– 5.47 ± 0.05	– 5.48 ± 0.01	– 5.47 ± 0.01	– 5.46 ± 0.02
4	9.97	– 6.22 ± 0.01	– 6.28 ± 0.03	– 6.27 ± 0.03	– 6.19 ± 0.02

*After filtration at 0.22 μ.

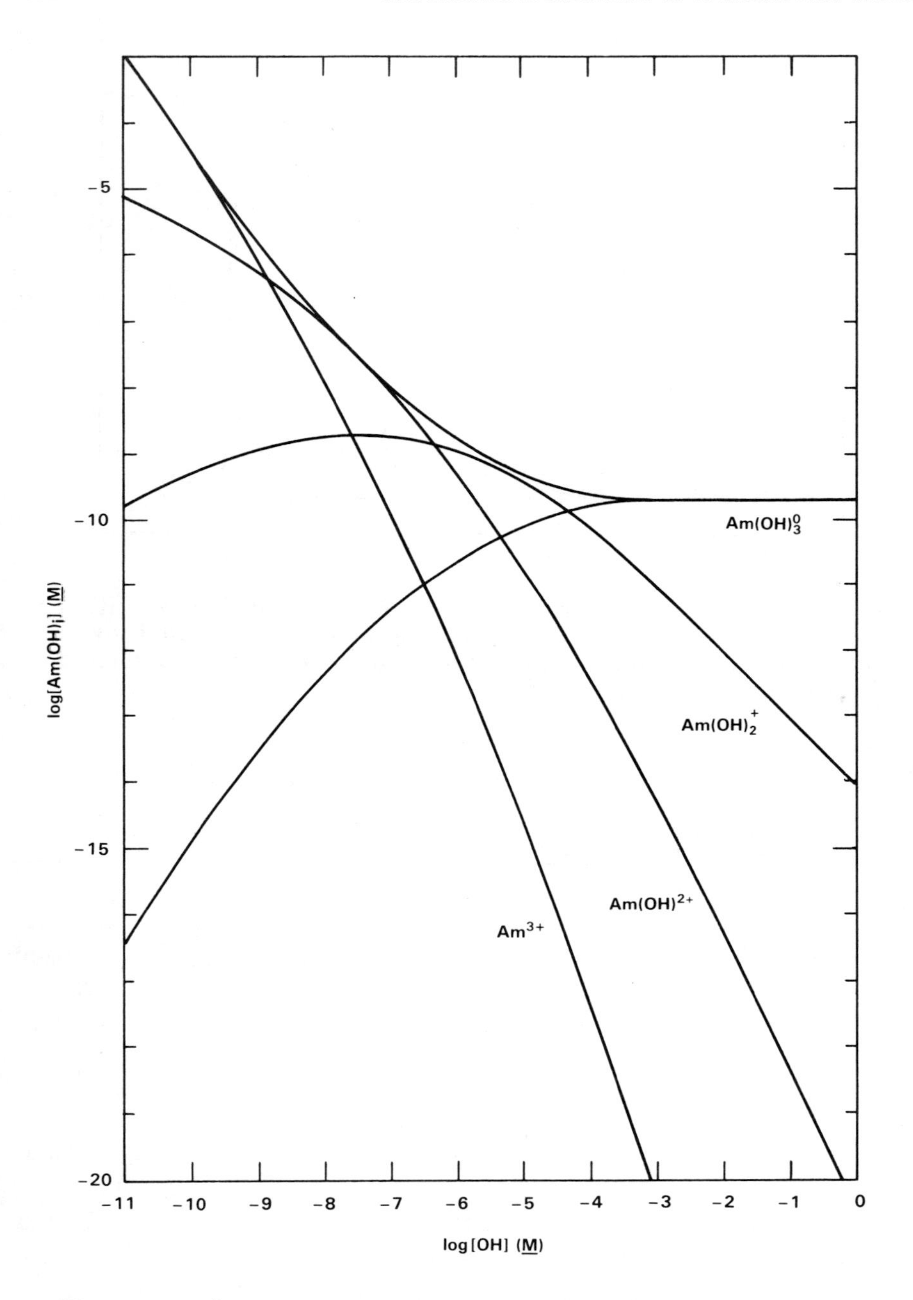

Figure 6. Concentrations of Am(III) Hydrolysis Products as a Function of the OH^- Concentration in AmO_2 Dissolution.

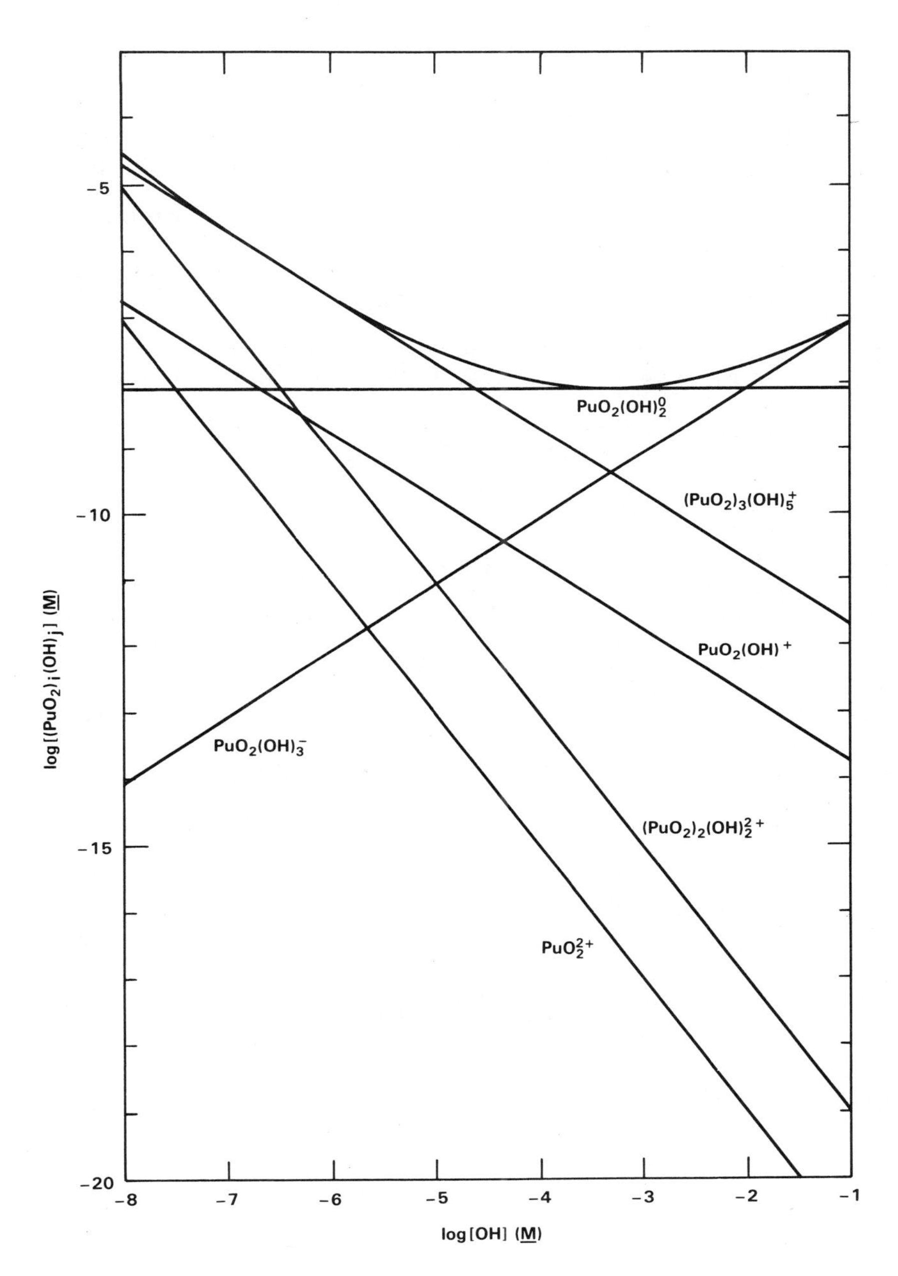

Figure 7. Concentrations of Pu(VI) Hydrolysis Products as a Function of the OH^- Concentration.

Literature Cited

1. Baes, C. F.; Mesmer, R. E. "The Hydrolysis of Cations"; John Wiley and Sons: New York, 1976.
2. Allard, B. in "Actinides in Perspective"; Ed. Edelstein, N. M.; Pergamon Press: Oxford, 1982, pp. 553-580.
3. Smith, R. M.; Martell, A. E., "Critical Stability Constants"; Plenum Press: New York, 1976.
4. Högfeld, E. "Stability Constants of Metal-Ion Complexes: Part A: Inorganic Ligands"; Pergamon Press: Oxford, 1982.
5. Rai, D.; Stricket, R. G.; Moore, D. PNL-SA-10635, 1982.
6. Shalinets, A. B.; Stepanov, A. V. Radiokhimiya 1972, 14, 280.
7. Marin, B.; Kikindai, T. C. R. Acad. Sci. Ser. 1969, C 268, 1.
8. Nair, G. M.; Chander, K.; Joshi, J. K., Radiochim. Acta, 1982, 30, 37.
9. Desire, B.; Hussonois, M.; Guillaumont, R., C. R. Acad. Ser. 1969, C 269, 448.
10. Bidoglio, G. Radiochem. Radioanal. Lett. 1982, 53, 45.
11. O'Connor, P. R. Report CN-2083, September, 1944.
12. Kraus, K. A.; Dam, J. R. Natl. Nuclear Energy Ser., Div. IV, 14B, Paper 4.18, 1949, 528-549; USAEC Reports CN-2831, October 1946, CL-P-432, July 1945; CI-P-449, October 1945.
13. Krevinskaia, Y. Ye.; Nikol'ski, V. D.; Pozharski, B. G.; Zastenker, Ye. Ye. Radiokhimiya 1960, 1, 238.
14. Moskvin, A. I.; Zaitseva, V. P., Radiokhimiya, 1962, 4(1), 73.
15. Perez-Bustamante, J. A. Radiochim. Acta 1965, 4(2), 67.
16. Schedin, U. Acta. Chem. Scand. 1971, 25(2), 747.
17. Cassol, A.; Magon, L.; Portanova, R.; Tondello, E. Radiochim. Acta. 1972, 17(1), 28.
18. Musante, Y.; Porthault, M. Radiochem. Radioanal. Lett. 1973, 15(4-5), 299.
19. Schedin, U. Acta. Chem. Scand. 1975, A29, 333.
20. Brückl, N.; Kim, J. I. Z. Phys. Chem. N. F. 1981, 126, 133.
21. Kim, J. I.; Bernkopf, M.; Stumpe, R. to be published.

RECEIVED October 31, 1983

8

Effect of Aging on the Solubility and Crystallinity of Np(IV) Hydrous Oxide

RICHARD G. STRICKERT, DHANPAT RAI, and ROBERT W. FULTON

Pacific Northwest Laboratory, Richland, WA 99352

Neptunium(IV) hydrous oxide ($NpO_2 \cdot xH_2O$) was aged in aqueous suspensions with pH values ranging from 4 to 12. Initially amorphous $NpO_2 \cdot xH_2O$ developed strong X-ray diffraction peaks within one month that corresponded to crystalline NpO_2. The oxidation state analyses of the 1.8 nm filtered solutions showed the Np to be primarily present as Np(V). The solubility of $NpO_2 \cdot xH_2O$ in suspensions at pH values <8.5 followed the $NpO_2 \cdot xH_2O \leftrightharpoons NpO_2^+ + e^- + xH_2O$ reaction with an apparent log K^o (after 200 days) of -10 ± 1. The results of the aqueous suspensions with pH values >8.5 were complicated by the partial conversion of the $NpO_2 \cdot xH_2O$ to an amorphous, oxidized Np solid, and possibly by Np aqueous speciation other than NpO_2OH^o.

Currently proposed licensing regulations for geologic nuclear waste repositories require a performance assessment involving long-term predictive capabilities. Previous work (1-5) has shown the importance of solubility controls for modeling maximum actinide concentrations in repository groundwaters. However, until reliable data are available on the actinide solid phases that may be present or that may precipitate in the environment, the solubility of solid phases such as hydrous oxides that have fast precipitation kinetics can be used to initially set maximum solution concentration limits.

In general, metal hydrous oxides in aqueous suspensions gradually age to form crystalline metal oxides. These changes in crystallinity of solid phases are accompanied by decreases in the standard free energy of formation. However this crystallization process is opposed by radiolytic effects such that the properties of the steady-state radioactive solid will be between those of the hydrous oxide and the crystalline oxide (6). For example, Rai and Ryan (6) have shown that over a 3.5-year period $^{239}PuO_2$

0097-6156/84/0246-0135$06.00/0

(crystalline) was altered to a less crystalline PuO_2 phase and that amorphous ^{239}Pu(IV) hydrous oxide was converted to the same less crystalline phase with a corresponding decrease in solubility. They also showed that, due to its greater alpha activity, crystalline $^{238}PuO_2$ was converted to a steady-state solid with properties, including solubility, similar to Pu(IV) polymer.

Because of the longer half-life of ^{237}Np compared with ^{239}Pu, the alpha radiation effects should be significantly less. Therefore, Np(IV) hydrous oxide is expected to rapidly develop crystallinity resulting in a decrease in solubility and thus a decrease in the maximum predicted solution concentration. The objectives of this study were to determine the solubility of Np(IV) hydrous oxide and to determine the effect of aging Np(IV) hydrous oxide on its solubility and crystallinity.

Materials and Methods

A ^{237}Np(IV) stock solution was prepared from an anion-exchange purified and filtered (1.8 nm) solution of 4.3×10^{-2}M Np (in 1M HCl) that was bubbled with H_2 through a black platinum gauze for two days. At the end of bubbling the solution appeared to be light brown due to some formation of Np(III) but changed to green indicative of Np(IV) within one day. Spectrophotometric measurements made on the reduced stock solution confirmed that the solution contained Np(IV) with less than 4% Np(V). Another oxidation state analysis of the Np solutions used the TTA extraction technique (7) which distinguishes the (IV) oxidation state from the oxidized states (V and VI). The uncertainty of the TTA extraction procedure was estimated to be about 5%. The TTA extraction analysis of the reduced stock solution also indicated that the Np was in a reduced state >90% TTA extractable as (IV).

All experiments were done in an atmospheric control chamber (initially nearly 100% Ar, <60 ppm O_2, and undetectable levels of CO_2). Aliquots of the Np(IV) stock solution were adjusted with NaOH to an approximate pH value of 12 to precipitate the brown Np(IV) hydrous oxide(11). These suspensions were centrifuged and the supernatants were discarded. The Np solids were resuspended and the pH of the suspensions (approximately 10 mg Np solid/30 ml solution) were adjusted with HCl or NaOH to pH values between 4 to 12. No further adjustments in the pH values were made. Some of the freshly precipitated Np(IV) hydrous oxide was redissolved in acid and the Np oxidation state was analyzed by the TTA extraction procedure. As before, approximately 90% of the Np was measured to be in the reduced state.

Visible amounts of the Np solid were present in the tubes during the study. Periodically, the pH and the redox potential of the various Np suspensions were measured with calibrated glass and

Pt electrodes, respectively (5). Voltage readings from the Pt electrode in the Np suspensions appeared stable within 15 minutes. The redox measurements were converted to pe values [pe = 16.9 • Eh (in volts) at 25°C] because it is convenient to discuss the redox potential of an aqueous solution in terms of the pe (defined as $-\log_{10}$ of the electron activity)(8).

Centrifugation alone is not sufficient to separate Np solid material from the solution (9). Thus, the Np solution concentrations and oxidation state analyses were determined from aliquots filtered through approximately 1.8 nm pore-size Centriflo® membrane cones (Amicon Corp., Lexington, MA). Each filter cone was pretreated with an initial aliquot of the suspension to be filtered. Tests using successive filtration confirmed that no significant amounts of Np were being sorbed by the filter cones. The solutions were alpha counted by liquid scintillation techniques. Major cations in the solutions were determined by inductively coupled plasma spectrometry (ICP) analyses.

X-ray diffraction (XRD) patterns of small portions of the solids were periodically obtained with a Gandolfi camera using CuK_α radiation (10) to monitor crystallinity.

Results and Discussion

Although no redox buffers were added to the suspensions, the pe+pH sums were initially between 8 and 10 (Figure 1). However, most of the pe+pH sums appeared to reach a value of about 12 within 70 days. Although not shown in the figure, the pH value of each suspension remained fairly constant over time, while the corresponding pe value tended to increase with time especially for suspensions with pH values greater than 7.

Available thermodynamic data (11,12) indicate that the dominant oxidation state of Np in the measured pe+pH range is (V). The Np oxidation state analyses of the solutions at different times (Table I) confirmed that the Np was present as Np(V). The solubility results for suspensions of pH values <8.5 and >8.5 are discussed separately as the nature of the aqueous Np species in these pH regions are expected to be different.

Low (<8.5) pH Solutions. The Np solution concentrations are plotted as a function of measured pe in Figure 2 for various equilibration periods. The data points for the low pH (<8.5) suspensions were fitted by a linear least squares method to a line of fixed unit slope. When the slopes were allowed to vary, their values ranged from 0.8 to 1.1.

Because the Np in solution is predominantly (V) and because the Np solution concentration increases with pe along a unit

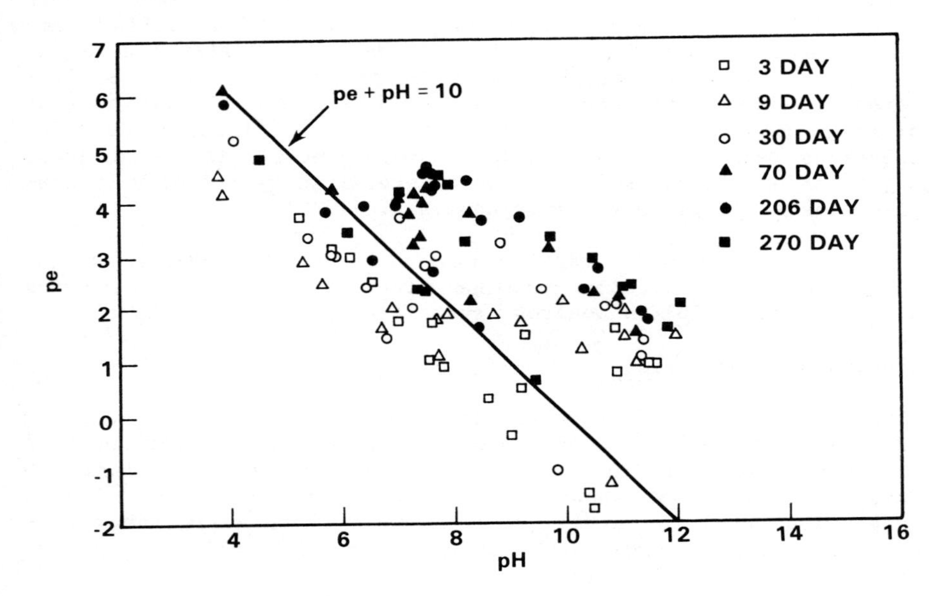

Figure 1. Measured pe and pH values of Np(IV) hydrous oxide aqueous suspensions for various equilibrium times. The line corresponds to a pe + pH value of 10.

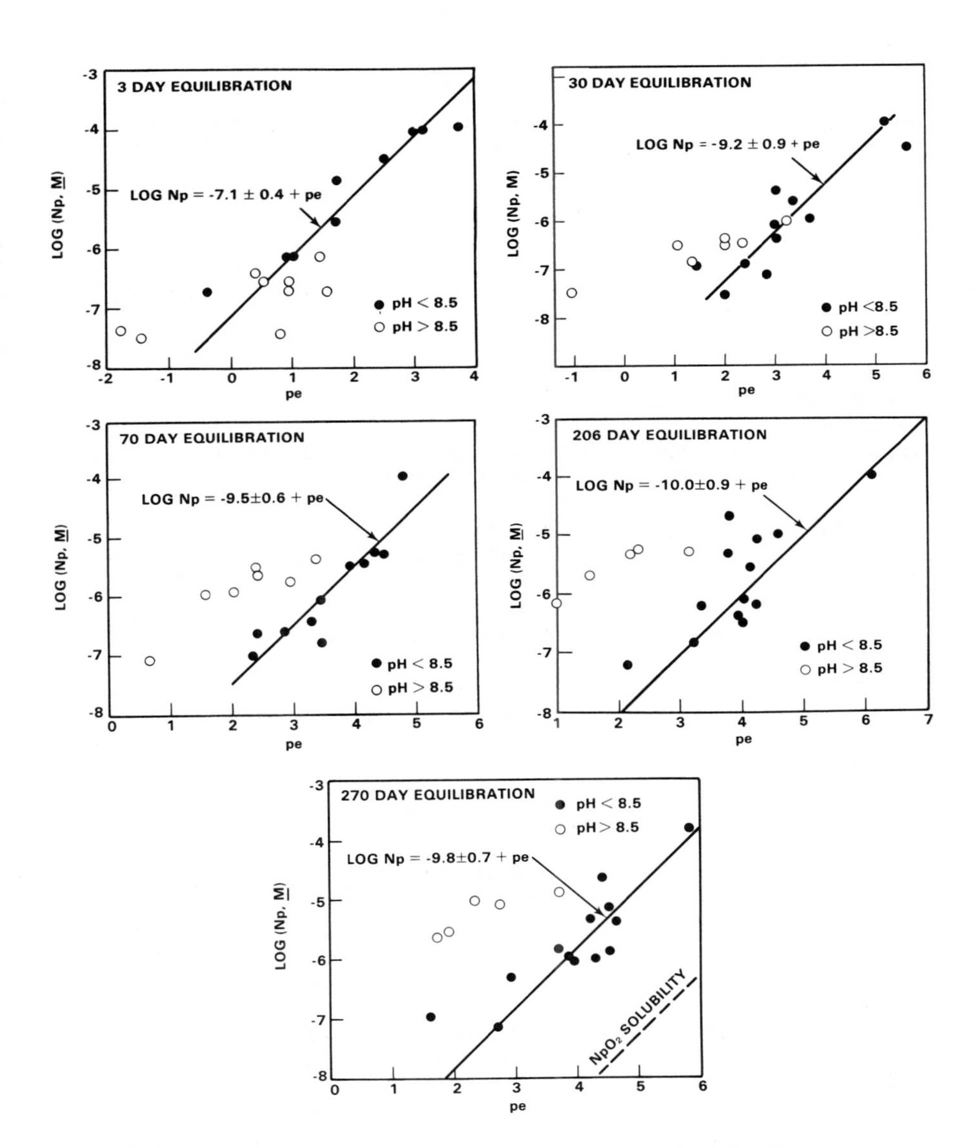

Figure 2. Log Np concentrations in filtered solutions contacting Np(IV) hydrous oxide for various equilibration times as a function of pe. Solid lines are fitted to data from pH<8.5 solutions. Dashed line is based on NpO_2 solubility data from Reference 5.

Table I. Neptunium Oxidation State Analyses*

Solution Equilibration (days)	% Extractable in TTA Np(IV)	% Non-extracted in TTA Np(V and VI)#
70	7 ± 4	101 ± 11
110	5 ± 2	92 ± 4
206	5 ± 2	92 ± 5
270	6 ± 3	88 ± 4

* The standard deviation of the average % Np of all solutions is included.

Expected to be Np(V) only as Np(VI) stability fields exist outside the experimental pe and pH range.

slope, the reaction in Equation (1) can be used to describe the Np solution behavior in the low pH range (<8.5), where no Np(V) hydrolysis is expected to occur.

$$NpO_2 \cdot xH_2O(s) \rightleftharpoons NpO_2^+ + xH_2O + e^- \qquad (1)$$

When K^m is a constant involving Np(V) concentrations and electron activities in Equation 1 (assuming a solid phase activity of one), the K^m is related to the Np(V) solution concentration as shown in Equation 2.

$$\log [NpO_2^+] = \log K^m + pe \qquad (2)$$

The NpO_2^+ activity coefficient (γ_{Np}) is needed to determine the thermodynamic equilibrium constant (K^o) for Equation 1. Analysis of major cations by ICP showed that the predominant cation in a filtered Np solution was Na (130 ppm) with less than 6 ppm of B, Ca, K, or Si. Less than 1 ppm amounts of Ba, Li, Mg, Sr, and Zn were also detected. These data were used to calculate the ionic strength. Because the ionic strength of the solutions was less than 0.01 and the Np species is singly charged, maximum activity corrections are small ($-\log\gamma_{Np} \simeq 0.05$). Based on these data, the log K^o values for Equation 1 would be less than 0.05 units lower than the log K^m values. However, this correction is relatively insignificant compared to an order of magnitude larger uncertainty in the fitted intercept values (Figure 2).

The apparent log K^o values (based on the intercept values) along with the uncertainties of the fit are plotted as a function of time in Figure 3. The apparent log K^o rapidly decreases during the first thirty days and then show a much slower decrease out through 270 days.

From these results (Figures 2 and 3) one can conclude that, in aqueous solutions contacting freshly prepared Np(IV) hydrous oxide, Np solution concentrations reach steady-state values within 70 days and show an expected pe dependence as described by Equation 2 for pH values less than approximately 8.5. Furthermore, with the aging of the Np solid phase, there is a corresponding change in the solubility such that the calculated log K^o value decreases from approximately -7 to -10 ($\pm$1) within 200 days. It is not possible at this time to determine whether the solubility will continue to slowly decrease to lower values.

Using estimated thermodynamic data from Allard et al. (12) the calculated log K^o value for the reaction, $Np(OH)_4(s) \leftrightharpoons NpO_2^+ + e^- + 2H_2O$, is -9$\pm$2. This estimated value appears to be in good agreement with the experimental value reported here. The experimental log K^o value for Equation 1 is still 2 units greater than the log K^o value estimated for the corresponding reaction in aqueous suspensions of initially crystalline NpO_2 (5). Because Np(V) was the only aqueous Np oxidation state that was possible to measure accurately, it was not feasible to calculate the redox potentials from activities of different Np species. Therefore to calculate the log K° values, it was assumed that the Pt electrode measured potentials truly reflected the equilibrium potentials. This assumption needs to be investigated further, although it appears to be valid because the log K° value measured in this study is similar to values reported in the literature (12).

High (>8.5) pH Solutions. The Np concentrations for the higher pH solutions were initially close to the fitted unit-slope line (Figure 2), but with time increased to values around 10^{-5}M. All of the high pH solutions contained Np in the (V) oxidation state. Neptunium(V) should not complex readily with the small amount of chloride (introduced through the use of HCl in adjusting some of the pH values)(11). Neptunium(V) is expected to be hydrolyzed (11) and its concentration in solution contacting Np(IV) hydrous oxide could be described by:

$$NpO_2 \cdot xH_2O(s) \leftrightharpoons NpO_2OH^o + H^+ + e^- + (x-1)H_2O \tag{3}$$

and

$$\log[NpO_2OH^o] = \log K^m + (pe + pH) \tag{4}$$

However, in contrast to Equation 4, the Np solution concentration versus pe+pH (Figure 4) appears to decrease rather than increase with the pe+pH sum. This suggests that at pH values

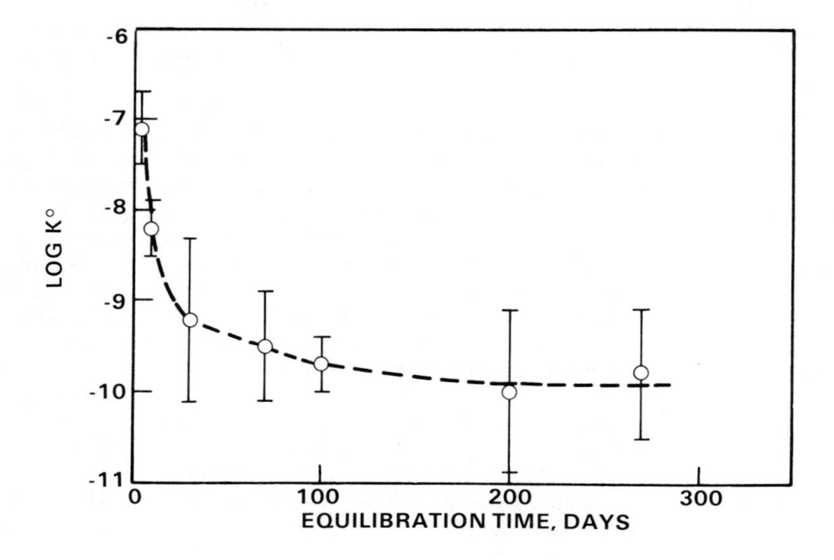

Figure 3. Apparent log K° for the reaction $NpO_2\ xH_2O \leftrightarrows NpO_2^+ + e^- + xH_2O$, for various equilibration times. Error bars based on fitted intercept values (see text).

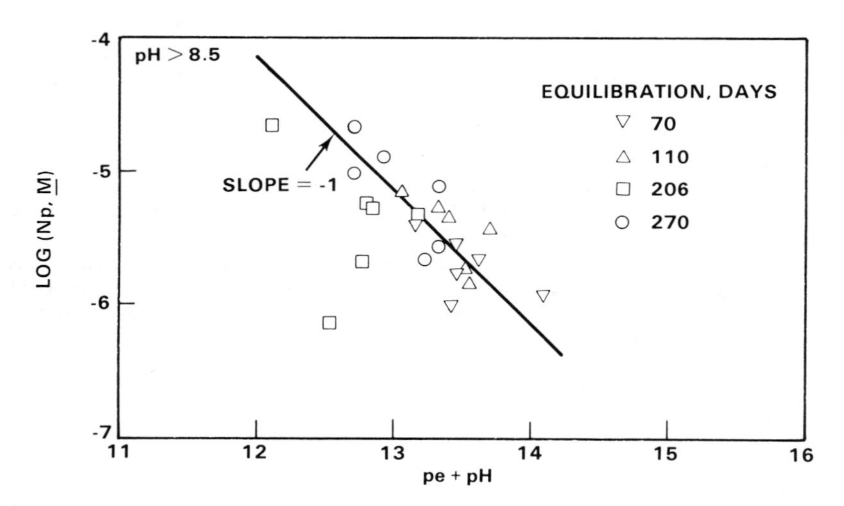

Figure 4. Log Np concentrations in filtered solutions ($pH \geq 8.5$) contacting Np(IV) hydrous oxide for various equilibration times as a function of pe + pH. The solid line is drawn to indicate a slope of -1.

greater than 8.5 the solubility-controlling solid phase is not Np(IV) hydrous oxide alone. This conclusion is supported by Np oxidation state analyses of the solid phases from various suspensions (discussed in the following section). There is also the possibility that the Np solution species may not exist as the simple mono-hydroxy Np(V) complex in the pH range 8.5 to 12.

Solid Phase Analyses. Portions of the Np solids from several suspensions were removed after 200 days, rinsed and dissolved in HCl. The Np oxidation state analyses of the dissolved solids (Table II) indicate that while the solids from various low pH suspensions contain predominantly Np(IV), high pH suspensions contain Np solids of mixed-Np oxidation states. Even though the TTA values are lower, the sum of the extracted and non-extracted neptunium (relative to the total Np solution concentration) does not account for all the Np. This may indicate that the Np(IV) values could be higher. Recoveries in predominantly Np(V) solutions were nearly 100%.

Table II. Np Oxidation State Analyses of Dissolved Np Solids*

Suspension	% Extractable in TTA Np (IV)	% Non-extracted in TTA Np(V and VI)
Fresh ppt	88	8
Low pH	73 ± 4	11 ± 1
High pH	43 ± 3	42 ± 2

* Standard deviation based on multiple samples.

The results of XRD analysis (Table III) show that freshly prepared Np(IV) hydrous oxide is amorphous. However, within one month several samples from both low and high pH suspensions showed X-ray diffraction peaks that corresponded to NpO_2(c). No other crystalline structure was observed in the solids examined. Therefore, the Np concentrations in high pH solutions may be controlled by an amorphous, oxidized Np solid phase. A similar study (6) of ^{239}Pu(IV) hydrous oxide demonstrated that crystallinity similar to PuO_2 did not begin to appear until after 3 and 1/2 years. The rapidity of the crystallization of Np(IV) hydrous oxide compared to Pu(IV) hydrous oxide is no doubt associated with the difference in half-lives between the two alpha-emitting isotopes.

The results presented in this paper along with other work (6) show that alpha radiation affects the nature of the equilibrium

Table III. Summary of X-Ray Diffraction Analyses

Equilibration Time, Days	Solution pH	d Spacing (Å)					
Fresh ppt	--	--	--	--	--	--	--
41	7.7	3.16	2.69	1.92	1.65	1.36	1.24
112	6.55	--	--	--	1.64	--	1.24
289	5.75	3.17	--	1.93	1.64	1.36	1.22
31	9.6	3.17	2.70	1.94	1.66	1.38	1.24
126	12.05	--	--	1.92	1.64	1.35	1.25
289	11.5	3.19	2.74	1.92	1.65	1.37	1.23
NpO_2^*		3.14	2.72	1.92	1.64	1.36	1.24

* From Reference 13.

solid phase and thus the maximum solubility-controlled concentration. Such results indicate that modeling for repository performance assessments must include the effects of alpha radiolysis on the properties of actinide solid phases.

Acknowledgments

This research was performed for the U.S. Department of Energy under Contract No. DE-AC06-76RLO 1830. We thank Robert Fulton for his efforts in the experimental measurements, Harvey Tenny for the X-ray diffraction analyses, and Frank Hara for ICP analyses. The preparation and presentation of this paper was supported through the Materials Characterization Center.

Literature Cited

1. Rai, Dhanpat; Serne, R. J. "Solid Phases and Solution Species of Different Elements in Geologic Environments." U.S. Department of Energy Report PNL-2651, Pacific Northwest Laboratory, Richland, Washington, 1978.

2. Rai, Dhanpat; Strickert, R. G.; Swanson, J. L., "Actinide Solubilities in the Near-Field of a Nuclear Waste Repository." In Workshop on Near-Field Phenomena in Geologic Repositories, (August 31 - September 3, 1981, Seattle, Washington), Nuclear Energy Agency of OECD, Paris, France, 1981 pp. 13-20.
3. Wood, B. J.; Rai, Dhanpat, "Nuclear Waste Isolation: Actinide Containment in Geologic Repositories." U.S. Department of Energy Report PNL-SA-9549, Pacific Northwest Laboratory, Richland, Washington, 1978.
4. Allard, B. "Solubilities of Actinides in Neutral or Basic Solutions" In Proceedings of the Actinides 81 Conference, N. Edelstein (ed.), Pergamon Press, Oxford, 1982.
5. Strickert, R. G.; Rai, Dhanpat, "Solubility-Limited Neptunium Concentrations in Redox-Controlled Suspensions of NpO_2", U.S. Department of Energy Report PNL-SA-10590, Pacific Northwest Laboratory, Richland, Washington, 1982.
6. Rai, Dhanpat; Ryan, J. L., "Crystallinity and Solubility of Pu(IV) Oxide and Hydroxide in Aged Aqueous Suspensions." Radiochim. Acta. 1982 30, 213-216.
7. Foti, S. C.; Freiling, E. C., "The Determination of the Oxidation States of Tracer Uranium, Neptunium, and Plutonium in Aqueous Media." Talanta 1964 11, 384-392.
8. Stumm, W.; Morgan, J. J., "Aquatic Chemistry"; 2nd Ed., John Wiley and Sons; New York, 1981.
9. Rai, Dhanpat; Strickert, R. G.; McVay, G. L., "Neptunium Concentrations in Solutions Contacting Actinide-Doped Glass." Nuclear Technology 1982, 58, 69-76.
10. Tenny, H., "Single Particle Powder Patterns." Microchem. J. 1979, 24, 522-525.
11. Burney, G. A.; Harbour, R. M., "Radiochemistry of Neptunium" NAS-NRC Nucl. Sci. Ser., NAS-NS-3060, 1974.
12. Allard, B.; Kipatsi, H.; Liljenzin, J. O., "Expected Species of Uranium, Neptunium, and Plutonium in Neutral Aqueous Solutions" J. Inorg. Nucl. Chem. 1980, 41, 1015-1027.
13. Joint Committee on Powder Diffraction Standards (1601 Park Lane, Swarthmore, Pennsylvania 19801), Card No. 23-1269, 1973.

RECEIVED October 31, 1983

9

Geochemical Controls on Radionuclide Releases from a Nuclear Waste Repository in Basalt

Estimated Solubilities for Selected Elements

T. O. EARLY, G. K. JACOBS, and D. R. DREWES

Rockwell International, Energy Systems Group, Richland, WA 99352

Two basalt flows within the Grande Ronde Basalt at the Hanford Site in southeastern Washington are candidates for a high-level nuclear waste repository. In order to determine the anticipated rate of release and migration of key radionuclides from the repository, solubility controls must be determined. Solubilities, solids controlling solubility, and aqueous speciation in groundwater have been determined from available thermodynamic data for a variety of actinides and fission products. Groundwater compositions used include all available analyses from the Grand Ronde formation determined by the Basalt Waste Isolation Project (BWIP). Over the range of conditions investigated, solids predicted to control solubility for the selected radionuclides include hydroxides and hydrous oxides (palladium, antimony, samarium, europium, lead, and americium), oxides (nickel, tin, thorium, neptunium, and plutonium), elements (selenium and palladium), and silicates (zirconium and uranium). Dominant soluble species include hydroxy complexes (zirconium, palladium, tin, antimony, samarium, europium, thorium, uranium, neptunium, and plutonium) and carbonate species (nickel, samarium, europium, lead, uranium, neptunium, plutonium, and americium). In addition to the limitations in completeness and accuracy of thermodynamic data, solubility estimates of the radionuclides are sensitive to the following: (1) Eh and the degree of redox equilibrium, (2) temperature, (3) formation of metastable solid phases, and (4) coprecipitation. The Eh effects have been evaluated for each radionuclide and are significant for selenium, palladium, and tin, and possibly uranium and neptunium. Solubility estimates also have been made at ambient temperature (~55 ± 5°C) for Grande Ronde

0097–6156/84/0246–0147$06.25/0

basalts for those nuclides for which sufficient data exist. Effects of metastability and coprecipitation cannot be treated quantitatively, but their contributions have been estimated in reference to available experimental data.

The Basalt Waste Isolation Project (BWIP), under the direction of the U.S. Department of Energy, is investigating the feasibility of locating a nuclear waste repository in the deep basalts beneath the Hanford Site, Richland, Washington. The repository is designed to be constructed in one of four basalt flows (Rocky Coulee, Cohassett, McCoy Canyon, or Umtanum) in the Grande Ronde formation (Figure 1).

The National Waste Terminal Storage Program has adopted the approach of utilizing a series of engineered barriers to complement the geologic site in providing for the safe isolation of nuclear waste. The engineered barriers include the waste package [e.g., waste form, canister(s), and backfill] and the repository seal system (e.g., shaft, borehole, room, and tunnel seals).

Under current regulatory criteria (1-2), performance may be apportioned among the isolation components in order to demonstrate compliance with the proposed criteria and to ensure the safe isolation of the waste. Data from site characterization activities and information regarding the geochemical behavior of radionuclides may be used to determine the isolation capacity of the geologic site component for a range of expected conditions. From this information, the performance of the engineered components of the isolation system, required to complement the site for the safe isolation of nuclear waste, may be determined. From such an analysis, the required performance for the engineered components may be allocated among the waste package and repository seal systems so as to allow designs to be optimized for both performance and cost effectiveness, while at the same time providing for an adequate factor of safety in overall isolation capacity.

In order to evaluate the rate and cumulative release of radionuclides from a repository in basalt to the accessible environment, a necessary first step is the identification of the important natural and engineered isolation parameters. They include the following:

- Waste package containment time
- Radionuclide release rates from the waste package
- Site-controlled radionuclide solubilities
- Groundwater flow through the repository
- Groundwater travel time to the accessible environment
- Near- and far-field radionuclide sorption.

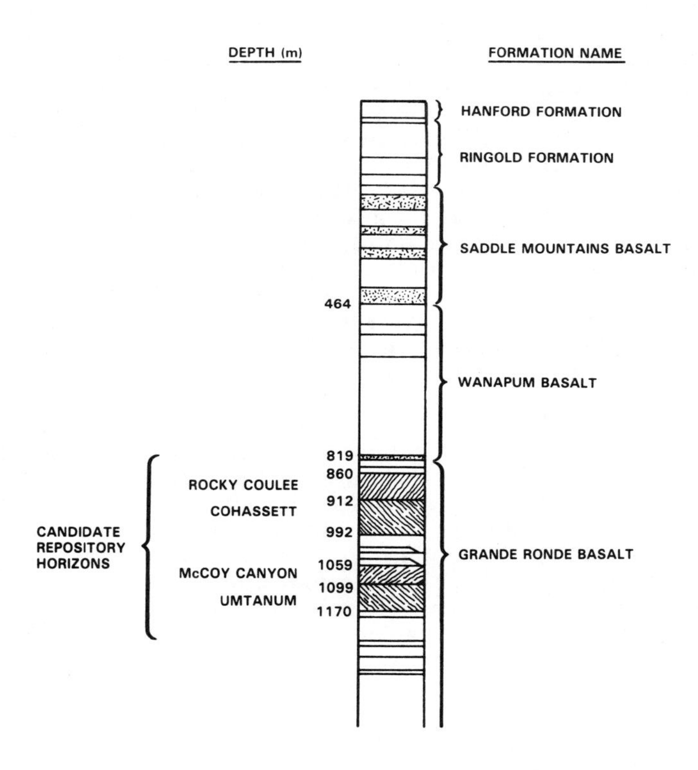

Figure 1. Simplified Stratigraphy of the Pasco Basin, Washington.

Of these parameters, radionuclide solubilities that are addressed in this paper represent an important control on the concentration of specific radionuclides in groundwater available for transport to the accessible environment.

Radionuclide concentrations in groundwater outside the waste package will be controlled by the solid phases and aqueous species forming in the basalt geochemical environment. For a given radionuclide, one can predict via thermodynamic arguments the most stable solid phase which should form and thus control the concentration of the radionuclide in this system. Then incorporating the effect of complexing agents present in the groundwater, one may calculate the theoretical, maximum possible concentration of all dissolved radionuclide species in equilibrium with the stable solid phase. The solubilities, associated solid phases, and aqueous complexes, while thermodynamically predicted, may not actually occur in the natural basalt geochemical system. Deviations from these predictions could result from an incomplete thermodynamic data base, uncertainties in the thermodynamic data, kinetic inhibitions to the formation of the most stable phases, and/or the formation of colloidal particles. However, for the basalt site-specific case, it is useful to delineate baseline calculations from available thermodynamic data at this time, with the understanding that the limitations noted above may be very important in determining the final controls on the concentrations of radionuclides in the basalt geochemical system. Confirmation of these calculations undoubtedly requires careful laboratory tests addressiong solubility controls as well as kinetic and colloidal transport factors.

Method

Solubility Relationships. Estimates of the solubility of radionuclides in characteristic groundwaters for a nuclear waste repository in basalt (NWRB) were made by considering the thermodynamic data available for solid phases and aqueous species. For each radionuclide, published $\Delta G°_{f,298}$ values for individual solids and aqueous species in conjunction with log K_{eq} values for appropriate chemical equilibria were used to derive a set of mathmetical expressions that relate the activity of all dissolved species in solution to the solid phase(s) calculated to be the most stable under a given set of environmental conditions (pH, Eh, groundwater chemistry). A detailed summary of the theoretical approach used can be found in standard texts dealing with chemical thermodynamics of heterogeneous geologic systems (3).

Assumptions and Limitations

Thermodynamic Data. There are two key considerations in evaluating the effect of available thermodynamic data on calculated solubilities. The principal limitation is the lack of a complete thermodynamic data base containing all important solid and aqueous species pertinent to the basalt geochemical system. It is possible that thermodynamic data may not exist for solids that are more stable than those currently available, resulting in high solubility estimates. Alternatively, thermodynamic data may not be available for aqueous complexes of radionuclides that are more stable than those in the current data base, resulting in solubility estimates that are too low. Therefore, one must recognize the potential limitations of the thermodynamic data base in making solubility estimates.

The reactions with associated solid and aqueous species and log K_{eq} values used in this study come from published compilations (4-6) and more recent experimental data (7-8). No attempt was made to critically evaluate the available data for accuracy, although comparison among different compilations was made as a check on typographical errors. Additionally, no attempt was made to evaluate the effect of data uncertainties on the resulting computations. These uncertainties will be addressed in future studies.

Temperature. The estimated solubilities presented in this paper are strictly applicable only at 25°C. At elevated temperatures, thermodynamic data exist for only a few radionuclides and much of the data is estimated. This limitation should not significantly affect the applicability of present calculations to a repository in basalt because measured temperatures in the candidate horizons, the Rocky Coulee, Cohassett, McCoy Canyon, and Umtanum flows, are 49°C ± 1°C, 51°C ± 2°C, 56°C ± 1°C, and 58°C ± 2°C, respectively.

From data provided in the literature (9-10), it is possible to estimate solubilities for uranium and plutonium at a temperature which closely approximates that of the candidate repository horizons. These estimates are discussed in a later section of this paper.

Eh-pH Conditions. Measured pH values for Grande Ronde groundwater are found to be 9.5 ± 0.9. Hydrothermal experiments on the system basalt-groundwater (11) have shown that during the thermal period, pH could be depressed as much as 1 to 2 pH units. However, since significantly elevated temperatures will not extend far into the host rock (12), this study treats only ambient conditions and is applicable to the far-field or the near-field after the thermal period.

Unlike pH, Eh in the basalt-groundwater system is difficult to determine quantitatively. Therefore, since some radionuclide solubilities are highly sensitive to Eh, results are presented for a range of Eh conditions encompassing values expected in the candidate horizons. This approach offers two advantages: (1) the sensitivity of radionuclide solubilities to Eh may be evaluated, thus defining the degree to which Eh must be known in the basalt geochemical system for establishing baseline solubilities and (2) key radionuclides and environmental conditions for experimental confirmation of calculated solubilities are defined.

Measured Eh values for Grande Ronde groundwater, although having associated uncertainties, typically range from +0.35 to -0.2 V. Based upon an indirect approach utilizing mineralogical and aqueous speciation information (13), it is estimated that the Eh conditions in the basaltgroundwater system are -0.40 ± 0.05 V at T = 57°C ± 2°C and pH = 9.5 ± 0.05. A detailed analysis is now underway to evaluate ambient Eh conditions utilizing groundwater chemistry data, laboratory experiments at low and high temperatures, and geochemical modeling.

Other Effects. In the very near-field, dissolution of the waste form may lead to local changes in pH and Eh. In addition, radiolysis effects in this region also may be important. The consequences of these factors are not addressed in this study because their effects are extremely localized and such a detailed analysis of solubility and speciation is beyond the scope of this paper.

Sensitivity Analysis

Two considerations: (1) kinetic limitations in aqueous sulfate/sulfide species equilibration and (2) observed lateral and vertical differences in Grande Ronde groundwater chemistry, were evaluated for their effect on calculated solubilities and associated aqueous speciation.

The presence of sulfate in Grande Ronde groundwater and the uncertainty of the actual Eh conditions require consideration of sulfate/sulfide equilibration kinetics in the calculation of solubilities and aqueous speciation of radionuclides. Under strongly reducing conditions, aqueous sulfide species can be very important complexing agents for some radionuclides, and the precipitation of very insoluble sulfides can control solubilities to very low values. To date, aqueous sulfide species have not been detected in Grande Ronde groundwaters (14). This suggests that either Eh conditions are not reducing enough to stabilize reduced aqueous sulfur species or that equilibrium has not been reached. It has been shown recently (15) that the kinetics of equilibration between aqueous sulfate and sulfide species are very slow without the presence of a mediating component (e.g., sulfate-reducing bacteria or reactive mineral surfaces) at the ambient temperature of the candidate repository horizons (~50°C to 60°C).

For these reasons, it is useful to evaluate the effects of forming or not forming aqueous sulfide complexing agents and sulfide precipitates on radionuclide solubilities and aqueous speciation. Therefore, calculations were performed for key radionuclides for two cases: (1) reduction of sulfate to sulfide and bisulfide was allowed to occur as Eh was lowered and (2) sulfate reduction as a function of Eh was prohibited so that sulfate was the only sulfur complexing agent available at all Eh values.

Most solubility computations were made using the results from 29 natural Grande Ronde groundwaters collected by BWIP hydrologists from five wells on the Hanford Reservation and chemically analyzed in the Basalt Materials Research Laboratory over the past 2 yr. The rationale for using all available analytical data results from the fact that significant vertical and lateral variation in Grande Ronde water chemistry exists and using an average analysis probably is not justified. Sensitivity analyses for the effects of sulfate/sulfide equilibration kinetics in this paper were performed on a single groundwater composition (16). This groundwater has been chosen by the BWIP as a standard reference Grande Ronde groundwater for use in experimental hydrothermal, sorption, and solubility studies. Consequently, its inclusion in this study will permit our conclusions to be related to experimental studies now in progress. An analysis of the reference groundwater is listed in Table I.

Results

Figures 2 through 7 show the results of solubility calculations for a selected group of radionuclides considered in this study as a function of Eh. The bold, solid line in the figures delineates the total solubility of each element in the reference groundwater. In addition, a range of solubilities for each radionuclide is shown by the patterned zones. The range in solubilities results from calculations using the 29 Grande Ronde groundwater analyses. In all calculations the concentration of each aqueous species was computed from its respective activity coefficient. Activity coefficients were calculated by the Güntelberg approximation (17) which is useful for aqueous systems containing mixtures of electrolytes of unlike charge.

Of the elements considered in this study (see Table II), nickel, palladium, antimony, and lead are particularly sensitive to the presence of reduced sulfur species (S^{2-}, HS^-) in the groundwater. For each of these radionuclides, if sulfur speciates under thermodynamic equilibrium conditions, solid sulfide phases will control their solubility at low Eh values. The implication of this fact is illustrated in Figure 1 by a bold, dashed line that corresponds to the solubility of nickel in the reference groundwater and a patterned zone representing the total range

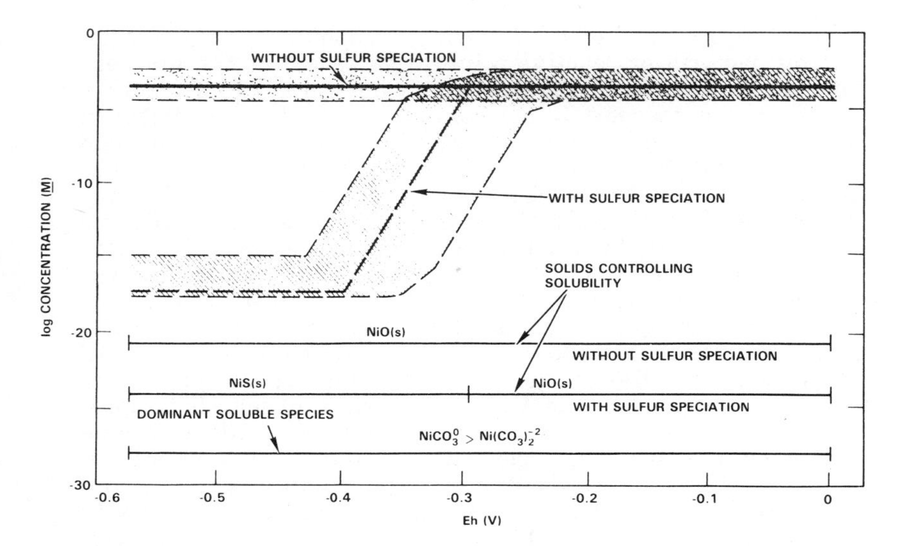

Figure 2. Solubility of Nickel as a Function of Eh at 25°C. The horizontal lines in the lower part of the figure show the range of dominance of solids controlling solubility and aqueous species as functions of Eh. Figures 3 through 7 are interpreted similarly. See text for further explanation.)

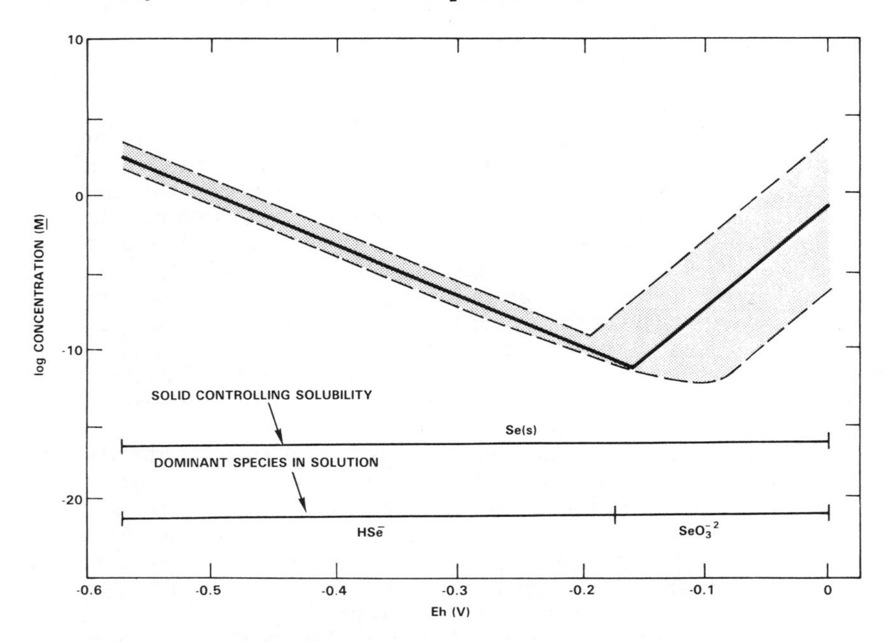

Figure 3. Solubility of Selenium as a Function of Eh at 25°C.

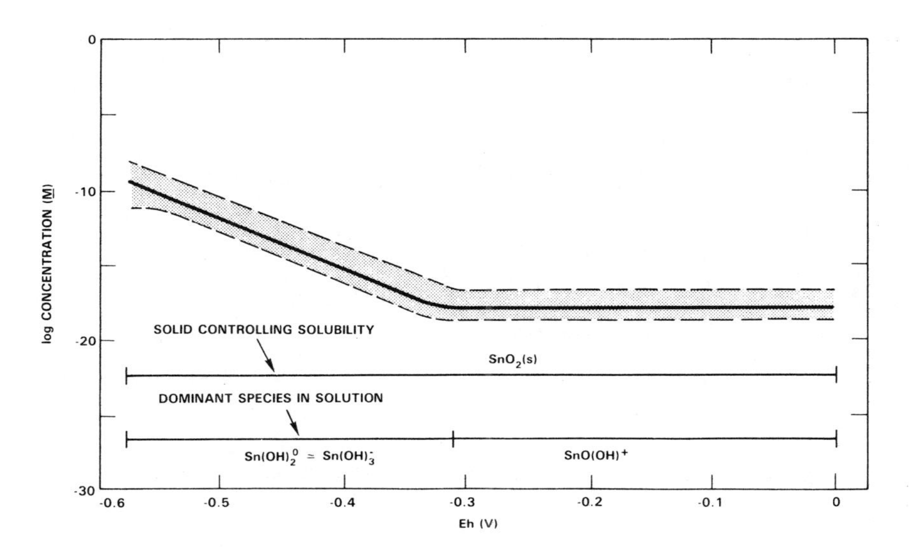

Figure 4. Solubility of Tin as a Function of Eh at 25°C. [Sulfur speciation has only a minor effect on tin solubility (not shown in the figure) for Eh <0 to -.55 V].

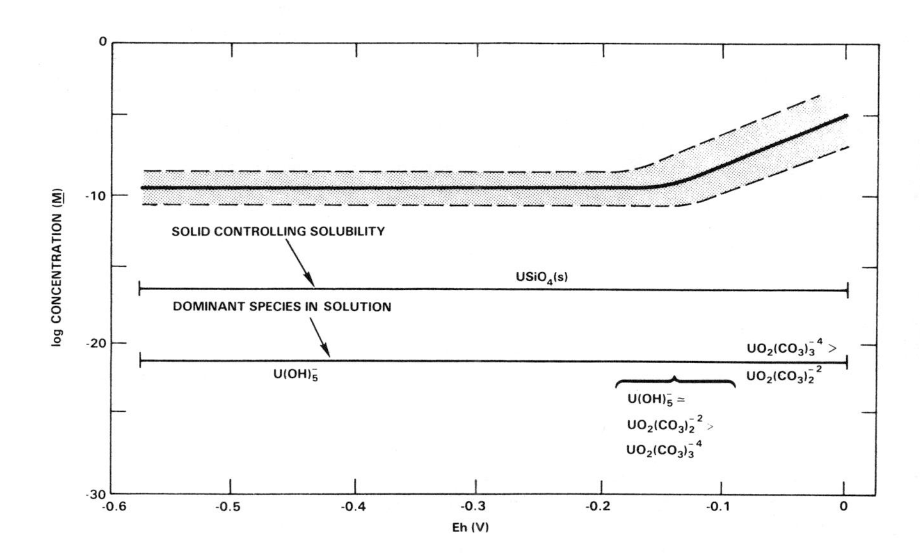

Figure 5. Solubility of Uranium as a Function of Eh at 25°C.

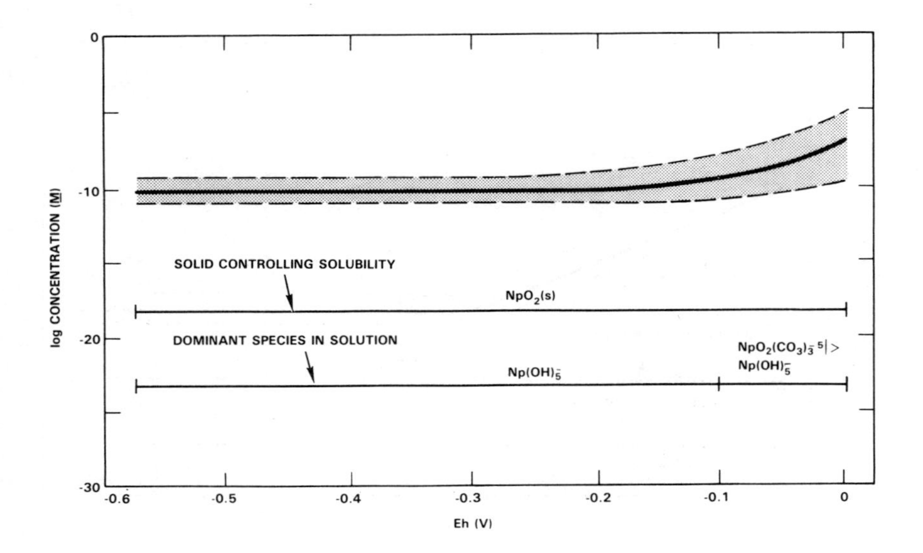

Figure 6. Solubility of Neptunium as a Function of Eh at 25°C.

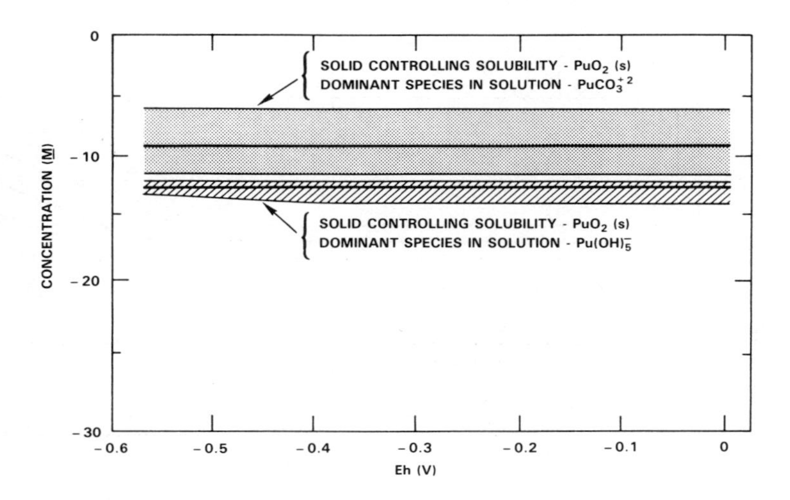

Figure 7. Solubility of Plutonium as a Function of Eh at 25°C. (The upper patterned area represents plutonium solubility in Grande Ronde groundwaters using the published value for the stability constant of $PuCO_3^{2+}$. The lower patterned area represents plutonium solubility in the absence of $PuCO_3^{2+}$ species.)

Table I. Reference Grande Ronde Groundwater Composition at 25°C *

Chemical Species	Concentration (mol/L)
H	1.82×10^{-10}
Na	1.56×10^{-2}
K	8.78×10^{-5}
Ca	6.9×10^{-5}
Mg	1.55×10^{-7}
Si (as SiO_2)	1.27×10^{-3}
F	1.76×10^{-3}
Cl	8.81×10^{-3}
SO_4	1.81×10^{-3}
C (inorganic)	8.96×10^{-4}

*Reference (16).

of its solubility calculated for 29 groundwater analyses with the effects of sulfur speciation included. As noted in the previous section of this paper, analytical data from these groundwaters and kinetic studies (15) suggest that equilibrium sulfur speciation probably does not occur under the conditions encountered in the Grande Ronde formation. Consequently, it will be assumed that this type of equilibrium is not a significant factor in this study and that the solubilities for nickel, palladium, antimony, and lead will not be controlled by sulfide phases at low Eh.

Included in the figures are the specific solid phases that control solubility and those dissolved species that dominate the aqueous solution as a function of Eh for each radionuclide. The criterion used for listing soluble species is that, collectively, they account for $\geq 90\%$ of the total solubility of the element over the range of Eh values considered.

A complete summary of the results of the solubility computations for all radionuclides considered is found in Table II. The solubility values listed assume no sulfur speciation and are for a reference Eh of -0.3 V. This value is approximately midway between estimates of Eh from basalt-mineral equilibria and direct measurements of Grande Ronde groundwaters.

Table II. Solubilities of Radionuclides in Grande Ronde Groundwater at Eh = -.0.3 V

Element	T = 25°C; Eh = -0.3 V		Computed Solubility (T = 25°C; Eh = -0.3 V)		Solubility estimates from experiments and natural waters (mol/L)
	Solids controlling	Dominant species in solution	Range in Grande Ronde groundwaters (mol/L)	Reference Grande Ronde groundwaters (mol/L)	
Ni	NiO	$NiCO_3^0$ $Ni(CO_3)_2^{2-}$	4×10^{-5} - 3×10^{-3}	2×10^{-4}	$\geq 10^{-6}$ [a]
Se	Se	HSe^-	1×10^{-8} - 5×10^{-7}	6×10^{-8}	$\sim 10^{-6}$ [b]
Zr	$ZrSiO_4$	$Zr(OH)_5^-$	1×10^{-10} - 2×10^{-8}	2×10^{-9}	
Pd	Pd	$NiCO_3^0$	2×10^{-12} - 4×10^{-9}	2×10^{-10}	
Sn	SnO_2	$SnO(OH)^+$ $SnO(OH)_2^0$, $Sn(OH)_3^-$	4×10^{-19} - 2×10^{-17}	2×10^{-18}	
Sb	$Sb(OH)_3$	$Sb(OH)_3^0$	1×10^{-7} - 2×10^{-7}	1×10^{-7}	
Sm	$Sm(OH)_3$	$Sm(OH)_3^0$, $Sm(OH)_4^-$ $Sm(OH)_3^{+3}$, $Sm(OH)_2^+$	2×10^{-9} - 6×10^{-8}	2×10^{-9}	
Eu	$Eu(OH)_3$	$Eu(OH)_4^-$, $Eu(OH)_3^0$ $EuCO_3^+$, $Eu(OH)_2^{+3}$	3×10^{-9} - 5×10^{-8}	3×10^{-9}	
		$PbCO_3^0$	7×10^{-9} - 7×10^{-7}	3×10^{-8}	$\geq 10^{-7}$ [a]

Pb	$Pb(OH)_2$	$Pb(CO_3)_2^{2-}$			
Th	ThO_2	$Th(OH)_4^0$	8×10^{-15}	8×10^{-15}	$\leq 10^{-8}$ [b]
U	$USiO_4$	$U(OH)_5^-$	$2 \times 10^{-11} - 2 \times 10^{-9}$	2×10^{-10}	$\geq 10^{-9}$ [a] 10^{-6} [b]
Np	NpO_2	$Np(OH)_5^-$	$2 \times 10^{-11} - 7 \times 10^{-10}$	1×10^{-10}	4×10^{-5} [b]
Pu	PuO_2	$Pu(OH)_5^-$ [c] $PuCO_3^{2+}$ [d]	$5 \times 10^{-12} - 1 \times 10^{-6}$ [c] $4 \times 10^{-14} - 2 \times 10^{-12}$ [d]	8×10^{-10} [c] 3×10^{-13} [d]	
Am	$Am(OH)_3$	$Am(CO_3)_2^-$, $Am(CO_3)_3^{3-}$ $AmCO_3^+$	$4 \times 10^{-9} - 1 \times 10^{-6}$	2×10^{-8}	$10^{-9} - 10^{-8}$ [e]

[a] From maximum concentration levels observed in Hanford groundwaters.
[b] From sorption experiments.
[c] Solubilities based upon published stability constant for $PuCO_3^{2+}$ complex.
[d] Solubilities based upon absence of $PuCO_3^{2+}$ complex.
[e] Edelstein et al. (1983).

In view of the fact that the basalt flows that currently are being considered for a repository at Hanford are at an ambient temperature between 50°C and 60°C, the effect of temperature on radionuclide solubility should be evaluated. As noted previously, only limited thermodynamic data exist with which to make such extrapolations. Estimates of thermodynamic parameters for uranium and plutonium are available (9-10). These estimates suggest that at temperatures of 50°C to 60°C the solubilities of these elements in Grande Ronde groundwater will be $\leq$ one order of magnitude greater than their respective solubilities at 25°C. Analogous estimates for the other radionuclides considered in this study are even more uncertain, but increases in solubility due to temperature of about one order of magnitude seem reasonable.

Examination of the results in the figures and Table II leads to several significant conclusions. Solids that control solubility for these radionuclides include hydroxides and hydrous oxides (palladium, antimony, samarium, europium, lead, and americium), oxides (nickel, tin, thorium, neptunium, and plutonium), elements (selenium and palladium), and silicates (zirconium and uranium). This rather narrow range of structural/chemical types is due largely to the high pH and low Eh of the groundwater. Similarly, the dominant soluble species are predicted to be hydroxy complexes (zirconium, palladium, tin, antimony, samarium, europium, thorium, uranium, neptunium, and plutonium) and carbonate species (nickel, samarium, europium, lead, uranium, neptunium, plutonium, and americium). These complexes result from the high pH of the solution and the apparent lack of other ligands in significant quantities in the groundwater that are known to form strong complexes with many radionuclides. For example, phosphates have not been detected in these groundwaters (detection limit ~<1 mg/l).

The range of observable compositions for Grande Ronde groundwater generally leads to calculated ranges in solubilities of about two to three orders of magnitude. Because many of the dominant soluble species are either hydroxy or carbonate complexes, much of the variation in solubility for these radionuclides is due, either directly or indirectly, to the range of pH values for the groundwaters. Measured differences in concentrations of F^-, Cl^-, SO_4^{2-}, or SiO_2^0 do not seem to alter significantly the computed solubility results for these radionuclides. However, recent experimental studies (18) suggest that plutonium solubility in groundwater from a flowtop of the Grande Ronde basalt in the Pasco Basin is enhanced by the high fluoride concentration of the water. The implications of this result have not been fully evaluated but it might suggest inadequacies in the thermodynamic data base.

The potential significance of organic complexing agents in these groundwaters must not be ignored. Approximately 0.3 mg/L of fulvic acids were found in a single sample of Grande Ronde groundwater from a borehole at Hanford (19). A literature survey to document available information relative to complexing of actinides with naturally occurring humic and fulvic acids has been compiled (20). Apparently both humic and fulvic acids can form very strong complexes with all oxidation states of these radionuclides. Even if strong actinide-organic acid complexes form in an NWRB their mobility through bentonite backfill material may be reduced significantly due to the strong sorptive capacity of clays for humic and fulvic acids. Nevertheless, an evaluation of the relative importance of these acids in the basalt geochemical environment awaits future, planned characterization studies.

A number of studies have been reported in the literature in which theoretical aqueous speciation of key radionuclides has been determined from available thermodynamic data (4,9,21-25). In general, where comparisons are possible, one finds reasonably close agreement in the calculated dominant dissolved species, the solids controlling solubility, and computed total solubilities for this study and those listed above.

Few direct, experimental determinations of radionuclide solubilities applicable to this study are available in the literature. However, solubility estimates for some radionuclides can be extracted from the results of conventional sorption experiments (26). In addition, natural levels of some elements listed in Table II that are observed in Hanford groundwaters provide crude estimates of minimum solubility limits for them. Finally, measured solubility results for americium in aqueous solutions have been reported (27).

Solubility estimates made by the techniques discussed above are reported in the last column of Table II. In addition to the limited number of such measurements, the results do not compare favorably in all cases with the theoretical values listed. This fact is hardly surprising considering the recognized limitations in the thermodynamic data base and difficulties encountered in interpreting results of solubility experiments. Furthermore, the theoretical estimates are based on the assumption that the thermodynamically most stable solid for a radionuclide controls its solubility. The effects of metastability are not included and, in this sense, theoretical solubility estimates are not conservative. A series of sorption-type experiments designed to yield solubility estimates for a number of the radionuclides included in this paper is in progress, and the results will be reported at a later date.

Discussion

One application of the results from the present study to practical problems of waste management addresses the dissolution of a spent fuel waste form and transport of radionuclides into the geologic environment. According to present plans, an NWRB at the Hanford Site will be designed to receive at least a portion of its high-level waste as spent fuel, while the remainder will be in the form of borosilicate glass. A glass waste form has not been well characterized structurally, but radionuclide releases from it probably will be controlled by competing rates of glass dissolution and subsequent reprecipitation reactions, perhaps involving some of the solid phases treated in this study as solubility controls. Discussions of the chemical and structural characteristics of spent fuel are available (28-30). In general, spent fuel consists of a matrix of UO_2 in which the actinides such as plutonium, neptunium, thorium, and americium are in solid solution in the UO_2. Some uranium may be in the form of uranate compounds such as $CsUO_4$. Of the remaining elements considered in this study, europium, samarium, lead, nickel, zirconium, and antimony probably will be in the form of oxides. Tin and palladium will be found as metallic alloy inclusions, and selenium will form various selenide compounds. A bundle of fuel elements will be inserted into metal canisters and emplaced into boreholes in the basalt. Currently, a mixture of about 75% basalt and 25% sodium-montmorillonite clay is planned to be used to backfill around the canisters within the boreholes. Radioactive heating of the waste during the early stages of storage is expected to produce a thermal gradient around the repository extending tens of meters from the repository and having a duration of up to hundreds or thousands of years.

After breaching of the canisters by corrosion, dissolution of the spent fuel waste form will begin. The dissolution process will be controlled by the solubility of the UO_2 matrix and the other discrete phases present. These crystalline phases will have suffered thermal and radiation damage and will be more soluble than predicted by thermodynamics. Additionally, the elevated temperatures and radiolysis effects likely to exist during the early stages of dissolution may lead to enhanced solubilities of most radionuclides. However, it is expected that reprecipitation of more stable solid phases will occur within the waste package backfill and ultimately control the solubility of the radionuclides to lower levels. One can argue effectively that kinetic effects may prohibit the most stable phases for some radionuclides from forming in this environment. However, no data exist at present to evaluate the importance of kinetic factors. In addition, one must consider the effects of solid solution in determining radionuclide solubilities. The rare earth elements (europium and samarium) may coprecipitate as a hydroxide solid solution. Similarly, the actinides and zirconium may well coprecipitate in a common phase (oxide or hydroxide solid solution). Coprecipitation may occur even if a thermodynamically more table phase exists for one or more of the nuclides. These coprecipitation effects tend to raise the observed solubilities of those nuclides for which more stable solid phases exist under the environmental conditions considered.

Transport of the dissolved radionuclides will occur down a thermal gradient in the backfill and repository host rock. As temperature decreases with increasing distance from the waste package, solubilities for most nuclides should decrease, resulting in precipitation. As before, the effects of coprecipitation may be significant. As transport processes carry remaining dissolved radionuclides beyond the region of thermal perturbation around the repository, further precipitation reactions probably will continue only in response to local variations in groundwater chemistry, or if a significant level of supersaturation exists for specific radionuclides.

One can attempt to overlay the effects of sorption on this scenario. It is apparent that strong sorption of radionuclides will tend to depress the equilibrium concentration of the element in groundwater, and the precipitation "front" described here may be restricted to areas very near the repository when compared to radionuclide transport in the absence of sorption.

Conclusion

This study has resulted in the following conclusions:

- In general, oxide and hydrous oxide solid phases provide solubility control for most of the radionuclides considered

- The predominant soluble species for the radionuclides considered are predicted to be neutral or anionic hydroxide and carbonate complexes

- Selenium, palladium, and tin solubilities appear to be strong functions of Eh. The solubility of selenium and palladium may reach very high levels under some Eh conditions. This result suggests that future experimental studies should include solubility determinations for these elements under appropriate T, pH, and Eh conditions.

It has been the underlying theme of this paper that the theoretical approach to solubility and speciation of key radionuclides adopted here can be helpful in identifying the general types of species likely to be dominant in aqueous solutions. At present, these estimates are useful in identifying those radionuclides that require special attention in repository performance assessment studies. However, sophisticated experimental studies are necessary to provide important confirmatory data for nuclides of critical importance to the safe isolation of nuclear waste in a repository in basalt.

Literature Cited

1. EPA, "Environmental Standards and Federal Radiation Protection Guidance for Management and Disposal of Spent Nuclear Fuel, High-Level, and Transuranic Radioactive Wastes," 40 CFR, 191, working draft No. 20, U.S. Environmental Protection Agency, 1981.
2. NRC, "Disposal of High-Level Radioactive Wastes In Geologic Repositories: Technical Criteria," 10 CFR Part 60, U.S. Nuclear Regulatory Commission, 1982.
3. Garrels, R. M.; Christ C. L. "Solutions, Minerals, and Equilibria"; Freeman, Cooper and Company, San Francisco, California, 1965.
4. Rai, D.; Serne R. J. "Solid Phases and Solution Species of Different Elements in Geologic Environments," PNL-2651, Pacific Northwest Laboratories, Richland, Washington, 1978.
5. Benson, L. V.; Teague L. S. "A Tabulation of Thermodynamic Data for Chemical Reactions Involving 58 Elements Common to Radioactive Waste Package Systems," Topical Report No. 4 for Rockwell Hanford Operations, LBL-11448, Lawrence Berkeley Laboratory, University of California, Berkeley, California, 1980.
6. Phillips, S. L. "Hydroloysis and Formation Constants at 25°C," LBL-14313, Lawrence Berkeley Laboratory, University of California, Berkeley, California, 1982.
7. Rai, D.; Strickert, R. J.; Moore, D. A. "Am(III) Hydrolysis Constants and Solubility of Am(III) Hydroxide," PNL-SA-10635, Pacific Northwest Laboratories, Richland, Washington, 1982.
8. Nair, G.; M., Chander, K.; J. K. Joshi, J. K. "Hydroloysis Constants of Plutonium(III) and Americium(III)," Radiochimica Acta 1982, 30, 37-40.
9. Langmuir, D. "Uranium Solution-Mineral Equilibria at Low Temperatures with Applications to Sedimentary Ore Deposits," Geochimica et Cosmochimica Acta 1982, 42, 547-569.
10. Lemire, R. J.; Tremain, P. R. "Uranium and Plutonium Equilibria in Aqueous Solutions to 200°C," J. Chem. and Engr. Data 1982, 25, 361-370.
11. Apted, M. J.; Meyers, J. "Comparison of the Hydrothermal Stability of Simulated Spent Fuel and Borosilicate Glass in a Basaltic Environment," RHO-BW-ST-38 P, Rockwell Hanford Operations, Richland, Washington, 1982.
12. King, I. P.; McLaughlin, D. B.; Norton, W. R.; Baca, R. G.; Arnett, R. C. "Parametric and Sensitivity Analysis of Waste Isolation in a Basalt Medium," RHO-BWI-C-94, Rockwell Hanford Operations, Richland, Washington, 1982.

13. Jacobs, G. K.; Apted, M. J. "Eh-pH Conditions for Groundwater at the Hanford Site, Washington: Implications for Radionuclide Solubility in a Nuclear Waste Repository Located in Basalt," Trans. Am. Geophys. Un.1981, 62, 1065.
14. Gephart, R. E.; Arnett, R. C.; Baca, R. G.; Leonhart, L. S.; Spane, Jr. F. A. "Hydrologic Studies within the Columbia Plateau, Washington: An Integration of Current Knowledge," RHO-BWI-ST-5, Rockwell Hanford Operations, Richland, Washington, 1979.
15. Ohmoto, H.; Lasaga, A. C. "Kinetics of Reactions Between Aqueous Sulfates and Sulfides in Hydrothermal Systems," Geochimica et Cosmochimica Acta 1982, 46, 1727-1745.
16. Jones, T. E. "Reference Material Chemistry, Synthetic Groundwater Formulation," RHO-BW-ST-37 P, Rockwell Hanford Operations, Richland, Washington, 1982.
17. Stumm, W.; Morgan, J. J. "Aquatic Chemistry"; 2nd ed., John Wiley & Sons, New York, New York, 1981.
18. Cleveland, J. M.; Reese, T. F.; Nash, K. L. "Groundwater Composition and Its Relationship to Plutonium Transport Processes" Amer. Chem. Soc. Fall Annual Mtg. Kansas City, Sept. 1982, Abstracts of Papers (NUCL 37).
19. Means, J. L. "The Organic Geochemistry of Deep Groundwaters," ONWI-268, Battelle Columbus Laboratories, Columbus, Ohio, 1982.
20. Olofsson, U.; Allard, B. "Complexes of Actinides with Naturally Occurring Organic Substances - Literature Survey," KBS Technical Report 8309, Stockholm, Sweden, 1983.
21. Langmuir, D.; Herman, J. S. "The Mobility of Thorium in Natural Waters at Low Temperatures," Geochimica et Cosmochimica Acta 1980, 44, 1753-1756.
22. Allard, B.; Kipatsi, H.; Liljenzin, J. O. "Expected Species of Uranium, Neptunium, and Plutonium in Neutral Aqueous Solutions," J. Inorg. and Nuclear Chemistry 1982, 42, 1015-1027.
23. Goodwin, B. W. "Maximum Total Uranium Solubility Under Conditions Expected in a Nuclear Waste Vault," TR-29, Atomic Energy of Canada Limited, Pinawa, Manitoba, Canada, 1980.
24. Newton, T. W.; Aguilar, R. D.; Erdal, B. R. "Estimation of U, Np, and Pu Solubilities vs Eh and pH," Laboratory Studies of Radionuclide Distributions Between Selected Groundwaters and Geologic Media, Jan. 1, Mar. 31, 1980, Erdal, B. R. (compiler), LA-8339-PR, Los Alamos Scientific Laboratory, Los Alamos, New Mexico, 1980.
25. Schweingruber, M. R. "Evaluation of Solubility and Speciation of Actinides in Natural Groundwaters," TM-45-82-11, Swiss Federal Institute for Reactor Research, Würenlingen, Switzerland, 1982.

26. Salter, P. F.; Ames, L. L. "The Sorption Behavior of Selected Radionuclides on Columbia River Basalts," RHO-BWI-LD-48, Rockwell Hanford Operations, Richland, Washington, 1981.
27. Edelstein, N.; Bucher, J.; Silva, R.; Nitsche, H. "Thermodynamic Properties of Chemical Species in Nuclear Waste," LBL-14325, Lawrence Berkeley Laboratory, Berkeley, California, 1983.
28. Bramman, J. I.; Sharpe, R. M.; Thom, D.; Yates, G. "Metallic Fission-Product Inclusions in Irradiated Oxide Fuels," Jour. Of Nuclear Materials 1968, 25, 201.
29. Davies, J. H.; Ewart, F. T. "The Chemical Effects of Composition Changes in Irradiated Oxide Fuel Materials," Jour. of Nuclear Materials 1971, 41, 143-155.
30. Bazin, J.; Jouan, J.; Vignesoult, N. "Compartment et Etat Physico-Chimique des Produits de Fission dans les Elements combustibles pour Reacteur a Eau Pressurisee," Bulletin d' Information Scientifique et Technique 1974, 196, No. 55.

RECEIVED January 12, 1984

10

Radionuclide–Humic Acid Interactions Studied by Dialysis

LARS CARLSEN, PETER BO, and GITTE LARSEN

Riso National Laboratory, Chemistry Department, DK-4000 Roskilde, Denmark

Dialysis has been used to study the interaction between radionuclides and humic acid. The technique gives information on 1) the complexing capacity of the humic acid samples, i.e. the concentration of complexing sites, and 2) the radionuclide - humic acid complexes by a) its stoichiometry (i.e. metal ion/ligand ratio), and b) interaction constants. The applicability of the technique is illustrated by studies on the interaction between a humic acid and $^{134}Cs^{+}$, $^{85}Sr^{2+}$, $^{60}Co^{2+}$, and $^{154}Eu^{3+}$.

In the course of time it has been unambiguously demonstrated that humic- and fulvic acids interact with metal cations by forming rather stable, and often soluble complexes(1,2). The increasing awareness of a possible pollution of the environment, e.g. in connection with the disposal of nuclear waste, emphasizes the need for additional knowledge about the interaction between relevant metal ions, e.g. radionuclides commonly present in nuclear waste, and humic substances. The possible presence of soluble and rather stable complexes may play an important role in determining the migration behavior of the metal ions under shallow land burial conditions. The influence of humic- and fulvic acids on the migration behavior of metal ions has been discussed previously (2-6).

A variety of techniques, including ion-exchange equilibrium (7), potentiometric titration(8,9), application of ion-selective electrodes(9,10), spectroscopic measurements(11), liquid-liquid partition(12), and gel filtration(13), have applied to studies on metal ion - soil organic matter interactions. The former of these was originally developed by Schubert (7) and later was successfully modified by Ardakani and Stevenson (14). Together with the application of ion-selective electrodes, the ion-exchange equilibrium technique is, for the time being, the more applied method. However, a general disadvantage

0097–6156/84/0246–0167$06.00/0

using these techniques is the lack of immediate information about the complexing capacity of the ligand. In the past, the effective ligand concentration [A] has been expressed in terms of molar units (11, 15), amount of ligand material (16), and concentration of complexing sites (17, 18); the latter of these terms seems to be the more appropriate as it characterizes the humic acid samples more satisfactorily than an average molecular weight.

Recently, Weber and co-workers reported on the application of dialysis to the determination of the complexing capacity of fulvic acid for a series of metal cations (19, 20). In the present paper we report a further elaboration on the dialysis technique to allow simultaneous determination of complexing capacity for the humic acid sample, and interaction constants for the metal ion - humic acid complexes, as well as the stoichiometry of the latter. The technique furthermore opens up a possibility for an indirect determination of apparent ionization constants for the humic acids.

Experimental

Samples (10 mL) of humic acid in the appropriate buffer (μ = 0.05-0.1), typically containing between 0.1 and 2.0 grams of humic acid per litre (determined by electronic absorption spectroscopy), were placed in regenerated cellulose dialysis bags. The closed bags were placed in 100 mL bottles, previously filled with 40 mL of the buffer containing varying concentrations (10^{-7} - 10^{-4} M) of the metal ions Cs^+, Sr^{2+}, Co^{2+}, or Eu^{3+}, and trace amounts of $^{134}Cs^+$, $^{85}Sr^{2+}$, $^{60}Co^{2+}$, or $^{154}Eu^{3+}$, respectively.

The present studies were carried out using commercially available humic acid, obtained as the corresponding sodium salt (EGA H1,675-2) and dissolved in a phosphate buffer (pH = 6.99), and an acetate buffer (pH = 4.47), respectively. The humic acid solutions were dialyzed against pure buffer solution prior to their use in the complex formation experiments in order to remove any low molecular fractions that could pass through the dialysis membrane.

The closed bottles were agitated gently for approximately 48 hours, the temperature being kept at 25.0 ± 0.1°C by a Heto 02 PT 923 thermostat. After equilibration, samples mL of the solutions outside the dialysis bag were withdrawn, and the contents of metal ions, $[M]_o$, were determined by γ counting (Kontron MR 252 Automatic Gamma Counting System) and comparison with standard solutions. The relatively high ionic strength of the buffer solutions (μ = 0.05-0.1) ensures an equal distribution of low molecular species that pass the dialysis membrane, i.e. $[M]_o$ measured in the solution outside the dialysis bag equals that inside the bag. Blind-tests, i.e. using dialysis bags containing pure buffer solution, revealed that equilibrium was in general obtained after 25 - 30 hours. Furthermore the blind-tests afforded information on the possible sorption of metal ions on the dialysis bags as function of metal ion concentration: $Z([M]_o)$.

Dialyses are normally carried out with four different humic acid concentrations, each of which is combined with seven different metal ion concentrations (up to 2×10^{-5} mol/L). For the determination of the complexing capacity a further series of three metal ion concentrations (up to ca. 10^{-4} mol/L) was added in the case of the lowest humic acid concentration, in order to occupy all sites available.

Theory

Metal ions, M, react with humate anions, A, to form complexes of the general type MA_j, taking only mononuclear species into account.

$$M + jA \rightleftharpoons MA_j \tag{1}$$

$$I = \frac{[MA_j]}{[M][A]^j} = \frac{[M]_c}{[M]_f\,[A]^j} \tag{2}$$

The latter relation being valid since $[MA_j] = [M]_c$.

Since the metal ions interact with the humate anions to form the complexes the dissociation of the humic acid should be taken into account:

$$HA \rightleftharpoons H + A \tag{3}$$

$$\frac{[HA]}{[A]} = \frac{[H]}{K_A} \tag{4}$$

As $[A]$ in the present context is the concentration of humate anions, which may possibly participate in the complex formation K_A has to be regarded as an apparent ionization constant, not of nessessity equal to that obtained by acid-base titration.

The total concentration of complexing sites available is given by the expression (5), the number of ligands being occupied in the complex formation being equal to $j[MA_j] = j[M]_c$.

$$[A]_t = [HA] + [A] + j[M]_c = \left(\frac{[H]}{K_A} + 1\right)[A] + j[M]_c \tag{5}$$

It is important to note that $[A]_t$ cannot be regarded as a universal constant for a given humic acid sample, as it may be dependent of the nature of the participating metal ions, M (12).

The total amount of metal ions, M_t, can be expressed as a sum of the amounts of metal ions in the dialysis bag, in the solution outside the latter, and the amount possibly sorbed on the bag (cf. Fig. 1).

$$M_t = V_i\,([M]_c + [M]_o) + V_o\,[M]_o + Z([M]_o) \tag{6}$$

A rearrangement of the relation (5) gives the following expression of the actual free ligand concentration [A].

$$[A] = \left([A]_t - j[M]_c\right) \Big/ \left(1 + \frac{[H]}{K_A}\right) \tag{7}$$

Combining the equations (2) and (7) affords the expression (8) for the so-called pH dependent Metal ion - Humic acid Interaction constant β.

$$\frac{[M]_c}{[M]_f\left([A]_t - j[M]_c\right)^j} = \frac{I}{\left(1 + \frac{[H]}{K_A}\right)^j} = \beta \tag{8}$$

where $[M]_c$ is given by

$$[M]_c = \frac{M_t - Z([M]_o)}{V_i} - \left(1 + \frac{V_o}{V_i}\right)[M]_o \tag{9}$$

It shall be remembered that part of the metal ions, which are not engaged by the humic acid complexation, may interact with the buffer solution, containing a certain concentration of 'buffer - ligands', [L], i.e. $[M]_f < [M]_o$. The metal ion - buffer-ligand interaction is given by the following set of equations.

$$M + L \rightleftharpoons ML \tag{10a}$$

$$ML + L \rightleftharpoons ML_2 \tag{10b}$$

$$\vdots$$

$$ML_{i-1} + L \rightleftharpoons ML_i \tag{10c}$$

$$\vdots$$

$$[ML_i] \big/ [M]_f [L]^i = \beta_{ML_i} \equiv \prod_{i=1}^{n} K_{ML_i} \tag{11}$$

Since $[M]_o = [M]_f + \sum[ML_i]$ we obtain the rather simple relation, (12), between $[M]_f$ and $[M]_o$.

$$[M]_f = [M]_o \Big/ \left(1 + \sum_{i=1}^{n} \beta_{ML_i} [L]^i\right) = [M]_o \big/ Q \tag{12}$$

In cases where $[M]_c \ll [A]_t$ eqn's (8) and (9) can be simplified as follows, incorparating (12)

$$\beta [M]_f [A]_t^j = \frac{\beta [M]_o [A]_t^j}{Q} = \frac{M_t - Z([M]_o)}{V_i} - \left(1 + \frac{V_o}{V_i}\right)[M]_o \tag{13}$$

which easily is rearranged into the expression (14):

$$[M]_o = \frac{M_t - Z([M]_o)}{V_i}\left(\frac{\beta}{Q}[A]_t^j + \left(1 + \frac{V_o}{V_i}\right)\right)^{-1} \tag{14}$$

It is seen that for fixed values of β and Q (i.e. fixed pH),

V_o, and V_i plots of $[M]_o$ vs. $(M_t - Z([M]_o))/V_i$ will result in straight lines, the slope being dependent of $[A]_t$ only, i.e. the amount of humic acid. It is noted that an increasing amount of humic acid causes a decrease in the slope α (cf. Fig. 2).

$$\alpha^{-1} = \frac{\beta}{Q}[A]_t^j + \left(1 + \frac{V_o}{V_i}\right) \quad (15)$$

Introducing the term $y = \alpha^{-1} - (1 + V_o/V_i)$ we have

$$y = \beta[A]_t^j/Q \quad (16)$$

which gives

$$\log y = \log\beta + j\log[A]_t - \log Q \quad (17)$$

By means of the complexing capacity, w (eq/g), $[A]_t$ is expressed in terms of the amount (in g/L) of humic acid, or sodium humate, $(HA)_w$, present in the solution.

$$[A]_t = w(HA)_w \quad (18)$$

Accordingly (17) can be rearranged into

$$\log y = (\log\beta + j\log w - \log Q) + j\log(HA)_w \quad (19)$$

A plot of log y as function of log $(HA)_w$ represents a straight line, the slope being equal to j, the number of ligands pr. metal ion (cf. Fig. 3). The intercept, $\log\beta + j\log w - \log Q$, may in principle be used for the calculation of β, if w is known, since $\log\beta = (\log\beta + j\log w - \log Q) - j\log w + \log Q$. However, since determinations of intercepts in general may be rather defective, it seems more reasonable to use the dialysis results directly, calculating β according to eqn. (8). In general up to 20-30 individual sets of $[A]_t$ and $[M]_c/[M]_f$ will be available for this purpose vide supra). Determination of the complexing capacity is carried out by increasing the metal ion concentration to a level, where all sites available in the humic acid, under the actual pH-condition, will be occupied. For convenience a solutions with low humic acid concentration, $(HA)_w$ (0.1 - 0.2 g/L) are used for this purpose. A plot of $[M]_f$ ($\equiv [M]_o/Q$) as a function of $[M]_c$ will feature a vertical asymptote at $[M]_c = [M]_{c,max}$ (cf. Fig. 4), the complexing capacity, w, being determined by

$$[A]_t = j[M]_{c,max} = w(HA)_w \quad (20)$$

which gives

$$w = j[M]_{c,max}/(HA)_w \quad (21)$$

The total concentration of complexing sites (in eq/L) is extracted from eqn. (20), since w and $(HA)_w$ are both known. Finally β can, as mentioned, be calculated by eqn. (8).

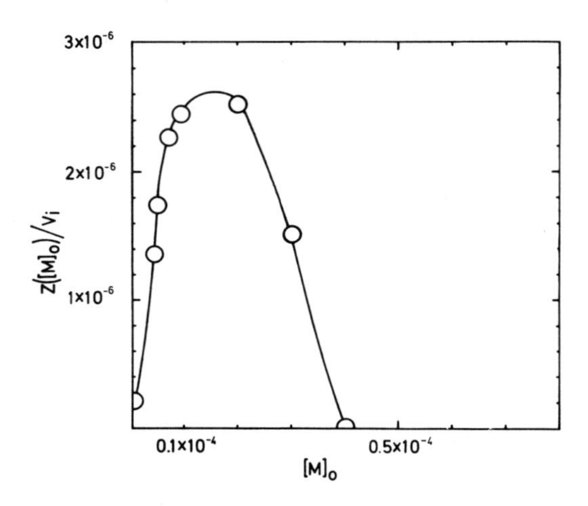

FIGURE 1. Sorption of europium ions on dialysis bags as function of europium concentration.

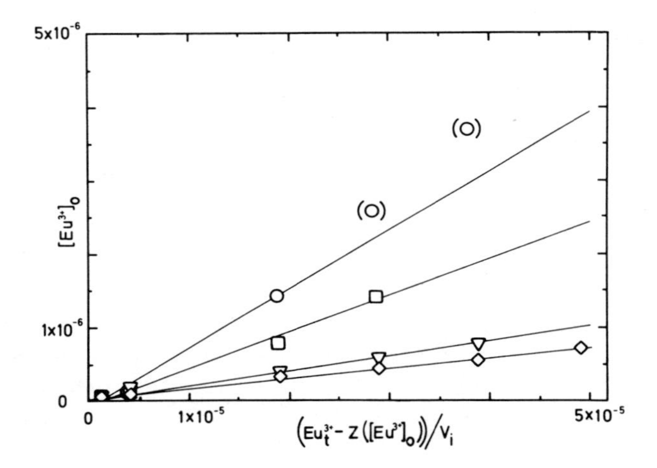

FIGURE 2. Variation in europium concentration outside the dialysis bag as function of the total amount of europium.

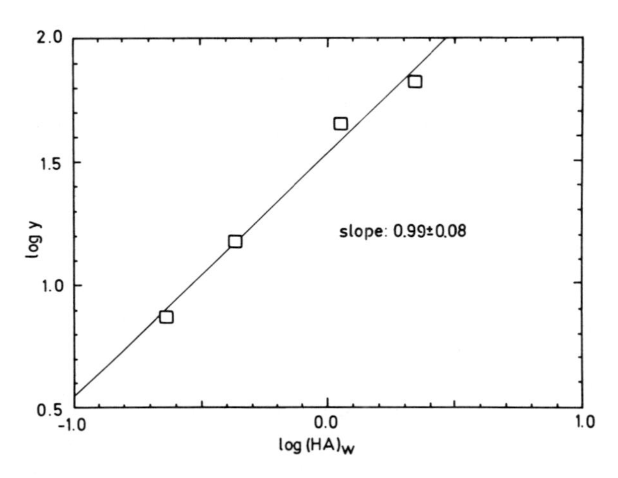

FIGURE 3. Determination of the number of ligands pr. europium ion (j) in Eu-HA complexes.

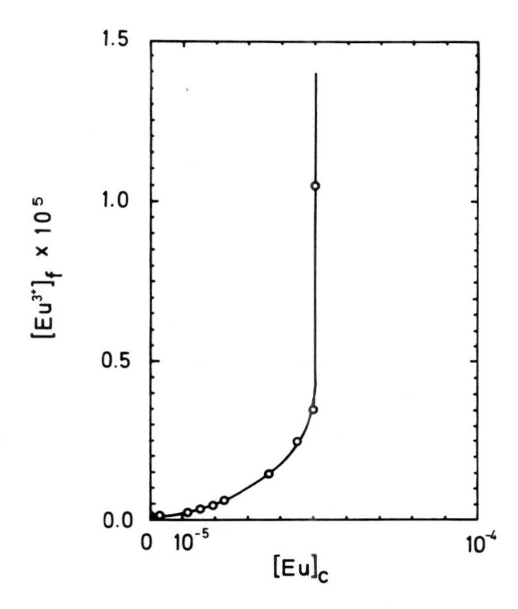

FIGURE 4. Free europium ion concentration as function of the concentration of complexed europium.

Results and Discussion

Based on previously reported results, it is expected that Eu^{3+} ions will interact strongly with the humate anions, leading to complexes exhibiting rather high stability constants. Bertha and Choppin (18) reported simultaneous formation of 1:1 and 1:2 complexes, the corresponding stability constants being log β^1 = 7.8 and log β^2 = 10.7, respectively (pH = 4.50). A somewhat lower stability constant for a 1:1 Eu^{3+} - fulvic acid complex was found (log β^1 = 6.90)(18) in close agreement with results reported by Marinsky et al. (log β^1 = 6.36)(21). On the other hand it is expected that Co^{2+} ions and humate anions will lead to considerably less stable complexes. Chimielewska (17) reported on Co-HA complexes in the pH-range 4.7 -5.8, the corresponding variation in log β ranged from 4.65 to 5.49 and 5.65 to 6.29, respectively, for two different humic acid samples. The corresponding variations in the number of ligands per metal ion, j, were 1.08-1.3 and 1.4-1.59, respectively. Adhikari et al. (22) reported on a 1:1 Co-HA complex, the stability constant being log β = 3.01 (pH = 4, T = 30°C). Additionally some results on Co-FA complexes are available (11, 15). In the cases of strontium and cesium no reports on the actual interaction with humate anions in solution have appeared. However, it has been reported that both types of ions sorb onto solid humic acid (24, 25).

An introductory series of experiments on the possible sorption of the metal ions on the dialysis bags revealed that only europium, to any significant extent, sorbed to the cellulose bags. Figure 1 depicts the sorption of Eu^{3+} by displaying the correction term $Z([Eu^{3+}]_o)$ as function of $[Eu^{3+}]_o$.

In Figure 2 the relation between $[M]_o$ and $(M_t - Z([M]_o))/V_i$ is visualized for the humic acid interaction with Eu^{3+} ions in an acetate buffer (pH = 4.47, μ = 0.05, T = 25.0°C). It should be remembered that the relation given in eqn. (14) is valid for $[M]_c \ll [A]_t$ only. In cases where the latter condition is not satisfied, strong deviations from the straight lines are observed, as indicated in Figure 2 by the points in parentheses. The slopes of the lines were determined by a least square procedure to be 8.04×10^{-2}, 5.00×10^{-2}, 2.03×10^{-2}, and 1.40×10^{-2}, for $(HA)_w$ equal to 0.23, 0.43, 1.13, and 2.21 g/L, respectively. The corresponding log y-values (cf. eqn. (17)) are 0.871 1.18, 1.65, and 1.82, respectively.

In Figure 3 the plot of log y as a function of log $(HA)_w$ is shown, the slope being determined to be j = 0.99 ± 0.08.

In order to calculate $[Eu^{3+}]_f$ (eqn. (12)) three different europium - acetate complexes have to be considered: $EuAc^{2+}$, $EuAc_2^{+}$, and $EuAc_3$, the corresponding stability constants being $10^{1.79}$, $10^{3.11}$ and $10^{4.24}$, respectively, (26) corresponding to a correction factor Q = 8.48.

In Figure 4 the dependence of $[Eu^{3+}]_f$ as function of $[Eu]_c$ is depicted, the plot featuring the vertical asymptote at $[Eu]_c$ = 5.1×10^{-5} mol/L. Since the corresponding amount of sodium humate is 0.23 g/L, the complexing capacity can be calculated, according to eqn. (21), to be $(2.20 \pm 0.18)x10^{-4}$ eq/g.

Based on four different ligand concentrations $[A]_t$ $5.06x10^{-5}$ $9.46x10^{-5}$, $2.49x10^{-4}$, and $4.86x10^{-4}$ eq/L and seven different concentrations of europium ions, ranging from $2.66x10^{-7}$ to $2.01x10^{-5}$ mol/L, corresponding to 28 mutually connected sets of data, the overall Eu-HA interaction constant at pH = 4.47, log β, is calculated (eqn. (8)) to be 5.86 ± 0.15.

The results obtained for the differential radionuclides are summarized in Table I.

TABLE I. Interaction Between Cs^+, Sr^{2+}, Co^{2+}, and Eu^{3+} and Humic acid in Different Buffer Systems (Ph: phosphate buffer, Ac: acetate buffer)

M	pH	μ	Q	w x10^4 (eq/g)[a]	j	log β
Cs	7.00(Ph)	0.1	1.0[b]	-	-	<1.2
Cs	4.49(Ac)	0.05	1.0[b]	-	-	<1.2
Sr	6.99(Ph)	0.1	1.56	2.16 ± 0.40	0.82 ± 0.15	3.32 ± 0.23
Sr	4.49(Ac)	0.05	1.13	0.34 ± 0.02	0.71 ± 0.04	3.12 ± 0.04
Co	7.01(Ph)	0.1	6.49	1.36 ± 0.23	1.06 ± 0.18	5.68 ± 0.16
Co	4.49(Ac)	0.05	1.97	0.51 ± 0.01	1.00 ± 0.01	4.53 ± 0.14
Eu	4.47(Ac)	0.05	8.48	2.20 ± 0.18	0.99 ± 0.08	6.15 ± 0.15

[a] complexing capacity of sodium humate
[b] assumed values; no data on Cs-Ph and Cs-Ac interactions available

For rather weak complexes, e.g. Cs-HA, we are not able to derive interaction constants properly by the dialysis technique. Decreases in $[M]_o$ by less than 1%, relative to $[M]_t$, will be insignificant. Thus, it is seen that (cf. eqn.'s 8 and 13, $[M]_c \ll [A]_t$) $\log \beta < -(1.30 + j \log [A]_t)$, since $[M]_f > 0.99[M]_t$ and (cf. eqn. 14) $[M]_c < 0.01[M]_t(1+V_o/V_i)$.

In the case of cesium no significant variations in $[M]_o$ were observed for $(HA)_w$ at least up to 1 g/L, neither in the acetate nor in the phosphate buffer system. Assuming the existence of 1:1 complexes only and taking the original sodium content in the sodium humate (27) as a measure for sites available for cesium complexation, we find log β < 1.2.

In the case of europium a direct comparison between the above given data and those reported previously (18) is possible, since both studies are carried out at pH=4.5 although the ionic strength in the present study is 0.05 only, whereas Choppin used 0.1 (18).

The most striking difference between the two sets of data is the apparent discrepancy in the type of complexes found. We find the 1:1 complex (j=0.99) is formed only, in complete agreement with the results of Ardakani and Stevenson (14), who argued that j should be an integer. In contrast to this Choppin (18) reported that formation of both 1:1 and a 1:2 complexes, since j=1.6 was found corresponding to a 2:3 ratio between the two types of complexes. However, the study by Choppin was carried out using trace amounts of metal ions only, in contrast to the present study as well as that by Stevenson, which is suggested to account for the apparent discrepancy. This may well be due to a presence of a small number of sites, exhibiting very high affinities towards europium ions, which will influence the results in the case of ultra-low metal ions concentrations only, as used by Choppin (18).

Secondly, it is noted that the present study reveals a Eu-HA interaction constant which is ca. 1.5 orders of magnitude lower than that reported for the 1:1 complex by Choppin (18). However, the origin of the humic acid may in this case play a crucial role. As indicated above, the studies on Co-HA interactions strongly suggest that interaction constants may vary up to several orders of magnitude for different samples of humic acid. Hence, the here reported Eu-HA interaction constant is simply to be regarded as a reflection of a relatively low complexing ability of the humic acid used. Similar indications have been obtained in parallel studies (28). Analogously the here derived Co-HA interaction constants are relatively low, although the same order of magnitude compared to those reported by Chimielewska (17).

In contrast to the Co-HA system, where the expected decrease in log β with decreasing pH is observed, the Sr-HA interaction constant appears to be only slightly pH-dependent. However, small effects due to differences in ionic strength have to be taken into account.

Also the composition of the Sr-HA complexes apparently differs somewhat from those of Co-HA and Eu-HA, as the latter two types exhibit pure 1:1 character, whereas a tendency to a metal-deficient complex is found in the strontium case.

The different behavior of strontium and cobalt ions is not surprising. The former is an ordinary second-row element with strong tendencies to salt formation, whereas the latter is a transition metal with partly filled 3d-orbitals, which significantly will be involved in the complex formation.

Also the possible presence of mixed complexes, as M(OH)-A, M(Ph)-A, and M(Ac)-A, may play a role, as may differencies in hydration energies for the two types of ions (cf. ref. 29).

A deeper insight in the actual nature of the radionuclide - humic acid complexes requires further investigations, which, however, are outside the scope of the present work.

Finally, the complexing capacities, w, should be mentioned, Not surprisingly it is found that w decreases with decreasing pH in agreement with previously reported results (cf. ref.'s 19 and 20, as well as ref.'s cited therein). Additionally a pronounced dependence of the metal ion is seen. It is obvious that the complexing capacity, towards europium ions is significantly higher than for cobalt- and strontium ions, the two latter capacities being nearly identical, which strongly suggests that the ionic charge may play an important role in this context. However, differences in complexing capacities have also been explained in terms of aggregation (19, 20).

The present paper has demonstrated the versatility of the dialysis technique in studies of interactions between metal ions and humic acid samples. The method allows facile determination of interaction constants, as well as of complexing capacities of the humic acid samples. It is noteworthy that the method described here, without modifications, can be applied to other areas of complex chemistry involving macromolecular ligands.

Acknowledgment. The work has been partly financed by the Commission of the European Communities under Contract 194-81-6 WASDK(G)

Symbols

$[HA]$	:	concentration of undissociated humic acid
$[A]$	:	concentration of humate anions, which may act as ligands (eq/L)
$[A]_t$	:	total humate-ligand concentration (eq/L)
$(HA)_w$	:	total humic acid (or sodium humate) concentration (g/L)
w	:	complexing capacity of humic acid (or sodium humate) (eq/g)
$[L]$	:	concentration of 'buffer-ligand' (mol/L)
M_t	:	total amount of metal ions (mol)
$[M]_t$	:	total metal ion concentration ($\equiv M_t/(V_i+V_o)$) (mol/L)
$[M]_c$	:	concentration of metal ions complexed by humate anions (mol/L)
$[M]_f$	:	concentration of free (i.e. uncomplexed) metal ions (mol/L)
$[M]_o$	:	concentration of metal outside the dialysis bag (mol/L)
$[MA_j]$	:	concentration of metal - humate complex (mol/L)
$[ML_n]$	:	concentration of metal - buffer-ligand complex (mol/L)
μ	:	ionic strength of the buffer solutions
j	:	number of humate-ligands pr. metal atom in complex
β	:	metal - humate interaction constant
I	:	pH-independent metal - humate interaction constant
K_A	:	overall humic acid ionization constant
β_{ML_i}	:	stability constants for the metal - buffer complex
Q	:	$[M]_o$ to $[M]_f$ conversion factor
$Z([M]_o)$	:	correction of M_t due to metal ions sorbed on the dialysis bags
V_i	:	volume inside the dialysis bags (present study: 10 mL)
V_o	:	volume outside the dialysis bags (present study: 40 mL)

Literature Cited

1. Stevenson, F.J.; Ardakani, M.S. in "Micronutrients in Agriculture" Mortvedt, J.J.; Giordano, P.M., Lindsay, W. L., eds., Soil Sci. Soc. Am., Madison, 1972, chapter 5, and references therein.
2. Jackson, K.S.; Jonasson, I.R.; Skippen, G.B. *Earth-Science Rev.* 1978, *14*, 97-146.
3. Bolter, E.; Butz, T. R. Report PB-278 050 (Missouri Water Resources Research Center), Rolla, 1977.
4. Means, J.L.; Hastings, D.W. Report ONWI-84 (Battelle Columbus Laboratories), Columbus, 1979.
5. Beveride, A.; Pickering, W.F. *Water, Air, and Soil Pollut.* 1980, *14*, 171-185.
6. Halbach, P.; von Borstel, D.; Gundermann, K.D. *Chem. Geol.* 1980, *29*, 117-138.
7. Schubert, J. *J. Phys. Colloid Chem.* 1948, *52*, 340-350.
8. Stevenson, F.J. *Soil Sci.*, 1977, *123*, 10-17.
9. Takamatsu, T.; Yoshida, T. *Soil Sci.* 1978, *125*, 377-386.
10. Bresnahan, W.T.; Grant, C.L.; Weber, J.H. *Anal. Chem.*, 1978, *50*, 1675-1679.
11. Schnitzer, M.; Hansen, E.H. *Soil Sci.* 1970, *109*, 333-340.
12. Geering, H.R.; Hodgson, J.F. *Soil Sci. Soc. Am. Proc.* 1969, *33*, 54-59.
13. Hirata, S. *Talanta* 1981, *28*, 809-815.
14. Ardakani, M.S.; Stevenson, F.J. *Soil Sci. Soc. Am. Proc.* 1972 *36*, 884-890.
15. Schnitzer, M; Skinner, S.I.M. *Soil Sci.*, 1967, *103*, 247-252.
16. Tan, K.H.; Leonard, R.A.; Bertrand, A.R., Wilkinson, S.R. *Soil Sci. Soc. Am. Proc.* 1971, *35*, 107-120.
17. Chimielewska, B. *Pol. J. Soil Sci.* 1969, *2*, 107-120.
18. Bertha, E.L.; Choppin, G.R. *J. Inorg. Nucl. Chem.* *40*, 655-658.
19. Truitt, R.E.; Weber, J.H. *Anal. Chem.* 1981, *53*, 337-342.
20. Rainville, D.P.; Weber, J.H. *Can. J. Chem.* 1982, *60*, 1-5.
21. Marinsky, J.A.; Cramer, S.J.; Ephraim, E., unpubl. results.
22. Adhikari, M; Chakrabati, G; Hazra, G. *Agrochim.* 1977, *21*, 134-139.
23. Dunigan, E.P.; Francis, C.W. *Soil Sci.* 1972, *114*, 494-496.
24. Ibarra, J.V.; Osacacar, J.; Gavilan, J.M[a] *Fuel* 1979, *58*, 827-830.
25. Juo, A.S.R.; Barber, S.A. *Soil Sci.* 1969, *114*, 484-496.
26. "Critical Stability Constants," Smith, R.M.; Martell, A.E. eds., Plenum Press, New York, 1976, Vol. III.
27. The sodium content in the sodium humate was determined by AAS to be 8.15% corresponding to 3.4 x 10^{-3} eq/L.
28. Carlsen, L; Platz, D, to be published.
29. Saar, R.A.; Weber, J.H. *Environ. Sci. Technol.* 1982, *16*, 510A-517A.

RECEIVED December 5, 1983

HYDROTHERMAL GEOCHEMICAL REACTIONS

Preliminary Assessment of Oxygen Consumption and Redox Conditions in a Nuclear Waste Repository in Basalt

D. L. LANE, T. E. JONES, and M. H. WEST

Rockwell International, Rockwell Hanford Operations, Richland, WA 99352

During construction of a nuclear waste repository in basalt (NWRB), Eh conditions in the repository horizon will be perturbed as a result of air-saturation of groundwater, temporarily leading to redox conditions more oxidizing than in the undisturbed system. Performance assessment of an NWRB requires information on redox conditions, since they will greatly affect the corrosion rate of canisters and the solubility and transport of certain radionuclides. Experiments were conducted to evaluate rates of oxygen consumption and redox conditions in the basalt-water system under conditions expected in an NWRB. Two methods were used to obtain these data: (1) the As(III)/As(V) redox couple and (2) the measurement of dissolved oxygen levels in solution as a function of time. These experiments have provided evidence that basalt is effective in removing dissolved oxygen and in rapidly imposing reducing conditions on solutions. At 300°C, calculations showed that an upper limit on Eh of -400 ± 100 mV was attained in 11 days. The dissolved oxygen content of solutions from a 150°C experiment decreased from air-saturation (8.5-9 mg/L) to 0.4 mg/L after 8 days, while solutions maintained at 100°C for 130 days contained 1.8-1.9 mg/L dissolved oxygen.

The basalt flows underlying the Hanford Reservation near Richland, Washington, are being evaluated as a possible repository site for long-term storage of high-level nuclear wastes. Characterization studies (1) and calculations based on redox-buffering reactions (2) suggest that the in situ conditions of groundwaters within the deep basalt formations are low redox potential (Eh), moderate pH, and low ionic strength. The long-term performance of a nuclear waste repository in basalt (NWRB) is based on the ability of the engineered barrier and host rock systems to provide initial containment and subsequent retardation of radionuclide transport by maintaining these and other in situ conditions.

0097-6156/84/0246-0181$06.00/0

An important part of the engineered barrier system is the waste package, consisting of the waste form, canister, and backfill. The waste package backfill has several performance requirements, one of which is to impose and maintain low Eh conditions. These conditions will aid in minimizing canister corrosion and in limiting the dissolution rate, solubility, and transport of certain radionuclides (3). However, during repository construction and waste emplacement, the in situ low Eh conditions will be perturbed by entrapment of air, resulting in high dissolved oxygen contents (i. e., high Eh) of inflowing groundwater. Thus, it is necessary to evaluate the effectiveness of backfill components in re-establishing the low Eh conditions of the undisturbed basalt-groundwater system. Dissolved oxygen consumption will be the first step in this process.

Since crushed basalt has been recommended as a major backfill component (1), experiments were completed to evaluate the rate of dissolved oxygen consumption and the redox conditions that develop in basalt-water systems under conditions similar to those expected in the near-field environment of a waste package. Two approaches to this problem were used in this study: (1) the As(III)/As(V) redox couple as an indirect method of monitoring Eh and (2) the measurement of dissolved oxygen levels in solutions from hydrothermal experiments as a function of time. The first approach involves oxidation state determinations on trace levels of arsenic in solution (4-5) and provides an estimate of redox conditions over restricted intervals of time, depending on reaction rates and sensitivities of the analyses. The arsenic oxidation state approach also provides data at conditions that are more reducing than in solutions with detectable levels of dissolved oxygen.

An arsenic oxidation state experiment was conducted at 300°C and 300 bars pressure in the basalt-deionized water system, while dissolved oxygen experiments were performed at 100°C and 150°C and 300 bars in the basalt-synthetic Grande Ronde groundwater system. A control experiment consisting of synthetic Grande Ronde groundwater at 150°C and 300 bars was also conducted to evaluate dissolved oxygen levels in the absence of basalt. The synthetic groundwater composition was based on Hanford Site groundwater samples from the Grande Ronde Formation, Columbia River Basalt Group. Finally, this study does not address other processes in an NWRB which may affect redox conditions such as radiolysis of solutions.

Experimental

Materials. The basalt studied in these experiments was relatively unaltered tholeiite from the Umtanum flow entablature of the Columbia River Basalt Group, a repository host candidate for the Hanford Site (6). Phase and chemical characterization are discussed in Noonan et al. (7) and Palmer et al. (8).

The formulation of the synthetic Grande Ronde groundwater used as starting solution in the dissolved oxygen experiments is given in Jones (9). Table I provides an analysis of the starting solution.

Table I. Analysis of Starting Solution for Dissolved Oxygen Experiments

Component	Concentration[a] (mg/L)
Si	32
Na	340
Al	<0.08
K	3.8
Ca	2.8
Mg	0.3
Fe	<0.01
F^-	33
Cl^-	280
SO_4^{2-}	170
CO_{3_T}[b]	70
pH[c]	9.7

[a]Elements determined by inductively coupled plasma-atomic emission spectrometry, anions by ion chromatography, and carbon by total carbon analyzer.

[b]Total carbonate expressed as HCO_3^-.

[c]Measured at room temperature.

Sample Preparation. The basalt was crushed and sieved, and the -120 + 230 mesh fraction was used. The grains were ultrasonically washed in deionized water to remove very fine adhering particles. If these particles are not removed, they will preferentially dissolve under hydrothermal conditions, resulting in abnormally high rates of mineral-fluid reactions (10). Examination of samples of the basalt on a scanning electron microscope assured that all fines had been removed. Nitrogen B.E.T. specific surface area of the washed basalt was 2.7 m^2/g.

A solution of 125µg/L As(V) in deionized water was prepared by serial dilution of a 1,000 mg/L As(V) stock solution prepared from As_2O_5. This solution and the synthetic Grande Ronde groundwater were air-saturated.

Conditions. Table II provides temperature, pressure, and other conditions for the experiments. The surface area/volume ratio for all experiments was 2.7×10^3 cm^{-1}. The hydrothermal apparatus was a Dickson-type sampling autoclave with a gold-titanium reaction cell, a gold-lined sampling tube, and a titanium sampling valve block (11). Samples of the reacting fluid could be taken over time without disturbing the pressure-temperature conditions of a run. The autoclaves were rocked 180° at about 4 cycles/min.

Analytical. Arsenic oxidation state determinations were performed by hydride generation-flame atomic absorption spectroscopy (AAS) at the University of Arizona Analytical Center. The analytical procedures are discussed in Brown, et al. (12).

An Altex Model 0260 oxygen analyzer with Clark-type polarographic electrode was employed for most of the dissolved oxygen (DO) measurements. Air-saturated water calibration procedures followed the manufacturer's instructions. Colorimetric techniques employing CHEMetrics, Inc. kits were also used for several DO determinations (13). Prior to obtaining a solution sample from the autoclave for DO measurement, 1 mL of solution was taken to flush out stagnant fluid in the sampling tube. The sample was taken into a gas-tight syringe and cooled quickly to room temperature before DO measurement.

Results

Arsenic Oxidation States. A solution sample was taken 257 hr after initiation of the 300°C basalt + arsenic-doped deionized water experiment (Run D2-8, Table II). The data from arsenic oxidation state AAS analysis of the initial As(V)-doped water (0-hr sample) and of the 257-hr solution sample are given in Table III. All detectable arsenic was in the +3 oxidation state [As(V) <15µg/L] in the 257-hr sample. Standard additions of As(III) and As(V) to the 257-hr sample were quantitatively recovered. To desorb arsenic from particulates in this sample, an aliquot of the solution was treated with 5% hydrofluoric acid. The higher As(III) content of the treated 257-hr sample aliquot (110 vs. 61µg/L, Table III) demonstrates that sorption occurred. Scanning transmission electron microscopic (STEM) analysis of the particulates indicated the presence of poorly crystallized high-iron "illite".

Table II. Experimental Conditions

Run No.	System[a]	T(°C)[b]	P(bars)[b]	W/R[c]	Duration (hrs)	Data
D2-8	B+W	300	300	10	257	As(III)/ As(V)
D2-16	B+SW	150	300	10	357	DO[d]
D2-29	B+SW	100	300	10	3,139	DO
D1-32	SW	150	300	--	497	DO (control)

[a]B+W = basalt + arsenic-doped deionized water.
B+SW = basalt + synthetic Grande Ronde groundwater.
SW = synthetic Grande Ronde groundwater.
[b]Estimated two standard deviation uncertainties for temperature are ±5°C to 10°C and for pressure are ±30 bars; 300 bars = 4,350 psig = 30 MPa.
[c]W/R = water/rock mass ratio.
[d]DO = dissolved oxygen.

Table III. Arsenic Oxidation State Data (Run D2-8)

Component	Solution Data		
	0 hr[a]	257 hr	
		Untreated	Treated with 5% HF[b]
As(III) (μg/L)[c]	<5	61	110
As(V) (μg/L)	125	<15	<15
pH[d]	5.1	8.0	---
pH[e]	---	8.1	---

[a]Starting solution for Run D2-8 consisting of deionized water spiked with 125 μg/L As (V) as prepared by serial dilution of a 1,000 mg/L As (V) stock solution.

[b]Hydrofluoric acid treatment used to desorb As from particulates in 257-hr sample.

[c]Arsenic data determined by hydride generation-flame AAS.

[d]Measured at room temperature; uncertainty for 257-hr sample pH is ±0.1 (one standard deviation), but is larger for 0-hr sample since it is weakly buffered.

[e]Calculated at 300°C by D. E. Grandstaff of Temple University using computer code HIPH4.

Dissolved Oxygen. The DO data, measured at room temperature and pressure (RTP), for solutions from the 100°C and 150°C basalt + synthetic groundwater experiments are given in Tables IV and V. Figure 1 is a composite of all the DO data. Solution samples at RTP from the 150°C experiment showed a decrease in DO content from air-saturation to 0.4 mg/L after about 200 hr. Solutions from the 100°C experiment showed a decrease in DO content from air-saturation to 1.8-1.9 mg/L after about 3,000 hr. Results from the 150°C synthetic Grande Ronde groundwater control experiment are also given in Figure 1. Solution samples were taken at 19, 191, 334, and 497 hr. The DO measured in these control solutions consistently gave values from 6 to 7 mg/L. Thus, the DO content was interpreted to be constant with time in the absence of basalt.

Table IV. Dissolved Oxygen Data from 150°C Basalt + Synthetic Grande Ronde Groundwater Experiment (Run D2-16)

	Time (hr)					
	0	1	4	21	143	189[a]
DO[b] (mg/L)	8.8(0.2)[c]	7.5(0.8)	6.8(0.8)	4.8(0.5)	1.1(0.3)	0.4(0.1)
pH[d]	9.7	8.0	7.9	7.9	7.9	7.9

[a]Two samples were taken at this time.

[b]All DO measurements were made with an oxygen electrode except the 189-hr sample which was analyzed colorimetrically.

[c]One standard deviation in parentheses; 0 and 189 hr sample uncertainties were determined from replicate analyses; uncertainties of other data points are estimates based on experience with analytical technique.

[d]Measured at room temperature; one standard deviation uncertainty is ± 0.1 pH unit.

Table V. Dissolved Oxygen Data from 100°C Basalt + Synthetic Grande Ronde Groundwater Experiment (Run D2-29)

	Time (hr)					
	0	282	792	1,416	2,088[a]	3,139[a]
DO[b] (mg/L)	8.8(0.2)[c]	5.5(1.0)	4.0(0.5)	2.8(0.6)	2.3(0.5)	1.85(0.4)
pH[d]	9.7	7.9	7.9	8.0	8.0	8.1

[a]Two samples were taken at these times.

[b]All DO measurements were made with an oxygen electrode.

[c]One standard deviation in parentheses; 2088 and 3139 hr sample uncertainties were determined from replicate analyses; 282 and 792 hr sample uncertainties were determined from replicate tests.

[d]Measured at room temperature; one standard deviation uncertainty is ± 0.1 pH unit.

In Run D2-16 (150°C), a solution sample at 189 hr gave 0.5 ± 0.25 mg/L DO by the indigo carmine colorimetric method and 0.4 ± 0.2 mg/L by the rhodazine D method. A second sample was taken to test for reproducibility of the rhodazine D method, and 0.4 mg/L was again obtained. (Uncertainties are two standard deviation estimates.) In Run D2-29 (100°C), two samples were taken at 2,088 hr, and the results obtained by oxygen electrode agreed within 5 relative percent (0.1 mg/L). At 3,139 hr, two samples were again taken, one analyzed with the Altex meter and the other analyzed with a Wheaton meter employing an oxygen electrode requiring no stirring (14). These two DO values also agreed within 5 relative percent (0.1 mg/L).

Discussion

Arsenic Oxidation States. The reduction of As(V) to As(III) in the 300°C basalt + arsenic-doped water experiment clearly demonstrates the ability of basalt to rapidly impose reducing conditions on solutions. The primary reducing agent in basalt-water systems is considered to be ferrous iron derived from mesostasis (glass + microcrystalline phases) dissolution. The mesostasis makes up 45 to 50 vol% of the Umtanum flow entablature used in the experiments, and glass constitutes about half of this amount (8). The ferric iron produced during basalt-water interaction at 300°C was apparently incorporated in iron-bearing illite as determined by STEM analysis of alteration products.

Since rates of arsenic redox reactions are slow at room temperature (5), it is assumed that the oxidation state data represent adjustment of arsenic species to the electron activity of the solution at 300°C. A quantitative assessment of the Eh of the basalt-water system at 300°C requires high-temperature thermochemical data for aqueous arsenic species. Such data are not available and, therefore, approximations were used to calculate Eh at 300°C.

The appropriate reaction between the As(III)-As(V) species is

$$H_3AsO_3 + H_2O = HAsO_4^{2-} + 4H^+ + 2e^- \qquad (1)$$

Three approaches were used to estimate the equilibrium constant of reaction (1), K_1, at elevated temperatures: (1) an equation derived by Helgeson (15) valid to 200°C; (2) free energy of formation data to 200°C for As(III)-As(V) species in Naumov et al. (16); and (3) the Criss and Cobble (17-18) correspondence principle approach, applicable to 250°C. The effect of increasing temperature on the stability fields of As(III)-As(V) species is illustrated in an Eh-pH diagram in Figure 2. The pressure dependence of K_1 was not determined, but is assumed to be within the uncertainty of the elevated temperature K_1 estimates.

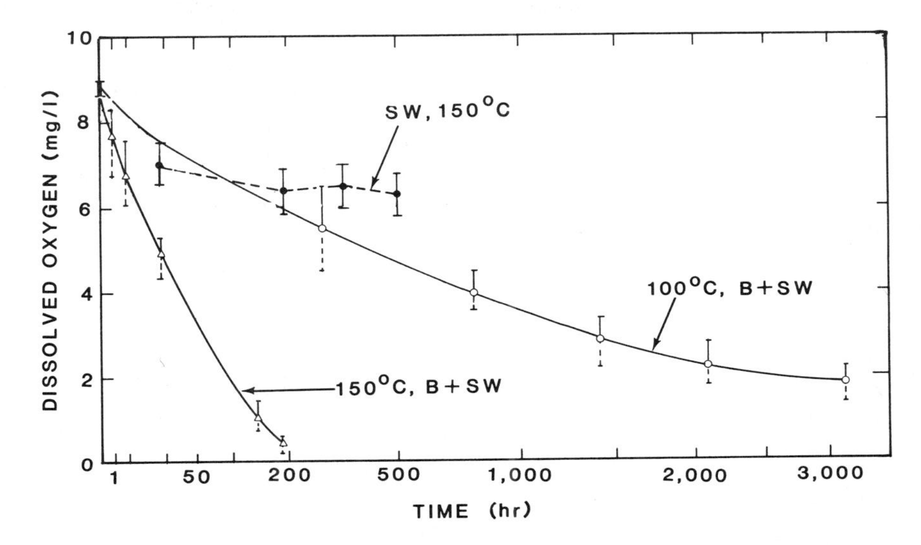

Figure 2. Portion of Eh-pH diagram for As-H_2O system, showing effects of increasing temperature from 25 to 100 °C. Free energy of formation data from Ref. 16 were used in the calculations.

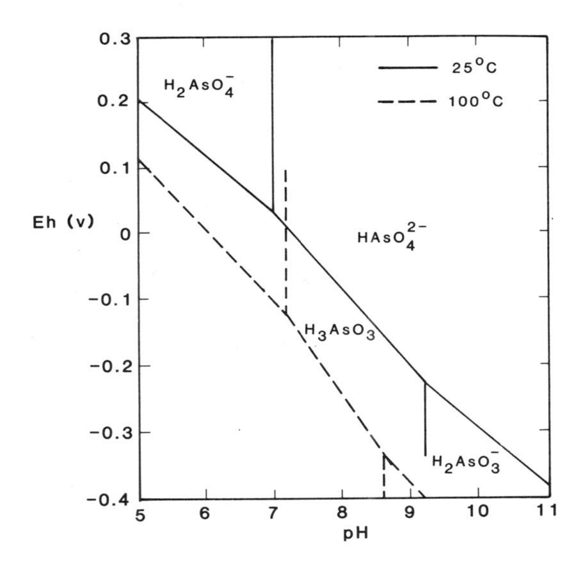

Figure 1. Dissolved oxygen vs time data. The experiments were basalt + synthetic Grande Ronde groundwater (B+SW) and synthetic Grande Ronde groundwater (SW) at 300 bars. Determination of uncertainties for B+SW data points is discussed in Table IV. Uncertainties for SW data were derived from replicable tests.

Averaged log K_1 values vary from -24 to -25 from 100°C to 250°C, suggesting little change with temperature. Extrapolation of the average heat capacity change for reaction (1) to 300°C following Criss and Cobble (18) yields log K_1 = -26 to -27. Considering the lower temperature log K_1 estimates and the potential errors in the extrapolation to 300°C, a range of log K_1 values from -24 to -27 at 300°C was assumed, and an equation for calculating Eh derived:

$$Eh\,(mV) = (1{,}450 \pm 100) - 227\,pH - 57\,\log \frac{a_{H_3AsO_3}\,a_{H_2O}}{a_{HAsO_4^{2-}}} \qquad (2)$$

The arsenic oxidation state data and the calculated pH at 300°C (see Table II) allow an upper limit on the Eh of the solution in the basalt-water experiment to be estimated from Equation (2). Assuming $a_{H_2O} = 1$ and As(V) = 15 μg/L, this upper limit Eh value is -400 ± 100 mV. The basalt-fluid redox buffer mechanism of Jacobs and Apted (2) gives an Eh of about -600 mV at 300°C and pH 7.8 (19). This mechanism involves ferrous iron-bearing basalt glass + water reacting to magnetite + silica.

Dissolved Oxygen. The experimental results demonstrate that in the absence of basalt, DO is maintained at high levels, while in the presence of basalt, oxygen is effectively removed (see Figure 1). Although the ferrous iron content of Umtanum basalt mesostasis is not well known, estimates from bulk ferrous/ferric iron data and from microcharacterization of mesostasis phases (7, 20) indicate that the amount of Fe^{2+} available for oxygen consumption by mesostasis dissolution is large relative to the amount of DO ($\geq$10 on a mole/mole basis).Thus, the available Fe^{2+} concentration should remain constant over the duration of the experiments.

The DO data from Runs D2-16 (150°C) and D2-29 (100°C) were fit to a number of rate equations, with linear correlation coefficients from 0.95-0.99 for first order kinetics. However, the uncertainties and lack of adequate data do not permit a rigorous statistical evaluation. Therefore, as a first approximation, a first order rate equation with respect to DO concentration was assumed. The data can then be expressed by the equation:

$$[O_2] = [O_2]_0 \exp(-kt) \qquad (3)$$

where $[O_2]_0$ is initial DO concentration, k is the apparent first order rate constant, and t is time.

An apparent first order rate constant of 1.5×10^{-2} hr^{-1} was derived from the DO data from the 150°C experiment. For the 100°C experiment, the rate constant is about 4.5×10^{-4} hr^{-1}. Further experiments are needed to determine the full rate law for oxygen consumption. Because reaction mechanisms and/or rates can change with time, extrapolation to conditions under which basalt controls Eh may not be justified.

The relationship of the stirring rate in these experiments to the rates of hydrolysis reactions of basalt phases is indicative of surface-reaction controlled dissolution (21). First order kinetics are not inconsistent with certain rate-determining surface processes (22). Approximate first order kinetics with respect to dissolved oxygen concentration have been reported for the oxidation of aqueous ferrous iron (23) and sulfide (24), and in oxygen consumption studies with roll-type uranium deposits(25).

Applications. In the following paragraphs, the conditions (temperature, time, water/rock mass ratio, surface area) and the results on closed system oxygen consumption and redox conditions of the basalt-water experiments are compared to expected conditions in the open system backfill and near-field environment of an NWRB. Crushing of basalt for pneumatically emplaced backfill could result in a substantial fraction of fine-grained basalt with a variety of active surface sites for reaction similar to the crushed basalt used in the experiments. The effects of crushing on rates of mineral-fluid reactions are well documented (10, 26).

Temperature-time calculations (27) indicate that, initially, the waste package will be at temperatures of $\geq$100°C for hundreds of years. Thus, the rates of oxygen consumption and establishment of reducing conditions, as determined in this study, are rapid compared to these times.

Preliminary transport model calculations (1) suggest that the fluid flux through a waste package integrated over 1,000 yr might be equivalent to fluid/rock mass ratios $<$10, which is consistent with modeling and estimates of integrated fluid fluxes in other basaltic systems (28-30). The combined effects of low integrated fluid fluxes, the probable large surface area of a crushed basalt backfill, and the experimental results on rates of redox reactions suggest that basalt should effectively remove dissolved oxygen and re-establish reducing conditions during the containment period of an NWRB.

With the addition of bentonite to a crushed basalt backfill, aqueous diffusion would be the most effective mass transfer process (31). Aagaard and Helgeson (32) state that at temperatures $\leq$200°C, aqueous diffusion rates are orders of magnitude greater than rates of silicate hydrolysis even in acid solutions. Therefore, the dissolution rate of backfill phases and the overall mass transfer process could be controlled by reactions at the mineral-fluid interface. As stated earlier, dissolution of basalt phases in the sampling autoclave experiments may also be controlled by interface reactions.

The experimental results of this study may also be applicable to processes occurring in the disturbed rock zones around a waste package. The surface area/volume ratio for these experiments is approximately equivalent to the solution in a 10-μm-wide planar fracture (33). With a fracture abundance of

20 m^{-1} (estimated for the Umtanum flow, 34), this fracture width is equivalent to an intrinsic rock permeability of about 1 millidarcy (35) or a hydraulic conductivity of about 10^{-8} m/sec. As a point of reference, hydraulic conductivity values for flow tops within the Grande Ronde basalt have been reported from 10^{-5} to 10^{-9} m/sec (1).

Conclusions

This study focused on the disturbance of the in situ redox conditions in the basalt-groundwater system during the early phases of an NWRB and on the ability of basalt and time-scale required to return the system to a state of redox control. Experiments in basalt-water systems, conducted under conditions expected in an NWRB, have provided evidence that basalt is effective in removing DO and in imposing reducing conditions on solutions at appreciable rates. Application of the experimental results to a crushed basalt backfill suggests that buffering of redox potentials at reducing conditions could occur rapidly relative to the containment period of an NWRB.

The usefulness of the As(III)-As(V) couple as a redox indicator in hydrothermal experiments is limited by the availability of thermochemical data and by the range of Eh dictated by analytical and reaction rate constraints. For low temperature ($\leq$150°C) short-term experiments, selenium oxidation state analysis may be more appropriate than arsenic, since the Se(VI)-Se(IV) reduction occurs at relatively high Eh (36).

Investigations of redox processes in natural water systems have emphasized the disequilibrium behavior of many couples (e.g., 37). The degree of coupling of redox reactions with widely varying rates, and its effect on radionuclide transport in an NWRB needs to be considered. Because of the generally slow kinetics of autoxidation reactions, the potential surface catalyzed reduction of a radionuclide at low temperatures in the presence of trace levels of DO may explain certain sorption data (e.g., 38).

Sufficient DO data were not obtained from basalt-synthetic Grande Ronde groundwater experiments to allow determination of a definitive rate law. A first order kinetic model with respect to DO concentration was assumed. Rate control by diffusion kinetics and by surface-reaction mechanisms result in solution composition changes with different surface area and time dependencies (32, 39). Therefore, by varying reactant surface area, determination of the proper functional form of the integrated rate equation for basalt-water redox reactions is possible.

Acknowledgments

We wish to express our indebtedness to the laboratory personnel who assisted in this research: M. O. Baechler, K. M. Cooley, A. P. Hammitt, C. W. Hobbick, J. R. Payne, and D. R. Schatz; and to our colleagues G. D. Aden and C. C. Allen for the STEM analyses. J. L. Moyers and G. Seplak at the University of Arizona Analytical Center kindly provided the arsenic oxidation state analyses without which this report would not be possible. We thank L. L. Ames and J. E. McGarrah of Battelle Northwest Laboratories for providing the B.E.T. surface area analysis of the basalt, and D. E. Grandstaff of Temple University for the high-temperature pH calculation. We appreciate the skills of D. B. Mill in drafting the figures. J. K. Haberstok spent long hours in typing the many versions of this paper. To her, our undying gratitude. Finally, we thank two anonymous reviewers whose comments were both useful and appreciated. This work was performed under U.S. Department of Energy contract DE-AC06-77RL01030.

Literature Cited

1. "Site Characterization Report for the Basalt Waste Isolation Project," DOE/RL 82-3, 3 vols., Rockwell Hanford Operations, for the U.S. Department of Energy, Washington, D.C., November 1982.
2. Jacobs, G. K.; Apted, M. J. EOS, Amer. Geophys. Union Trans. 1981, 62, 1065.
3. White, W. B.; Freeborn, W. P., in "Materials Characterization Center Workshop on Leaching Mechanisms of Nuclear Waste Forms, Summary Report"; Mendel, J. E., compiler; PNL-4382; Pacific Northwest Laboratories, Richland, Washington; pp. 247-254.
4. Shaikh, A. U.; Tallman, D. E. Anal. Chim. Acta 1978, 98, 251-259.
5. Cherry, J. A.; Shaikh, A. U.; Tallman, D. E.; Nicholson, R. V. J. Hydrology 1979, 42, 373-392.
6. "Subsurface Geology of the Cold Creek Syncline," RHO-BWI-ST-14, Rockwell Hanford Operations, Richland, Washington; Myers, C. W. and Price, S. M., Eds.; 1981.
7. Noonan, A. F.; Fredriksson, K.; Nelen, J., in "Scientific Basis for Nuclear Waste Management"; Moore, J. G., Ed.; Plenum Press: New York, 1981; Vol. 3, pp. 51-58.
8. Palmer, R. A.; Aden, G. D.; Johnston, R. G.; Jones, T. E.; Lane, D. L.; Noonan, A. F., "Characterization of Reference Materials for the Barrier Materials Test Program," RHO-BW-ST-27 P, Rockwell Hanford Operations, Richland, Washington; 1982.

9. Jones, T. E., "Reference Material Chemistry-Synthetic Groundwater Formulation," RHO-BW-ST-37 P, Rockwell Hanford Operations, Richland, Washington; 1982.
10. Schott, J.; Berner, R. A.; Sjöberg, E. L. Geochim. Cosmochim. Acta 1981, 45, 2123-2135.
11. Seyfried, W. E., Jr.; Gordon, P. C.; Dickson, F. W. Amer. Mineralogist 1979, 64, 646-649.
12. Brown, R. M., Jr.; Fry, R. C.; Moyers, J. L.; Northway, S. J.; Denton, M. B.; Wilson, G. S. Anal. Chem. 1981, 53, 1560-1566.
13. Gilbert, T. W.; Behymer, T. D.; Castaneda, H. B. Amer. Laboratory 1982, 14, 119-134.
14. Phelan, D. M.; Taylor, R. M.; Fricke, S. Amer. Laboratory 1982, 14, 65-72.
15. Helgeson, H. C. J. Phys. Chem. 1967, 71, 3121-3136.
16. Naumov, G. B.; Ryzhenko, B. N.; Khodakovsky, I. L., "Handbook of Thermodynamic Data"; Soleimani, G. J., translator; U.S. Geological Survey, Report No. USGS-WRD-74-001, 1974, pp.181,237.
17. Criss, C. M.; Cobble, J. W. Amer. Chem. Soc. J. 1964, 86, 5385-5390.
18. Criss, C. M.; Cobble, J. W. Amer. Chem. Soc. J. 1964, 86, 5390-5393.
19. Myers, J.; Apted, M. J.; Lane, D. L., in "Proc. of 1982 National Waste Terminal Storage Program Info. Meeting", DOE/NWTS-30, pp.36-40.
20. Allen, C. C.; Strope, M. B., in "Proc. of 17th Annual Meeting of Microbeam Analysis Society, Phoenix, Arizona"; San Francisco Press: San Francisco; in press.
21. Berner, R. A. Amer. J. Sci. 1978, 278, 1235-1252.
22. Dibble, W. E., Jr.; Tiller, W. A. Geochim. Cosmochim. Acta 1981, 45, 79-92.
23. Davison, W.; Seed, G. Geochim. Cosmochim. Acta 1983, 47, 67-79.
24. Hoffmann, M. R. Env. Sci. Tech. 1981, 15, 345-353.
25. Goddard, J. B.; Brosnahan, D. R. Mining Eng. 1982, 34, 1589-1596.
26. Petrovich, R. Geochim. Cosmochim. Acta 1981, 45, 1675-1686.
27. "Waste Package Concepts for Use in the Conceptual Design of the Nuclear Waste Repository in Basalt," AESD-TME-3142, Westinghouse Electric Co., Pittsburgh, Pennsylvania, for Battelle Project Management Division, Office of Nuclear Waste Isolation, Sept., 1982.
28. Norton, D.; Taylor, H. P., Jr. J. Petrol. 1979, 20, 421-486.
29. Taylor, H. P., Jr.; Forester, R. W. J. Petrol. 1979, 20, 355-419.
30. Cocker, J. D.; Griffin, B. J.; Muehlenbachs, K. Earth Planet. Sci. Lett. 1982, 61, 112-122.

31. Andersson, G.; Rasmuson, A.; Neretnieks, I., in "Scientific Basis for Nuclear Waste Management V"; Lutze, W., Ed.; North-Holland: New York, 1982; pp. 539-548.
32. Aagaard, P.; Helgeson, H. C. Amer. J. Sci. 1982, 282, 237-285.
33. Rimstidt, J. D.; Barnes, H. L. Geochim. Cosmochim. Acta 1980, 44, 1683-1699.
34. Long, P. E.; Davidson, N. J., in "Subsurface Geology of the Cold Creek Syncline"; Myers, C. W.; Price, S. M., Eds.; RHO-BWI-ST-14, Rockwell Hanford Operations, Richland, Washington, 1981; Chapter 5.
35. Norton, D.; Knapp, R. Amer. J. Sci. 1977, 277, 913-936.
36. Pourbaix, M. "Atlas of Electrochemical Equilibria in Aqueous Solutions"; Pergamon Press: Oxford, 1966; pp.555-559.
37. Emerson, S.; Cranston, R. E.; Liss, P. S. Deep-Sea Res. 1979, 26A, 859-878.
38. Bondietti, E. A.; Francis, C. W. Science 1979, 203, 1337-1340.
39. Lagache, M. Geochim. Cosmochim. Acta 1976, 40, 157-161.

RECEIVED October 20, 1983

12

Developments in the Monitoring and Control of Eh and pH Conditions in Hydrothermal Experiments

JONATHAN MYERS
Rockwell International, Rockwell Hanford Operations, Richland, WA 99352

GENE C. ULMER, DAVID E. GRANDSTAFF, and ROBERT BROZDOWSKI
Temple University, Philadelphia, PA 19122

MICHAEL J. DANIELSON and OSCAR H. KOSKI
Pacific Northwest Laboratory, Richland, WA 99352

In the design of a high-level nuclear waste repository it is essential to obtain accurate groundwater Eh-pH data. Design considerations such as the choice of matrix for the waste form, type and dimensions of canister material, use of buffers, and type and amount of backfill would all benefit from an exact knowledge of oxidation potentials (Eh) and acidity levels (pH) of the groundwater. For the radionuclides themselves, it cannot be overstated that the future containment, dissolution, migration, sorption, and precipitation of radionuclides all will be determined by the Eh-pH conditions existing in the repository environment. The Basalt Waste Isolation Project (BWIP) has initiated a research effort to develop sensors which can be mounted in autoclaves to provide constant monitoring of the Eh-pH conditions that exist during waste form/barrier material/groundwater hydrothermal interaction tests. Sensors must withstand temperatures up to 300°C and pressures up to 300 bars. This report considers Teflon* (*trademark of Dupont) hydrogen diffusion membranes and zirconia pH sensors. The development of these sensors represents a significant advance in the environmental monitoring of Eh and pH conditions at elevated temperatures and pressures.

The U.S. Government, through the National Waste Terminal Storage (NWTS) Program of the U.S. Department of Energy, is actively studying the technical feasibility of permanent disposal of high-level waste in repositories excavated in deep geologic formations. Geologic strata presently being considered include bedded salt, tuff, and Columbia Plateau Basalt. Individual waste packages would be emplaced in repositories mined in one or more of these strata in accordance with emerging Nuclear Regulatory Commission (NRC) and Environmental Protection Agency (EPA)

0097–6156/84/0246–0197$06.00/0

regulations. The component materials, or barriers, of a waste package provide initial containment and eventual slow release of radionuclides to meet these regulatory criteria.

Migration of contaminated groundwater has been identified as the principal mechanism for radionuclide transport from a repository to the biosphere. Over the lifetime of the repository, it is assumed that groundwater will become contaminated as the result of hydrothermal reactions and interactions within and near waste packages.

The two principal types of waste form expected to be emplaced in a repository are spent fuel and borosilicate glass. Each of these waste forms will have characteristic dissolution mechanisms and associated rates of release of radionuclides based on the solubility of primary and secondary solid phases. These mechanisms and rates are highly dependent on the Eh and pH of the groundwater (1). The rate of canister corrosion is also dependent on the Eh and pH of the groundwater solutions. Procedures for calculating the pH of a solution at elevated temperature and pressure from chemical analysis at 25°C and 1 atm have been developed (2). However, these calculations are based on thermodynamic data which are somewhat uncertain. The Eh conditions of a solution at elevated temperature and pressure can be inferred from chemical analysis at 25°C and 1 atm. These calculations are based on measuring ratios of dissolved redox-sensitive species such as SO_4^{-2}/HS^- or As^{+3}/As^{+5} (3). This type of analysis will not give direct Eh values, but will set limits on Eh conditions. A major drawback to this technique is that the quench effects on most of these redox-sensitive reactions are unknown. It is, therefore, essential that the Eh and pH of the solution be continuously monitored during waste form/barrier material/groundwater hydrothermal interaction tests.

Rockwell Hanford Operations has supported a research effort to develop sensors that can be incorporated into autoclaves to continuously monitor the Eh and pH conditions existing during hydrothermal interaction tests. These sensors must be mechanically stable at temperatures up to 300°C and pressures up to 300 bars and must be chemically stable in the presence of complex hydrothermal solutions. The sensors must also be easily incorporated into the standard autoclaves currently in use, and must be operable in a radiation environment.

Several parallel lines of research are underway to develop working sensors. Two techniques currently under consideration are Teflon hydrogen diffusion membranes and zirconia pH sensors.

At temperatures below 100°C, Eh values are usually measured directly using electrochemical techniques. At elevated temperatures, however, it is usually more convenient to measure hydrogen pressures in equilibrium with the system being investigated. The Teflon hydrogen diffusion membrane is a device which directly measures H_2 pressures in equilibrium with the solution being tested. If the pH of the solution is known, fH_2 values can be easily be converted to Eh. The zirconia device described in this report is an electrochemical sensor which develops a voltage which is proportional to the pH of the solution being tested. This report covers the theory of operation of the two types of sensors and provides details of the current status of design, testing, and calibration of zirconia pH sensors and Teflon hydrogen diffusion membranes.

Teflon Hydrogen Diffusion Membranes

Teflon, a fluorocarbon polymer, is well known for chemical inertness, thermal stability at temperatures up to 290°C, and excellent electrical insulating properties. Most inorganic and many organic compounds are insoluble in it. Teflon also exhibits a relatively large hydrogen gas permeability. It, therefore, has potential as a selective osmotic membrane for hydrogen.

Investigations performed by Dupont on the chemical stability of Teflon reveal that Teflon immersed in a 20% HCl solution at 200°C for extended periods will not absorb any chlorine within detectability limits. In addition, Teflon does not absorb any detectable sulfur from sulfuric acid or sulfur vapor. Therefore, a Teflon H_2 membrane may prove to be superior to a platinum-group metal membrane for use in sulfur- or chlorine-bearing systems. Investigations are currently being performed to determine the feasibility of employing a Teflon membrane as an H_2 monitor/controller in hydrothermal test apparatus.

The Arizona State University Chemistry Department, under contract to the BWIP, has designed a H_2 diffusion membrane consisting of a 3-in.-long,1/4-in.-OD, closed-end Teflon tube with a wall thickness of 0.03 in. The mechanical strength of the membrane has been tested to 260°C and 1,500 bars and was able to hold pressure and pass H_2 under these conditions.

Experimental Equipment and Procedures. The experimental work was carried out at Arizona State University in a small, high-frequency, low-angle agitation autoclave fitted with a wire-wound resistance furnace. Temperature was controlled within 2°C by means of a variable transformer. The pressure vessel employed was similar to a Dickson autoclave; however, it incorporated a Bridgeman, unsupported area seal in place of the usual delta ring seal. The sample was contained in a gold bag which was connected to a titanium plug, and the plug was fitted with a Teflon membrane. The membrane was connected to an electronic pressure transducer via a stainless steel capillary tube. The electronic pressure transducer was then connected to a strip-chart recorder to provide continuous monitoring of the hydrogen pressure in equilibrium with the experimental charge. The experimental charges were sealed along with two 6-mm-diameter titanium balls in the gold bag. The system was then assembled and pressurized to 350 bars to test for leaks. After this, the pressure was lowered to 300 bars and the furnace was switched on. Pressure was readjusted periodically to maintain 300 ± 17 bars while the temperature stabilized. Experiments took approximately 7 hr to reach the desired pressure and temperature and to stabilize.

Calibration. These experiments were carried out in order to test the efficiency of the Teflon membrane and to calibrate membrane pressure against known solid-solid buffer reactions. The reactions chosen for calibration were:

$$Co + H_2O = CoO + H_2 \quad (1)$$

and

$$3Fe + 4H_2O = Fe_3O_4 + 4H_2 \quad (2)$$

In both cases the method was to load a known quantity of pure metal (cobalt or iron) into the gold bag with a measured volume of boiled, distilled, deionized water.

The experimental results using the two buffer assemblages are given in Table I. After the runs were quenched, X-ray examination of the solids confirmed the presence of both product and reactant phases. As expected, the Co-CoO buffer yielded very low fH_2, and the Fe-Fe_3O_4 buffer yielded very high fH_2, close to total P. The discrepancies between observed and predicted fH_2 for Co-CoO are probably due to uncertainties in the thermodynamic data for CoO, on which the calculations were based. As can be seen from this data, the measured H_2 pressures agree with the predicted values within experimental uncertainty, thus demonstrating the utility of this technique.

A basalt plus groundwater oxygen buffering experiment was conducted at Arizona State University using Umtanum basalt plus synthetic Grande Ronde groundwater at 200°C in an autoclave equipped with the Teflon membrane previously described. The gas diffusing across the membrane initially rose to a maximum pressure of 10.62 bars after 242 hr. This was followed by a pressure decrease to a minimum of 4.4 bars after 358 hr. Thereafter the pressure steadily increased at a rate of 4.5×10^{-3} bars/hr. A possible explanation of the decrease in pressure from 10.62 to 4.4 bars is the reduction of sulfate:

$$SO_4^{-2} + 4H_2 = S^{-2} + 4H_2O \quad (3)$$

A sufficient quantity of sulfate was initially present in the groundwater to account for the calculated quantity of H_2 consumed. The steady increase in gas pressure toward the end of the experiment could be due to the slow diffusion of CO_2 or other gases such as H_2O, O_2, SO_2, H_2S, or CH_4 through the Teflon. This hypothesis was tested by an analysis of a small volume of gas

Table I. Results of H_2 Membrane Calibration Using Solid Buffer Assemblages

P_T (bars)	T(°C)	PH_2 (bars)	Predicted PH_2 (bars)[a]
Co-CoO			
172	103	0.21	0.008
207	106	0.31	0.008
Fe-Fe_3O_4			
166	216	159.6	156.6
172	219	171.7	163.1
193	220	178.6	180.1[b]

[a]From the data of (4-7) and assuming ideal mixing of H_2 and H_2O.

[b]At these conditions it is uncertain as to whether the fH_2 of the MI buffer or fH_2 defined by an invariant condition of the miscibility gap in H_2O-H_2 is being measured. In the latter case the value should be between 180 and 200 bars.

using gas chromatography. The analysis indicated a mole fraction of 12% CO_2. No gases other than H_2 and CO_2 were detected. The partial pressure of H_2 can easily be calculated from an analysis of the gas if it is assumed that the total pressure is equal to the sum of the partial pressures of the various components present.

An alternative technique was to use a small volume liquid nitrogen trap installed between the Teflon membrane and the pressure transducer. At the temperature of liquid nitrogen, all of the possible gas components except H_2 condense to a liquid or solid state. For the above experiment, a liquid N_2 cold trap was installed in the H_2 capillary line between the membrane and the transducer. The trap was evacuated, opened to the line, and cooled. The ratio of postcooling to precooling pressure was 0.92. Assuming ideal mixing of the gas species in the membrane, this implies that the membrane gas is 92 mol% H_2. There is agreement, within 4%, between the results of this method and the gas chromatography analysis.

Despite these problems, it can be appreciated from Table I that the Teflon membrane readily passes H_2 and the PH_2 (read at the gauge) is responding in the predicted manner to changes in fH_2 of the system. On present evidence, it is therefore concluded that PH_2 at the gauge is an accurate reflection of PH_2 in the system, after taking into account the presence of other gases using either gas chromatography or a cold trap.

Future Investigations. Investigations of the Teflon membrane redox sensor will concentrate on measurements of the diffusion rates of hydrogen through the sensor in the temperature range of 25°C to 250°C, optimization of the sensor design, and the use of gas chromatographic techniques to measure diffusion rates of other gases potentially present in hydrothermal experiments.

Zirconia pH Sensors

The zirconia pH sensor is based on a recent discovery by (8) that ZrO_2, stabilized with Y_2O_3, will develop an electromotive force (EMF) that is proportional to the difference in the pH of solutions contacting the inner and outer surfaces.

Both (8) and (9) have presented information on working ZrO_2 pH electrodes, thus establishing the potential for their use. However, both of these investigators had difficulty obtaining commercially available electrolyte material which developed a suitable response to pH. This difficulty in obtaining suitable electrolyte material prompted the BWIP to contract with Temple University, Pennsylvania State University, and Pacific Northwest Laboratory (PNL) to develop a program of research aimed at understanding the physical and chemical factors in ZrO_2 cells that correlate with good quantitive sensing of pH. Currently the research involves the determination of the effect of the following parameters on the pH response of commercially available and custom fabricated ZrO_2 material:

- Electrical resistance
- Chemical purity
- Concentration of dopant
- Texture
- Sintering time
- Sintering temperature
- Porosity (density).

The ultimate goal of this program is to develop a pH sensor that can be installed in an autoclave to provide constant monitoring of the pH conditions which exist during waste form/.barrier material/ groundwater hydrothermal interaction experiments. It is also hoped that a long-lived sensor could be developed for use in long-term field tests for monitoring repository conditions.

Expected Range of pH Values. Changes in solution pH in rock-water systems may result from two primary causes. The first cause is due to changes in equilibrium constants with variation in temperature and pressure. For example, the neutral pH of pure water changes from 7.00 at 20°C to approximately 5.6 at 200°C and 300 bars pressure due to changes in the value of the dissociation constant for water. Precipitation, dissolution, oxidation, or reduction of phases with consumption or generation of hydrogen ion is the second primary cause of pH variation.

In a nuclear waste repository located in basalt, solution pH is controlled by interactions between groundwater and the reactive glassy portion of the Grande Ronde basalt (10). In situ measurements and experimental data for this system indicate that equilibrium or steady-state solutions are saturated with respect to silica at ambient temperatures and above. Silica saturation and the low, total-dissolved carbonate concentration indicate the pH may be controlled by the dissolution of the basalt glass (silica-rich) with subsequent buffering by the silicic acid buffer. At higher temperatures, carbonate, sulfate, and water dissociation reactions may contribute to control the final pH values.

The pH of Grande Ronde solution as a function of temperature was calculated using the program HIPH4. The method of computation is described in (2) and utilizes data on the concentration of several dissolved acid-base pairs. The calculated pH value decreases with increasing temperature and passes through a shallow minimum near 250°C. The solution pH at 25°C is about 9.8; the calculated pH at 65°C is about 9.2, quite comparable with measured values; and the pH calculated at 300°C is about 8.2. The calculated solution pH is quite basic, even at higher temperature; for example, at 300°C the solution is about 2.5 pH units above neutral. Thus, the range of pH values expected in hydrothermal solutions is relatively restricted, spanning only two to three pH units. Therefore, electrode response to pH changes at high temperatures will need to be designed to give as large an EMF/pH unit as possible to achieve good resolution.

Construction. Two ZrO_2 pH sensor designs have been developed by Ulmer and Grandstaff at Temple University and by Danielson and Koski at Pacific Northwest Laboratories. The Temple University design consists of a 12-in.-long, 1/4-in.-OD, round, closed-end, zironcia ceramic tube as the membrane of the pH electrode (Fig. 1). A hollow copper plug, fitting the inside diameter of the tube, is inserted snuggly (-0.001 in.) into the bottom end of the zirconia tube. The inside cavity of the copper plug is filled with a mixture of Cu powder and Cu_2O powder in the weight ratio 1:10. Two long, 0.020 in.-diameter, pure copper wires are inserted through a pair of holes drilled at right angles to each other through the mid-point of the sides of the 1-in.-long hollow copper plug. The side walls of the plug are pregrooved with four slots 0.025-in. deep to receive the wires as they are bent upward from the holes and threaded through a four-hole alumina insulator rod which fits the inner diameter of the zirconia tube. The copper wires are then twisted to form a four-strand cable just above the hollow copper plug and at the top end of the alumina rod. The wire protruding above the alumina rod is threaded through a silicon rubber stopper which is put under compression in a Swagelok fitting. This serves to seal the top end of the zirconia tube, excluding air and preserving the Cu_2O assemblage. A platinum wire electrode mounted in the bottom of the test chamber forms the other half of the circuit in the aqueous fluid.

Based on the success of the Temple University design, PNL initiated an investigation of ZrO_2 pH sensors utilizing the Temple University basic design with a few minor modifications. An alternative method of holding and sealing the sensor to withstand operating temperatures of 300°C and pressures of 300 bars was suggested in the PNL design which may have some advantages. The illustrated sealing design (Fig. 2) is incorporated into the autoclave head. Other designs are possible, such as incorporating the tube seal within a threaded fitting which would facilitate sensor installation and removal, as well as simplify the autoclave head design.

The pressure seal is produced using a Teflon ring compressed around the zirconia tube. Because the Teflon is soft, it will evenly distribute the forces (to minimize tube breakage). Belleville spring washers will act as a seal-follower to maintain sealing forces since Teflon can extrude when subjected to directional stress. Boron nitride can also be used as a sealant and offers the advantage of being more resistant to radiation damage than Teflon. The electrical connection is made above the mechanical seal to keep solution from seeping through and shunting out the high impedance connection. Part of the tube may be glazed if the pH responding regions must be defined. Glazes can be created by sputtering high purity zirconia on the tube. In both the Temple University and the PNL design, the internal volume of the tube above the Cu-Cu_2O can be filled with zirconia cement. This will prevent a catastrophic blow out if the tube should develop a leak.

Resistivity Measurements and pH Response. The resistance of the ZrO_2 electrolyte material has been found at Pennsylvania State University to be a critical factor in the performance of a pH sensor. Therefore, an effort has been made to measure the resistivity of each cell, and to correlate this with cell performance, especially with respect to linearity of the EMF versus pH response.

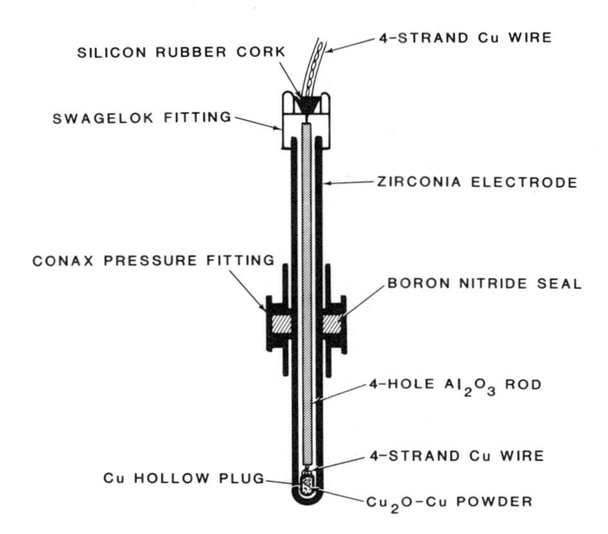

Figure 1. Temple University Design for a Zirconia pH Electrode.

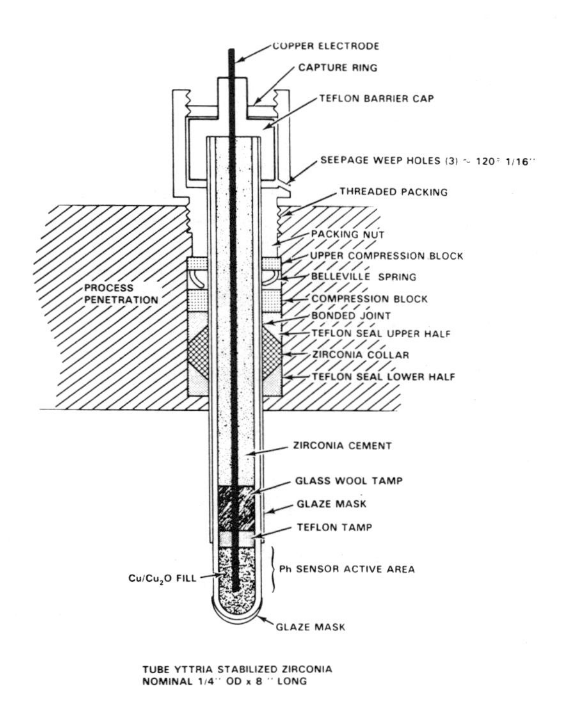

Figure 2. PNL Design for a Zirconia pH Electrode.

The pH response was initially tested at 90°C and one atmosphere pressure by immersing the sensor in standard pH buffer solutions. The EMF of the sensor was then compared with the EMF generated by a commercial Ross-type glass electrode. Figure 3 shows the EMF versus pH response of a representative tube, PSU-T1-18. The response over a pH range of 2.5 to 9.5 is linear and displays a slope which is 98% of the theoretical Nernstian slope for this temperature.

Resistance measurements were made by filling the zirconia tube with KCl solution and immersing it in a 90°C KCl bath. Resistance was then measured across platinum electrodes placed inside and outside the tube, using a high impedance electrometer. The measured resistivity was recast in ohm-cm units by recalculation using surface area, wall thickness, and depth of immersion. This Pennsylvania State University investigation determined that at a given temperature, a narrow range of resistivity values are displayed by those cells which have a Nerstian pH response. The corrected resistivities for tubes which displayed acceptable pH response at 90°C range from 9.51 to 9.97 log R, where R is in units of ohm-cm (Table II). Tubes having significantly higher or lower resistivities respond much less favorable, if at all. This testing procedure provides a rapid and simple screening process for identification of tubes that will display acceptable response.

Test Apparatus. A flow-through test vessel was constructed to allow the testing of any 1/4-in.-OD pH electrode at temperatures up to 300°C and pressures up to 300 bars. The system is illustrated schematically in Figure 4.

The test vessel has a 50-cm^3 cavity with a titanium lining set into a stainless steel body with a delta ring pressure seal. The "top hat" closure of the vessel has been designed to have the same configuration (with regard to diameters) as the Dickson-type rocking autoclave closure. This is to allow direct transfer of any evolved pH electrode design to the Dickson autoclaves and hot cell units with minimum redesign.

For the calibration and evaluation of kinetic response of pH electrodes, a Milton-Roy sapphire and Hastalloy pump (capable of 6,000 psi) were built into the pressure line in order to feed acid, base, or buffered solutions through the test vessel by way of stainless capillary tubing of 0.030-in. i.d. Outflow from this system can be controlled by means of two Hoke Micro-Metering valves which, when mounted in series, can provide an "engineered leak" with an outfall that can be matched by the pump to allow system flows from 0.6 cm^3/min to 16 cm^3/min. Such a pumpable system allows fresh reference solution to be continuously added to the system while maintaining constant pressure. This avoids a possible pH drift caused by reactions between the walls of the test chamber and the reference solution. This system also allows

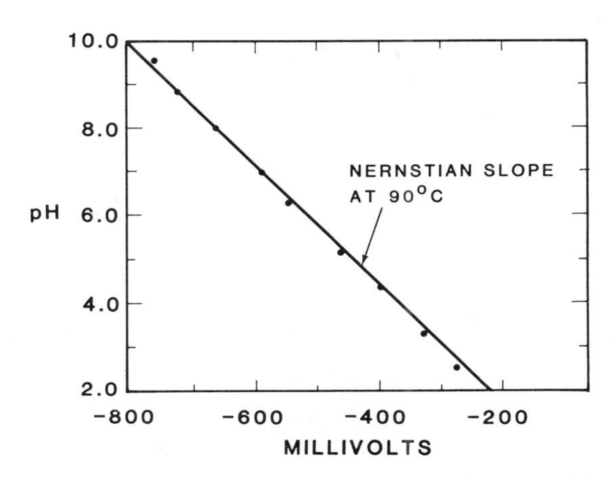

Figure 3. EMF versus pH Response of Pennsylvania State University Fabricated Tube Number PSU-T1-18.

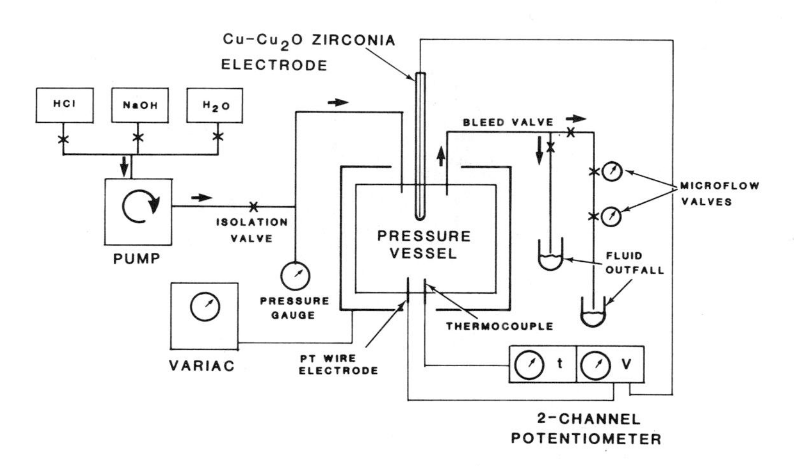

Figure 4. Pumpable pH Testing System.

Table II. Evaluation at 90°C of Resistivity and Nernstian pH Response of Various Suppliers of ZrO_2 Cells

Tube number (mol% Y_2O_3 in parentheses)	Log ohms, 90°C (R)	Log ohm-cm $\frac{(R \times Area)}{thickness}$ 90°C	pH response (% Nernstian in parentheses, 90°C)
(8) T 2-18	9.23	9.63	good (93%)
(4) T 3-18	9.11	9.52	none
(10) T 4-18	erratic (<7.17)	-	none
(6) T 5-18	erratic (<6.17)	-	none
(12) T 6-18	9.34	9.75	none
(8) ZT 2-14	9.04	9.61	good (89%)
(8) ZT 3-14	9.00	9.57	not yet tested
(8) ZT 6-14	9.08	9.65	not yet tested
(8) CT 3-14	8.11	9.52	good (89.3%)
(8) CT 1-38	8.25	9.77	fair (71.6%)
(8) CT 2-38	8.43	9.94	good (93.5%)
(8) CT 3-38	8.46	9.97	good (86%)
(8) CT 4-38	8.40	9.91	good (93.5%)

reference solutions of different pH values to be pumped into the test chamber to establish a calibration curve and to evaluate the response time of the sensor to changes in pH. As of this writing, the system has been used at pressures up to 135 bars and 200°C. The system can also be isolated from the pump and outflow valves by means of additional shutoff valves, which leave the pressure gauge functioning so that static testing can also be performed with the system. The pressure vessel is surrounded by a cylindrical furnace that is powered from a variable transformer as a resistive load. Temperatures are measured using an inconel-sheathed thermocouple (chromel-alumel) mounted through the vessel bottom by means of a split gland CONAX fitting. Through this same CONAX fitting (in its own insulated port) a 1-cm length of platinum wire of 0.020-in. diameter is introduced into the cavity. When the cavity is filled, this wire contacts the solution. The copper electrode (inside the ZrO_2 tube) is comprised of copper metal and cuprous oxide and forms the other half of the pH sensing circuit with the platinum wire. The circuit includes a high impedance potentiometer. The measured voltages can be calibrated in terms of pH as explained above.

Test Results. The pH response of the zirconia tubes was tested at elevated temperatures and pressures using the apparatus previously described. Standard HCl and NaOH solutions were pumped through the test chamber while the EMF of the sensor was monitored. The pH values at temperature and pressure were calculated from the room temperature pH measurement based on hydrogen ion mass balance using the data of (11).

Data from an early successful calibration test performed at Temple University is summarized in Figure 5. One manufacturer's tube was tested at 200°C and 50 bars pressure with unbuffered solutions of 0.01 N HCl and 0.001 N NaOH. The heavy dashed line in Figure 5 connects these two 200°C points while the thin dashed lines connect the equivalence values on the $pH_{25°C}$ (bottom) and $pH_{200°C}$ (top) scales of the graph. The solid line gives the calculated data from Equation 5 for theoretical Nernstian behavior as calculated from the literature data bases we have used. The heavy dashed line (2 points only) defines a slope of 91.9 mV per pH unit which is 97.8% of the predicted Nernstian slope. Thus, the tube exhibits a Nernstian slope; however, a +240 mV cell or blank correction would be necessary. The third data point (open circle) in Figure 5 is the result for a 0.01 N NaCl solution which, according to a Beckman (Zeromatic II) calibrated pH meter, had an input $pH_{25°C}$ of 5.6, but an outfall $pH_{25°C}$ of 9.0. This measured pH shift in the 0.1 N saline solution within the vessel would suggest that the new vessel was reacting with the hot saline solutions so that some pH shift was occurring. Note from Table III that both the NaOH and HCl were little changed by passage through the test vessel.

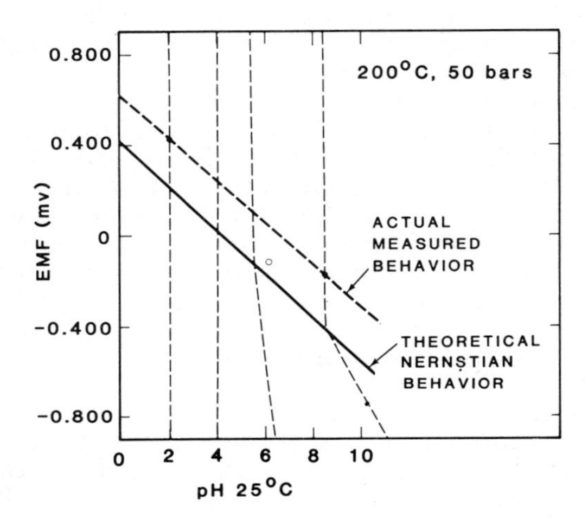

Figure 5. Response of Zirconia pH Sensor at 200°C, 50 bars.

Table III. Initial Results at High P and T for Electrode CT-TU-14-II-3

Beckman pH Meter Measured	Calculated $pH_{200°C}$	EMF (mV)	Solution chemistry
11.11 input	8.34	-138	0.001 N NaOH
11.11 outfall			
1.90 input	1.94	+450	0.01 N HCl
1.94 outfall			
5.6 input	6.28	-130	0.10 N NaCl
9.0 outfall			

*It would seem from this shift in pH that the solution may have been reacting with the new vessel since for 2-1/2 hr the outfall pH was steady but shifted from the 5.6 value at input.

Niedrach (1980) mentions that new vessels need time to passivate by redox reaction. We suspect, therefore, that the NaOH and HCl solutions have respectively high enough OH^- and H^+ concentrations to override the effects of the new vessel, whereas the NaCl solution may need replacing with a pH=7 buffer of higher ionic-strength in the vessel at least until the titanium metal develops a reaction-resistant oxide coating. Subsequent calibrations have been performed using well-buffered solutions with known pH at elevated pressures and temperatures. This has eliminated the uncertainties caused by unbuffered solutions reacting with the walls of the test chamber.

The ZrO_2 sensor investigated at Temple University responds to pH and is insensitive to the Eh of the solution. However, the bare platinum wire used to complete the circuit in the test chamber will respond to changes in the Eh conditions. It is believed that a linear relationship between pH and EMF was observed in this experiment because the test solutions employed were all air-saturated and therefore had similar Eh values. This hypothesis was tested by sequentially pumping three identical sets of pH buffer solutions through the test chamber. In the first calibration, the solutions were saturated with oxygen gas; in the second calibration, the solutions were saturated with hydrogen gas; in the third calibration, the solutions were saturated with air. There was a several hundred millivolt shift in the EMF in response to these three sets of solutions; however, in each calibration the pH versus EMF response was linear. Clearly, what is needed is a reference electrode substituted for the bare platinum wire used to complete the electrical circuit in the test chamber. The absence of a suitable reference electrode in the test circuit prevents an Eh-independent calibration of the pH sensor. However, the near perfect linearity of pH versus EMF measured between the bare platinum wire and the ZrO_2 sensor in the three calibration tests demonstrates the ability of the sensor to accurately measure pH once it is calibrated. Work is currently in progress at PNL to calibrate ZrO_2 sensors using a long-lived reference electrode in a large volume refreshed autoclave. Also, a project is currently in progress at PNL to design a long-lived reference electrode that will be compatible with the Dickson autoclave.

Effects of Chemical Purity. Zirconia tubes from five different sources were analyzed at Pennsylvania State University using scanning electron microscopy, plasma emission spectroscopy, energy dispersive X-ray spectroscopy, and electron beam microprobe analysis. The sources for the tubes included several commercially available tubes as well as tubes fabricated by the Pennsylvania State University Ceramics Department.

Tubes from one manufacturer have the highest concentration of calcium out of all the tubes tested (Table IV). X-ray mapping using the electron microprobe was used to determine whether the calcium is homogeneously distributed through the ceramic or occurs as discrete random or interstitial patches. The mapping technique reveals that the calcium is concentrated into patches in this manufacturer's ceramic, but is randomly distributed in other manufacturer's ceramic as well as Pennsylvania State University ceramics. X-ray mapping for magnesium, silicon, aluminum, and titanium all showed random patterns. If interstitial glass or crystalline phases were present, it is unlikely that they would be enriched only in calcium and not one or more of the other elements as well. Therefore, the heterogeneous distribution of calcium probably reflects calcium-enriched grains existing in the starting material. All of the commercially available tubes that functioned well as pH sensors are enriched in calcium relative to the other ceramics. However, the success of the calcium-poor Pennsylvania State University tubes as pH sensors, and the absence of an interstitial calcium-rich phase, makes it seem unlikely that enrichment in calcium is important in producing a pH response.

Test results at 25°C and 90°C have been interpreted at Pennsylvania State University as follows. The higher the purity of the ZrO_2, the more likely it is to behave well as a pH sensor. Exceptions to this rule are titanium, which is beneficial for thermal shock resistance, and yttria, which is necessary for polymorphic stabilization. The optimum Y_2O_3 content was investigated by testing ZrO_2 tubes with Y_2O_3 contents of 4, 6, 8, 10, and 12 mol%.

Table IV. Chemical Analyses of Various Zirconia Tubes *

Plasma Emission Spectroscopic Analysis (wt%)					
Oxide	CT-1-38	CT-2-38	CT-3-38	PSU-T1	ZT-3-14
Al_2O_3	0.78	0.78	0.78	0.78	0.78
TiO_2	0.11	0.09	0.07	0.12	0.06
Fe_2O_3	0.07	<0.05	0.06	0.06	0.11
MgO	0.09	<0.05	<0.05	<0.05	0.31
CaO	1.09	1.06	1.08	<0.05	0.31
HfO_2	1.53	1.51	1.53	1.45	1.84
Y_2O_3	8.74	9.38	9.39	7.76	7.83
Electron Microprobe Analysis (wt%)					
Oxide	ZT-NG-1	CT-3-38	ZT-1-14	PSU-T1	
Na_2O	0.03	0.09	0.10	0	
MgO	0.34	0.10	0.27	0.08	
Al_2O_3	0	0.14	0.01	0.16	
SiO_2	0.29	0.17	0.03	0.28	
Y_2O_3	9.95	10.54	8.94	5.11	
ZrO_2	87.69	82.37	85.29	87.91	
CaO	0.20	0.57	0.22	0	
TiO_2	0	0.09	0.08	0.31	
Total	98.5	94.6	94.9	93.8	

*Analyses performed at Pennsylvania State University.

Porosity could not be held constant. However, only tubes with 8 mol% Y_2O_3 responded to pH changes. The 6 and 10 mol% Y_2O_3 tubes sintered poorly, probably because of phase transitions during cooling. These tubes remained permeable, shorting out any potential developed by a pH difference. The 4 and 12 mol% Y_2O_3 tubes had resistivities in the working range; however, no pH response was observed. The effects of variations in the amounts of Y_2O_3 in the ceramic are still unclear, but the lack of success with tubes that have compositions differing from 8 mol% Y_2O_3 (which is also the composition having maximum conductivity at high temperatures), demonstrates that having a Y_2O_3 content in a narrow range near 8 mol% is important in generating pH response.

pH Sensing Mechanisms. A simple ion exchange-type reaction similar to that taking place on glass pH electrodes is likely for several reasons. The glass pH electrodes and zirconia electrodes are analogous in composition. A typical glass electrode contains primarily SiO_2 doped with 10 to 20 mol% of mono-, di-, or trivalent oxides. The most successful zirconia similarly contains primarily ZrO_2, a tetravalent oxide like silica, with 8 mol% Y_2O_3. In glass electrodes, the dopant cations are responsible for forming hydrogen-selective exchange sites on the hydrated surface of the glass. Similar sites may form on the hydrated zironia ceramic.

Normally, glass electrodes must be soaked in water for a few minutes to a few hours for the electrode to develop a pH response. This allows hydration of the outer surface to take place with the formation of the hydrogen ion-selective sites. Similarly, in 90°C pH tests at Pennsylvania State University, it has been found necessary to allow the zirconia electrodes to soak for several hours before a pH response was observed.

Preliminary results indicate that only zirconia tubes having a narrow range of yttria contents respond well as pH sensors. Compositions close to 8mol% yttria show the best response. Apparently, the existence of hydrogen ion-selective sites is dependent on the ceramic composition, as with the glass pH electrodes.

The pH response mechanism of the zirconia electrode is therefore most likely due to formation of hydrogen ion-selective sites, perhaps related to oxygen ion vacancies, developed on the electrode surface during hydration. Exchange of hydrogen ions between these sites and the hydrogen ions in solution generates an electrical potential across the electrode following the simple exchange reaction:

$$H^+(\text{at } O^{2-} \text{ site}) = H^+_{(aq)} \qquad (3)$$

This mechanism accounts for the lack of response of the electrode to redox conditions at temperatures below 400°C, the need for preliminary soaking and formation of an active, hydrated surface layer, and the influence of yttria content on pH sensitivity.

Future Investigations. Investigations are currently underway to determine the long-term stability of these sensors at temperatures as high as 300°C and pressures as high as 300 bars. Efforts are also underway to employ impedance spectroscopy techniques (12-13) to determine the charge-carrying mechanism in stabilized zirconia materials over a temperature range of 25°C to 300°C. These investigations will hopefully lead to an increased understanding of the pH-sensing mechanism so that variables such as concentration of dopant, grain size, sintering time, and sintering temperature can be optimized to produce an electrolyte material that can be reliably employed for extended periods at elevated temperatures and pressures.

Summary and Conclusions

Experience gained in using Teflon hydrogen diffusion membranes indicate that Teflon can be used to measure the hydrogen pressure in equilibrium with the fluid phase in hydrothermal experiments. During long-term experiments (months), interferences in H_2 measurements can be introduced by the diffusion of other gas species such as SO_2, H_2O, CO_2, and CH_4 through the membrane. However, the effect of the presence of these gases can be accounted for by gas chromatography analysis of the diffused gas mixture or by installing a liquid nitrogen cold trap between the membrane and the pressure transducer.

It has been demonstrated that the zirconia pH sensor can be used at temperatures at least as high as 300°C and at pressures at least as high as 300 bars. Only a small percentage of commerically available ZrO_2 tubes display a pH response, however. A simple screening technique has been developed to accurately predict which individual tubes will have a suitable pH response. If the measure resistance value at a given temperature lies outside of a certain range, the tubes will not display a pH response. Zirconia tubes were fabricated with Y_2O_3 contents ranging from 4 to 12 mol% and were tested for pH response. The ZrO_2 tubes stabilized with 8 mol% Y_2O_3 provide optimum response for the grain size and fabrication techniques used. Test results on the pH response of ZrO_2 tubes at 200°C and 50 bars indicate that some tubes display up to 98% of the theoretical EMF versus pH slope. Work is currently underway to develop a suitable reference electrode to allow accurate calibration of ZrO_2 pH sensor. These current investigations provide evidence that sensors capable of measuring Eh-pH conditions at elevated pressures and temperatures can be developed in a timely manner for use in hydrothermal experiments.

Acknowledgments

The authors wish to acknowledge contributions to this paper by J. R. Holloway and J. D. Clemens of Arizona State University and H. L. Barnes and W. L. Bourcier of Pennsylvania State University. They have agreed to the incorporation here of a summation of the results of their research performed under contract to Rockwell Hanford Operations.

Literature Cited

1. Early, T. O.; Jacobs, G. K.; Drewes, D. R. "Geochemical Controls on Radionuclide Releases from a Nuclear Waste Repository in Basalt: Estimated Solubilities for Selected Elements" RHO-BW-ST-39, Rockwell Hanford Operations, Richland, Washington, 1982.
2. Bischoff, J. L.; and Seyfried, W. E. American Journal of Science 1978 278, 838-860.
3. Cherry, J. A.; Shaikh, A. U.; Tallman, D. E.; Nicholson, R. V. J. Hydrology 1979 43, 373-392.
4. Eugster, H. P.; Wones, D. R. J. Petrol. 1962, 3, 82-125.
5. Burnham, C. W.; Holloway, J. R.; Davis, N. F. Geol. Soc. Am. Spec. Paper 1969, 131, 96.
6. Robie, R. A.; Hemingway, B. S.; Fisher, J. R. U.S. Geol. Surv. Bull. 1978, 1452, 456.
7. Ryzhenko, B. N.; Volkov, V. P. Geochemistry International 1971, 468-481.
8. Niedrach, L. Science 1980, 207, 1200-1202.
9. Tsuruta, T.; MacDonald, D. D. J. Electrochem. Soc. 1982, 129, 1221-1225.
10. Apted, M. J.; Myers, J. "Comparison of the Hydrothermal Stability of Simulated Spent Fuel and Borosilicate Glass in a Basaltic Environment" RHO-BW-ST-38, Rockwell Hanford Operations, Richland, Washington, 1982.
11. Helgeson, H. C.; Kirkham, D. H.; Flowers, G. C. Am. J. Sci. 1981, 281, 1249-1516.
12. Kleitz, M.; Bernard, H.; Fernandez, E.; Schouler, E. Impedance Spectroscopy and Electrical Resistance Measurements on Stabilized Zirconia in "Advances in Ceramics;" Vol. 3, SCIENCE AND TECHNOLOGY OF ZIRCONIA, ed. Hever, A. H., American Ceramic Society; Columbus, Ohio, 1981.
13. Bauerle, J. E. J. Phys. Chem. Solids 1969, 30, 2657-2670.

RECEIVED October 28, 1983

13

Surface Studies of the Interaction of Cesium with Feldspars

D. L. BROWN, R. I. HAINES, D. G. OWEN, F. W. STANCHELL, and D. G. WATSON

Atomic Energy of Canada Limited, Whiteshell Nuclear Research Establishment, Pinawa, Manitoba R0E 1L0 Canada

The interaction of cesium ions with feldspars at 150°C and 200°C has been studied in distilled water, granite groundwater and saline solution. Pollucite, $CsAlSi_2O_6$, was identified by infrared spectroscopy, and was formed as a cubic crystalline phase. Surface analytical techniques (XPS, SAM, SIMS and SEM/EDX) show Cs to be sorbed onto the mineral surfaces and alteration products. The mechanism of pollucite formation and its relevance to cesium transport/retardation in the near field of a nuclear waste-disposal vault is discussed.

In the Canadian Nuclear Fuel Waste Management program, disposal of radioactive waste is envisaged within an engineered multi-barrier vault, deeply embedded in a plutonic formation within the Precambrian Shield (1). The geologic formation is the final barrier in the scheme, and the interaction of radionuclides with rock-forming minerals is the important element of that barrier. This paper describes our research into the behaviour of cesium in contact with feldspars. Feldspars constitute a major portion of the igneous rock of a pluton. We have chosen microcline, $KAlSi_3O_8$ and labradorite, $CaAl_2Si_2O_8$ as representative feldspars of granite and gabbro rocks, respectively, for this study. We have included albite, $NaAlSi_3O_8$ for comparison. This feldspar occurs together with the former two minerals in nature. Cesium has been chosen for study for several reasons. Although the amount of cesium present in the fission products of used fuel is low (0.7%) (2), it's isotopes have relatively long half-lives ($\tau_{1/2}(^{137}Cs)$ = 30 years, $\tau_{1/2}(^{135}Cs)$ = 3 x 10 years), they emit penetrating radiation, and are deemed to be relatively mobile in aqueous solutions, percolating through rocks and soils. Hence, to be able to predict the movement of this element in the geosphere, it is necessary to understand its physical and chemical interactions with the geologic materials it is likely to contact.

0097–6156/84/0246–0217$06.00/0

Two extreme scenarios are plausible. Firstly, the waste vault remains intact for a long period, following which fission products escape into the immediate vicinity of the vault. Here, interactions will occur between the radionuclides and the geosphere, which has been hydrothermally altered by the groundwater, heated by the thermal energy released by radioactive decay. At the other extreme, we may consider the case where the vault is breached early in the thermal period, releasing radionuclides into a hydrothermal environment. It is this second case which is being considered here. We have treated feldspars with hydrothermal solutions of non-active cesium chloride in water, and selected groundwaters, and analyzed the resulting solutions and mineral surfaces, to elucidate the nature of the interaction of cesium with feldspar surfaces, and to determine whether this nuclide will be mobilized or retarded by the geosphere under these conditions.

Experimental

Microcline, albite and labradorite were obtained from Ward's Canada Limited. Cleaved crystal fragments and 60 mesh-sized samples were ultrasonically cleaned prior to use. Powdered samples (particle size $< 25\ \mu m$) were prepared by grinding in a tungsten carbide ball mill, wet-sieving and washing in water.

The water was doubly distilled and deionized before use. Granite groundwater (G.G.W.) and standard Canadian Shield saline solutions (SCSSS) were prepared according to the method of Vandergraaf (3).

Experiments were performed in Teflon-lined titanium autoclaves, submerged in a thermostated oil bath. The supernatant liquors were analyzed using atomic absorption spectrometry. Solid residues were washed with water, dried and examined by the following surface analytical methods:

1. SEM/EDX studies were performed with an ISI DS 130 scanning electron microscope, using gold-sputtered samples. XPS measurements were made using a McPherson ESCA-36 spectrometer, with an operating pressure of 5×10^{-7} Pa. Spectra were obtained using the Al K_{α} exciting radiation. Energy dispersion calibrations were made using the known energy difference of 1253.6 eV between the Mg (3p) and Mg(1s) lines for evaporated magnesium metal. An internal carbon standard was used for measuring binding energies.
2. Auger spectra of the mineral surfaces were obtained using a Physical Electronics Industries (PHI) Model 590A scanning Auger microprobe (SAM). Spectra were recorded in the pulse count (N(E)*E) mode at 0.6% resolution ($\Delta E/E$) using a 3 kV, 3 nA primary electron beam. Background instrumental pressure was maintained at 1.33×10^{-7} Pa during data acquisition. Secondary ion mass spectra (SIMS) were collected using a PHI Model 3500 SIMS II, attached to the SAM. A PHI Model 04-303

differentially pumped Ar^+ ion gun was used as the excitation source. The secondary ion spectra were excited by a 4.5 kV, 400 nA Ar^+ beam, rastered over a 1.5 mm-square area.

3. Fourier transform infrared (FTIR) spectra were obtained using a Nicolet 10 MX FTIR spectrometer, in the spectral range 4000 cm^{-1} to 270 cm^{-1}. Two hundred mg KBr discs, containing about 1 mg sample were used, in a nitrogen-purged system. Discs of reaction product were prepared by ultrasonically removing the product from the sample surface under acetone, decanting, centrifuging and drying the resultant product suspension, and mixing with an appropriate amount of KBr.

Results

Cesium-ion concentrations in distilled water and synthetic groundwaters were measured after contact with the feldspars for various periods of time, over the temperature range 150 °C to 200 °C. It was found that for short reaction times (< 5 days), there was little reduction in the concentration of cesium ion, i.e. little sorption of Cs^+ by the minerals. Removal of Cs^+ from solution was enhanced by increased mineral surface area, reaction temperature and time. It was observed that in the extreme case for powdered labradorite, 98% of an initial 10^{-2} mol dm^{-3} solution of Cs^+ was sorbed after 14 days at 200 °C in distilled water. The morphology, composition and chemical structure of the mineral surfaces were investigated by several analytical methods, as described below.

Scanning Electron Microscopy. Examination of the surface of cleavage fragments of microcline and albite, after reaction for 14 days at 200 °C with 10^{-2} mol dm^{-3} aqueous cesium chloride solution, revealed the formation of large (10 - 20 μm length) cubic crystals (see Figure 1(a)). Sub-micron crystals were observed covering the mineral surface surrounding the cubes, formed by hydrothermal alteration of the mineral. EDX spot analyses showed the cubes to be composed of Cs, Al and Si (elements lighter than F cannot be detected by this method).

Reaction times of $\leq$ 5 days produced no detectable (by SEM) cesium-containing phases. However, after treating the minerals with water at 200 °C for 14 days in the absence of cesium, then adding the appropriate amount of CsCl, we observed cesium aluminosilicate crystals within 42 hours of further hydrothermal reaction.

The cubic crystals were not observed in experiments involving low (< 10^{-4} mol dm^{-3}) cesium concentrations. For runs involving powdered and coarse-grained minerals, spherular and 30-sided polyhedral cesium alumino-silicate crystals were observed (Figure 1(b) and (c)). For powdered labradorite, rosette formations were seen (Figure 1(f)). Such morphologies are typical of analcime-type minerals, particularly pollucite, $CsAlSi_2O_6$ (4). EDX analysis of the alteration product surrounding the cesium aluminosilicate showed that Cs was also sorbed onto this material. Similar behaviour was observed for reactions at 150°C, although reaction times were much longer (47 days), and crystal growth much less (Figures 1(d) and (e)). Albite behaved in a similar fashion to microcline under all conditions. Experiments performed in G.G.W. also produced results analogous to those obtained using distilled water as the medium. This observation is not unexpected, since the GGW contains quite low concentrations of dissolved ions, and contact of distilled water with feldspars at high temperatures is expected to produce solutions of similar ionic strength (due to K^+ or Ca^{2+}, Na^+ and Si species) quite rapidly. For example, large cubes were produced by reaction of cleaved microcline with 10^{-2} mol dm^{-3} CsCl for 14 days in G.G.W. (Figure 2(a)). For experiments performed with SCSSS, no large cubes were detected. However, considerable alteration product was observed, as 1 - 2 μm-sized crystals of montmorillonite (Figure 2(b)). EDX showed them to consist of Mg, Ca, Al, Si (and O).

FTIR Spectra. After reaction, the mineral samples were subjected to ultrasonic treatment in acetone for periods of up to 1 hour. Transmission IR spectra of the released product fines from reactions of 60-mesh labradorite and microcline with 10^{-2} mol dm^{-3} CsCl in distilled water at 150°C for 47 days are presented in Figure 3. The spectrum obtained from of the labradorite reaction product (Figure 3(a)) is identical with that of a sample of natural pollucite (Figure 3(c)). The spectrum obtained from the microcline reaction product (Figure 3(b)) contains bands in the 500 cm^{-1} to 800 cm^{-1} region, due to unreacted microcline, in addition to the major bands of pollucite. Heating the samples overnight at 105°C resulted in the disappearance of the water absorption bands at about 3500 cm^{-1} and 1680 cm^{-1}.

XPS, SAM, SIMS Analyses. For reaction times < 5 days, or $[Cs^+] \leq 10^{-4}$ mol dm^{-3}, XPS results showed about 10 atomic % Cs to be present on the surface of the mineral samples. Depth profiling using SIMS indicated that in these cases, the sorbed Cs was present for only the first 200 nm.

Table I lists the binding energies for the major constituent elements present on the mineral surfaces, both before and after reaction. These values are, in general, in good agreement with literature values for aluminosilicates (5). The value for Cs corresponds to that for ionic Cs salts. No large shifts are expected in these numbers, since there is no oxidation state difference between reactants and products. A detailed XPS study of the interaction of cesium (and strontium) with feldspar surfaces is in progress, and will be the subject of a future paper.

The O/Cs region (500 - 600 eV) of the numerically differentiated Auger spectrum of one of the cesium aluminosilicate crystals, found on the face of microcline exposed to 10^{-2} mol dm^{-3}

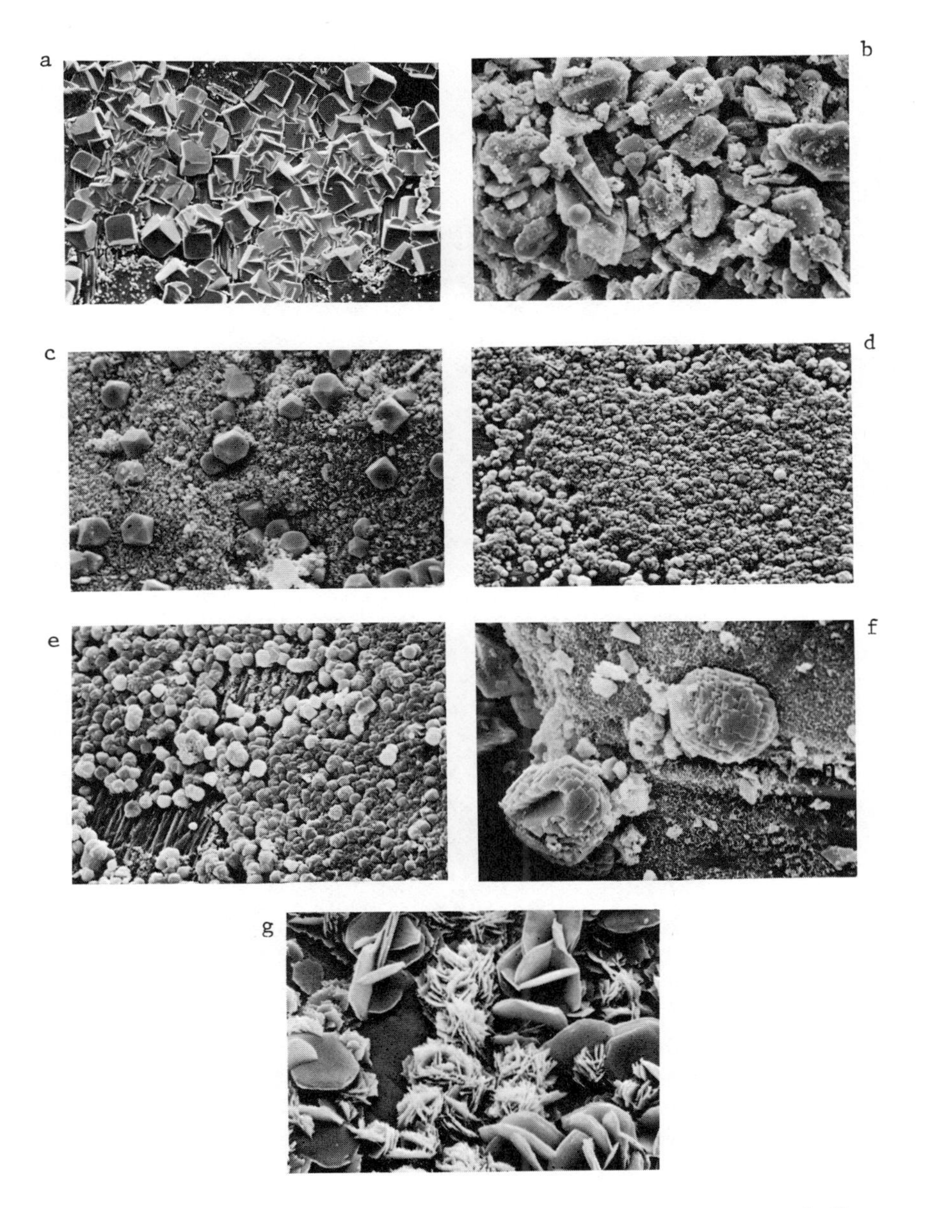

Figure 1. SEM photomicrographs of microcline (a, c and d), albite (b) and labradorite (e–g) after hydrothermal reaction with CsCl (see text).

Scale: 4.5 mm = 19.1 μm (a), 2.98 μm (b), 3.64 μm (c), 1.98 μm (d), 2.36 μm (e), 2.42 μm (f), and 2.34 μm (g).

Figure 2. SEM photomicrograph of (a) cleaved microcline surface after reaction with 10^{-2} mol dm^{-3} CsCl at 200°C for 14 days in G.G.W; (b) as (a), but in SCSSS.
Scale: 6 mm = 10.9 μm (a) and 5.62 μm (b).

Table I. Elemental Binding Energies (eV)[a] for Feldspars (i) Before and (ii) After Hydrothermal Reaction with Aqueous Cesium Solutions

Mineral	Al(2p)	Si(2p)	K(2p)	Na(1s)	Ca(2p)	Cs(3d)
(i) fresh minerals						
Microcline	74.50	102.7	293.65	1071.95		
Albite	74.65	103.0		1071.85		
Labradorite	74.60	102.50		1072.95	348.20	
Pollucite	76.10	102.80				724.6
(ii) after reaction with 10^{-2} mol dm^{-3} Cs^+ at 200°C for 14 days						
Microcline	75.45	103.10	293.50	1071.75		724.95
Albite	75.70	103.20				725.05
Labradorite	74.9	102.65		1072.2	348.45	724.4

[a] Error is ± 0.25 eV.

CsCl at 200°C for 14 days, is shown in Figure 4. Before this spectrum was collected, the cube was sputtered with Ar^+ ions to remove surface contamination. Two Cs MNN transitions arise at 554 and 566 eV (6) and the larger O KLL peak is at 506 eV. Superimposed on Figure 4 is a spectrum of natural pollucite, recorded under the same instrument conditions and corrected for a -4 eV positional difference observed in each of the three peaks. The same charging shift was observed in the Si peaks (not shown in Figure 4), which were found at 76 eV (cube) and 72 eV (natural pollucite). Transitions from four elements, Al, Si, O, and Cs, were observed in both spectra. Potassium was not detected in either spectrum.

The chemical composition of the cube was calculated from the Auger spectra using elemental sensitivity coefficients derived from the spectrum of natural pollucite, assuming the composition of the latter to be $CsAlSi_2O_6$. In this manner, an empirical composition of $Cs_{1.3}Al_{1.1}Si_{1.7}O_{5.9}$ was determined for the cube on the surface of the microcline. Several possible sources of error may contribute to the observed difference in composition. Other than the assumption that the natural pollucite was anhydrous, the most likely reason for the difference between the two is the electron-beam-induced reduction of Si in the matrix, resulting in a shift of the 76 eV peak (SiO_2) to 88 eV (Si), and partial desorption of O. Indeed, some reduction of SiO_2 to elemental Si was observed in the spectrum of the cube. The apparent high Cs concentration may be due to Cs adsorbed on the surface that was incompletely removed by sputtering. A similar quantitative analysis of a cubic crystal found on albite resulted in the empirical formula $Cs_{1.5}Al_{1.1}Si_{1.5}O_{5.9}$. According to the relative inten-

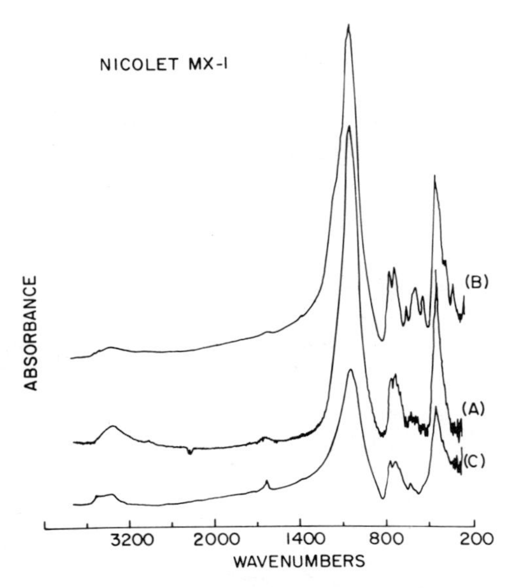

Figure 3. FTIR spectra of (a) fines removed ultrasonically from 60-mesh labradorite (see text); (b) fines removed ultrasonically from 60-mesh microcline (see text); (c) natural pollucite.

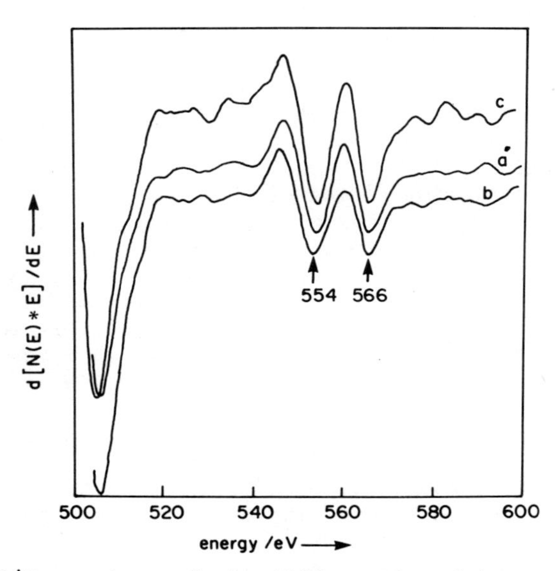

Figure 4. The O/Cs regions of the differentiated Auger spectrum of (a) a cubic pollucite crystal grown on microcline; (b) natural pollucite; and (c) background alteration product on microcline.

sities of the 76 and 88 eV peaks in this Auger spectrum, reduction of SiO_2 to Si occurred to a greater extent than previously observed on the microcline cube. The higher Cs concentration on the albite is likely due to adsorbed Cs, since this sample was not sputtered.

An Auger spectrum of the unsputtered surface of granulated microcline (soaked in CsCl as above), on which no cubic crystals were observed, is also included in Figure 4. The Cs peaks are, as before, at 554 and 566 eV. Oxygen falls at 505 eV and the Si line (not shown in Figure 4) is at 76 eV, in agreement with the other specimens analyzed. Cesium would not be expected to exhibit large chemical shifts, and none were observed.

Since the detection limit is relatively high ($\sim$ 0.6 atomic %), Cs was not detected by SAM on the samples treated with lower concentration CsCl solutions. However, Cs was detected by SIMS on all surfaces analyzed. Typical SIMS profiles showed Cs signals which slowly decreased to constant levels after 30-40 min sputtering at a rate of 6 nm/min. These results suggest Cs is adsorbed on the surface of, and incorporated in, the surface alteration product.

Discussion

Several studies (7) have shown that granites and granitic feldspars have a significant k_D value for cesium. The sorption phenomenon at ambient temperatures is considered to be due to ion exchange, with the k_D values for a series of minerals having been correlated with the cation-exchange capacities of the geologic materials. Under hydrothermal conditions (150-200^oC), significant sorption of cesium ion by the feldspars microcline, albite and labradorite occurs. Although monitoring of solution concentrations of Cs^+ shows negligible reduction for short reaction times, XPS analysis of the mineral surfaces showed about 10 atomic % Cs in at least the first 500 nm layer. The surface compositions indicated virtually complete replacement of the host metal cation. For longer reaction times, morphological studies of the mineral surfaces by SEM indicated crystallization of a secondary cesium phase in addition to mineral alteration products which have sorbed cesium. Coalesced, lens-shaped etch pits are apparent on the feldspar surfaces (Fig. 1(a)), a feature typical of mineral dissolution (8). Outcrops of microcrystalline alteration products proliferate around these highly etched regions. For mineral cleavage fragments, large (10-20 μm) cubic crystals are formed, also in the vicinity of etch pits. The size and purity of these crystals suggests slow growth, requiring a controlled rate of crystallization. These cubic crystals, shown by infrared spectroscopy to be pollucite, are formed from aluminum and silicon species which are released into solution by the feldspar dissolution. The concentration of these species in solution will be controlled by the rate of feldspar dissolution versus the rate of alteration product precipitation. The concentrations are expected to be low for cleavage fragments with low surface area. It may be noted that for powdered mineral reactants, greater removal of cesium from solution was evident, but no large pollucite crystals were seen, only small, spherular ones. This suggests rapid pollucite formation, consistent with rapid dissolution of the finely ground feldspars and subsequent reaction with the dissolved cesium ions present. These crystals are unlikely to have formed because of quenching of the reaction, since such a process (taking several minutes) would form more amorphous material. After rapid cooling of the reaction systems, the solid residues were washed thoroughly, thus tending to remove any loose, ultrafine quenched products.

There are no data in the literature on the solubility or stability of pollucite, so it is difficult to predict whether or not the solutions would be over-saturated by this phase at $\{Cs^+\} \leq 10^{-4}$ mol dm^{-3}. No pollucite was observed under these conditions, over the relatively short (14 days) reaction times at 200°C. Experiments are continuing to elucidate this question using XPS to look for the unique binding energy of Al(2p) (= 76.1 eV) of pollucite on the mineral surfaces, and hence provide information relevant to more realistic possible near-field conditions.

The mid-infrared spectrum of the cesium aluminosilicate unequivocally identified it as pollucite. Moreover, the SAM results indicate that the material is uncontaminated by foreign alkali metal ions, such as K^+ or Na^+ from the feldspars, or the aqueous media.

Earlier studies (9) have shown that kaolinite is a major alteration product of feldspars in acidic aqueous media. In SCSSS, montmorillonite appears to be formed readily under hydrothermal conditions. Strachan and Schulz (10) showed that pollucite is formed in hydrothermal reactions of montmorillonite with cesium, and pollucite was considered an excellent material for long-term storage of ^{137}Cs, because of its low leach rate (2×10^{-9} kg m^{-2} s^{-1}). Barney (11) has shown that kaolinite reacts with cesium under fairly mild conditions to form pollucite. Komarneni (12) has shown that pollucite resists ion exchange to some extent. Thus, should cesium be released from a vault into a hydrothermal environment, it may be expected to be considerably retarded via mineral precipitation and surface absorption by the geologic barrier. Further experiments are being performed to elucidate the kinetics of the cesium/ feldspar interaction, and to determine the rate and extent of reaction at temperatures closer to ambient. These data should lend insight into the behaviour of cesium in the near-field geosphere after the thermal period of the vault is over.

Acknowledgment

We are grateful to members of the Analytical Science Branch, WNRE, for the atomic absorption measurements.

Literature Cited

1. Boulton, J. (ed.), AECL-6314, Chalk River, Ontario, Canada, 1978.
2. Goodwin, B. W.; Johnson, L. H.; Wuschke, D. M. Proc. NEA Workshop, OECD, Near Field Phenomena in Geologic Repositories for Radioactive Waste, 1981, 33.
3. Abry, D. R. M.; Abry, R. G. F.; Ticknor, K. V.; Vandergraaf, T. T., TR-189, Chalk River, Ontario, Canada, 1982.
4. Barrer, R. M.; Kerr, S. I. *J. Chem. Soc.* 1963, 434.
5. Adams, I.; Thomas, J. M.; Bancroft, G. M. *Earth Planet. Sci. Lett.*, 1972, *16*, 429.
6. Thomas, S.; Haas, T. W. *J. Vac. Sci. Technol.* 1973, *10*, 218.
7. Torstenfelt, B.; Andersson, K.; Allard, B., Report Prav 4.29, Programrodet for radioaktivt avfall, 1981.
8. Berner, R. A.; Holdren, Jr., G. R. *Geochem. Cosmochim. Acta*, 1979, *43*, 1173.
9. Haines, R. I.; Owen, D. G., unpublished observations.
10. Strachan, D. M.; Schulz, W. W., ARH-SA-294, Atlantic Richfield Hanford Co., Richland, Washington 1977.
11. Barney, G. S., ARH-SA-218, Atlantic Richfield Hanford Co., Richland, Washington, 1975.
12. Komarneni, S.; McCarthy, G. J.; Gallagher, S. A. *Inorg. Nucl. Chem. Lett.*, 1978, *14* 173.

RECEIVED December 5, 1983

14

Interaction of Groundwater and Basalt Fissure Surfaces and Its Effect on the Migration of Actinides

G. F. VANDEGRIFT, D. L. BOWERS, T. J. GERDING, S. M. FRIED, C. K. WILBUR, and M. G. SEITZ

Argonne National Laboratory, Argonne, IL 60439

Experiments are being performed at Argonne National Laboratory (ANL) (1) to identify interactions of radionuclides and repository components that effect nuclide migration and (2) to assess changes in nuclide migration caused by modifications expected upon aging of the waste, backfill, and rock. This paper describes the philosophy of these experiments, the experimental set up, and some early results of this effort. The experiments are conducted with radioactive borosilicate glass, bentonite and mechanically fissured basalt rock in flowing water analogous to their configuration in a breach of a nuclear waste repository. Changes undergone by the groundwater as it passed through the fissure include: (1) drop in pH from 10 to 8, (2) loss of suspended particulate, and (3) loss of dissolved/suspended U, Np, and Pu. These effects, also studied as functions of radiation dose and of laboratory "aging" of the repository components, are related to the predicted long-term performance of a nuclear waste repository.

This paper is a progress report of an experimental study underway at ANL on the interaction of flowing simulated groundwater and the components that may be used in a nuclear waste repository constructed in basalt. The components are placed in the water stream analogously to the configuration that could occur from a breach of the repository; hence, the experiments

0097-6156/84/0246-0229$06.00/0

are designated as "analog experiments". In these experiments the movement of radionuclides by the flowing water is determined by the chemical composition of the groundwater and interactions with components at various positions along the flow path.

Radionuclide migration from a breached nuclear-waste repository by groundwater flow depends on the leaching of radionuclides from solid waste and on the chemical reactions that occur as a radionuclide moves away from the repository. Therefore, migration involves the interactions of leached species and the groundwater components with (1) the waste form and canister, (2) the engineered barrier, and (3) the geologic materials surrounding the repository. Some of these interactions would occur in the radiation and thermal fields centered on the solidified waste. Rather than trying to predict what the important interactions are and then to study them individually, we consider the combination of all potential interactions, using these analog experiments.

A schematic of the apparatus for an analog experiment is shown in Figure 1. In an experiment, groundwater is pumped through the system so that it passes through the first vessel, which contains basalt chips, bentonite, and the glass waste form; then through the second vessel, which contains more bentonite and basalt chips; and then through a narrow basalt fissure in the third vessel.

Five experiments have been initiated and a sixth experiment is planned; three of these experiments have been completed. Table I lists these experiments, shows how they have differed,

Table I. Analog Experiments Completed or Currently Underway

Experiment Number	Condition of Waste Form Backfill and Rock	Apparatus and* Experimental Condition	Status (3/22/83)
1	Unaltered	Hastelloy, HC-276	Completed
2	Unaltered	Monel-400	Completed
3	Unaltered	Hastelloy, HC-276, γ field	Completed
4	"1000 yr aged"	Hastelloy, HC-276	Near completion
5	"2000 yr aged"	Hastelloy, HC-276	On line
6	"1000 yr aged" bentonite and rock/unaltered waste form	Hastelloy, HC-276	Planned

*Apparatus constructed of the metals indicated; γ field indicates experiments carried out in the presence of a Co-60 gamma source.

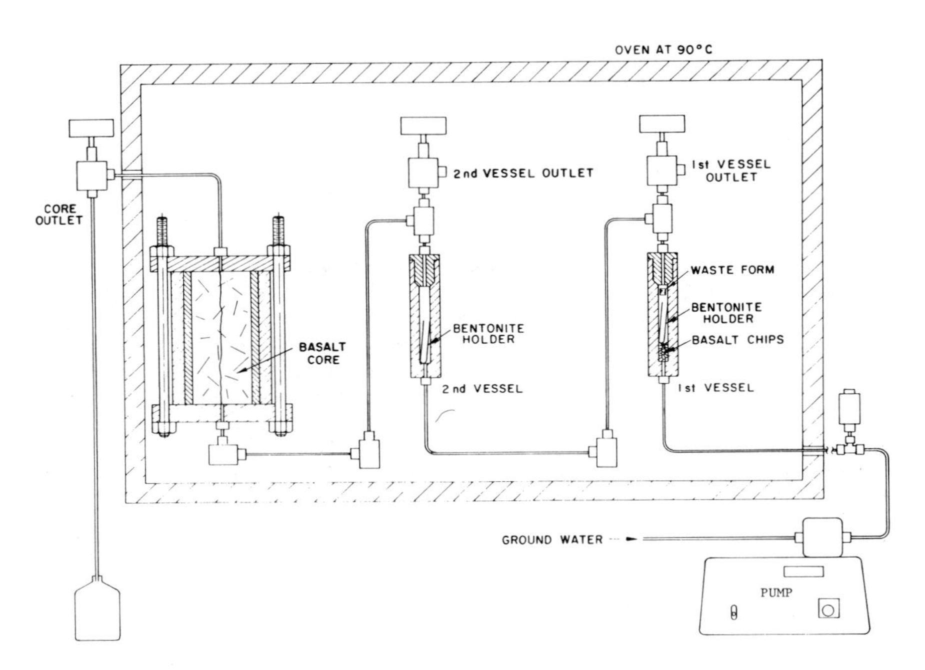

Figure 1. Apparatus for analog experiments.

and shows their current status. Experiments 4 and 5, labeled as "1000 yr aged" and "2000 yr aged", use radioactive solid, bentonite, and basalt rock that have been hydrothermally modified in the laboratory to hydrate the surfaces in a similar manner and to a degree that would occur over extended periods of time in the repository. "Unaltered" refers to materials receiving no hydrothermal treatment before their use in the experiments.

Experimental Materials

The simulated groundwater used in these experiments was prepared to represent the compositon of groundwater from Well DC-6, Grande Ronde Formation, which samples water primarily from 200 m below the Umtanum unit on the U.S. DOE Hanford site near Richland, WA (1). Its composition is shown in Table II. The groundwater used in these experiments was continually sparged with N_2 gas before it entered the apparatus to reduce the dissolved oxygen level and to limit CO_2 pickup.

Basalt cores were cut from rocks supplied by Basalt Waste Isolation Project (BWIP) personnel and ground to size (6.83-cm diameter by 14.60-cm long) on a lathe with distilled water coolant. These rocks were Pomona-flow basalt from the Pasco basin in the vicinity of the Near Surface Test Facility, U.S. DOE Hanford site. Characterization of this basalt can be found elsewhere (2).

Table II. Recipe for Simulated Groundwater Prepared for Analog Experiments[1]

Compound	Quantity Added to 20L of H_2O (g)	Concentration in m Mol/L
NaCl	3.51	2.99
Na_2SO_4	3.19	1.12
$Na_2B_4O_7 \cdot 10\ H_2O$	0.246	0.0331
NaF	1.64	1.95
Na_2CO_3	1.20	0.57
$NaHCO_3$	1.18	0.71
$Na_2SiO_3 \cdot 9\ H_2O$	11.46	2.01
K_2SO_4	0.085	0.024
$CaSO_4 \cdot 2\ H_2O$	0.111	0.032
$MgSO_4 \cdot 7\ H_2O$[2]	0.0036	7.3×10^{-4}
HCl (1M)[3]	61 mL	3.05

[1]The concentration of major constituent ions in ppm are presented in Table III.

[2]Added from a concentrated solution--1 mL to 20 L.

[3]Added to reduce the pH at 22°C of the groundwater to 9.9.

Because these core were not permeable to water even at high pressure, they were mechanically split to provide a path for water flow. To split the core, a sharp, hardened steel edge was pressed along the length of the core, using a hydraulic press, until the cores fractured. Details of the operation and the mounting of basalt cores in the core holder vessel can be found elsewhere (3).

The bentonite backfill material is sodium montmorillonite from Crook County, Wyoming. A 500 g sample of this material, SWy-1, was obtained from the Clay Mineral Society, Source Clay Minerals Repository, Department of Geology, University of Missouri, Columbus, MO.

The waste form was prepared by adding to 40 g of PNL 76-68 frit containing 3.4 wt % U_3O_8: 0.65 μCi of ^{137}Cs, 100 μCi of ^{85}Sr, 45 μCi of ^{133}Ba, 1640 μCi of ^{141}Ce, 2000 μCi of ^{152}Eu, 500 μCi of ^{241}Am, 23 mg (1400 μCi) of $^{239}PuO_2$, and 115 mg (80 μCi) of $^{237}NpO_2$. This mixture was melted in a 95% Pt/5% Au crucible with a procedure detailed elsewhere (3), and cut into 24 wafers with an average weight of 0.4 g and average dimensions of 1.1-cm x 0.90-cm x 0.13-cm. These waste-form wafers were characterized in an earlier publication (4).

Materials of contruction of vessels, tubing, and couplings for each experiment were either Hastelloy HC-276 or Monel-400.

Experimental Apparatus

Analog experiments are run in the apparatus shown in Figure 1 by flowing simulated groundwater through the apparatus at a rate of 0.5 mL/hr. Based on the geometric cross sectional areas of the first vessel and of the basalt fissure, this flow rate corresponds to a linear velocity of ∿30 m/y past the waste form and ∿500 m/y through the basalt fissure. The groundwater is initially at room temperature but is brought to the temperature of the apparatus, 90°C, in the length of tubing leading to the first vessel. There is an initial "equilibration" of the groundwater and the geologic components of the system before the waste form is loaded into the first vessel. As is evident from data to be presented later in this paper, an equilibration period of ∿30 days is sufficient for the groundwater to reach a steady-state condition with the repository components under the conditions of the first three analog experiments. This steady state condition, which is characterized by the exiting groundwater having a composition different from its entering composition, yet, having a constant composition with time, will be discussed at greater length in the results section.

During the first experiment it was found that to maintain flow through the basalt fissure it was necessary hold the fissure open by placing two strips of 0.14-mm thick gold ribbon along the length of the fissure. Before this was done, the pump pressure to maintain the 0.5 mL/h flow rose in several days from an initial 11 psig to one greater than 150 psig (the maximum transducer value). The core had become clogged by bentonite that had been transported by the flowing eluate from the first and second vessels. Even with the gold strips in place, the pump pressure rose steadily to ∿130 psig by the end of an experiment in all experiments with unaltered repository components (Table I; 1, 2, and 3). This same effect was not noted for analog experiments using hydrothermally-altered bentonite; no increase in pressure with time has been noted for these experiments (Table I, 4 and 5).

The special properties of bentonite clay (i.e., small particle size and swelling capability) made it impossible to localize the clay in the apparatus in a loose state. To maintain the bentonite in the system without obstructing the flowing water, a special dispersing system for bentonite was devised. This system comprised of a Hastelloy C-276 tube (2" long by 1/4" I.D.), crimped at one end, and 1 g bentonite packed into the tube to ∿1-inch from the top. When the tubes were contacted with groundwater, the bentonite swelled out of the tube and bentonite particles constantly dropped into the flowing groundwater.

The third analog experiment was performed in the presence of a cobalt-60 gamma ray source. The average dose rate to experimental components was 1×10^5 rad/h.

Experiment Shutdown

At the end of an experiment, the apparatus was disassembled beginning with the downstream end. Flow of water was continued throughout the shutdown phase. As a part of the apparatus was removed, it was checked for bentonite and radionuclide content. Large samples (150 mL) of eluate were taken at the second vessel outlet and then the first vessel outlet as the shutdown continued over a several week period.

Analytical Methods

Analog groundwater compositions were analyzed by emission spectroscopy (emission by inductively-coupled plasma, ICP) and by ion chromatography. Uranium concentrations were measured by laser fluorescent spectrometry and radioisotopes by gamma- and alpha-counting analyses. Because of the high dissolved solid content (832 mg/L) of the eluant groundwater, it was necessary to perform a chemical separation of ^{239}Pu and ^{237}Np by hexone extraction. These procedures are described elsewhere (5). Acid-base/pH titrations were performed on the simulated groundwater eluant samples to establish the carbonate plus bicarbonate concentrations. The initial pH value of samples was used to establish the concentrations of $[H_3SiO_4^-]$, $[H_2BO_3^-]$ and $[HCO_3^-]$ at 25°C. This procedures are also described in detail elsewhere (6).

Hydro-, Thermo-Alteration of Components

The fissured basalt cores and bentonite were altered by placing them in an autoclave under simulated groundwater at 320°C for 30 and 60 days to simulate 1000 y and 2000 y aging, respectively. The waste-form wafers for both experiments were aged in the same manner by treating them for 17 days in saturated steam at 340°C. The details of these procedures and the rationale for their use have been published previously (7). The effects of saturated steam on borosilicate glass were discussed in a recent publication (8).

Results

As the simulated groundwater passes through the three vessels of the analog expriment (Figure 1), it undergoes compositional changes which appear to affect its ability to transport actinide ions through the system. The following two sections describe separately (1) the changes in groundwater composition and (2) the actinide migration in the rock core fissure. Following these results is the discussion section, which briefly attempts to tie these two phenomena together.

Changes in Analog Groundwater Composition. The data in Figure 2 for the first three analog experiments (which used unaltered waste form, bentonite, and basalt cores) show that the groundwater exiting the basalt fissure reaches a steady-state composition for $[Na^+]$ and $[Ca^{2+}]$ soon after the start of the experiment. (There are three symbols for the three separate experiments, experiments 1, 2, and 3 in Table I, that are either empty or solid. The term "cold" in the figure legend refers to the equilibration period of the experiments, before the waste form was put into vessel 1. "Hot" refers to that part of the experiment where the waste form was in place.) The steady-state concentration of sodium ion in the groundwater eluant at the core's exit is ∿2/3 of its original groundwater concentration. The calcium-ion concentration is higher by almost an order of magnitude at the steady-state condition reached inside the core than at its original concentration.

An auxiliary experiment, performed at 90°C with groundwater and a column of crushed basalt chips and with a basalt surface area to groundwater volume ratio equivalent to that of the analog basalt fissure (2×10^2 cm^{-1}), has elucidated this phenomenon. Data from this experiment, which are presented in Figure 3, show that the sodium-ion concentration of the groundwater does not decrease gradually from its initial value of

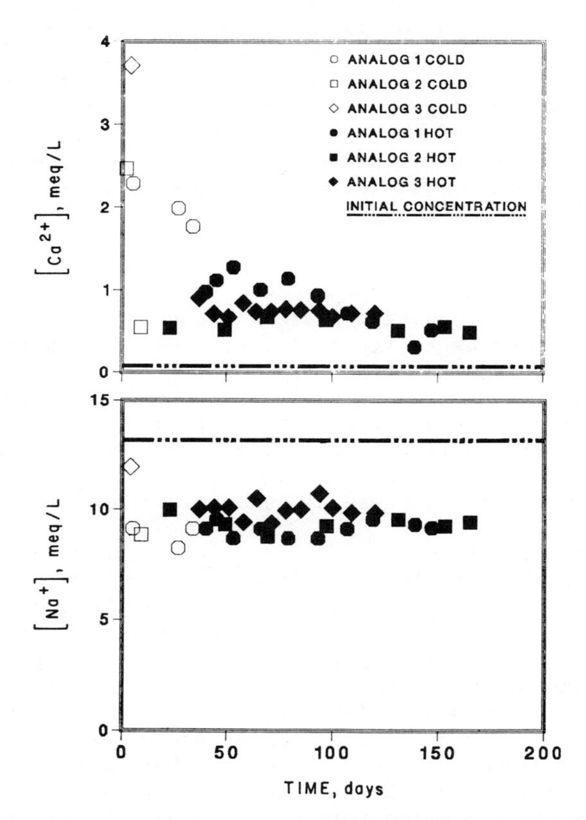

Figure 2. Attainment of steady-state concentration of sodium and calcium ions in the first three analog experiments.

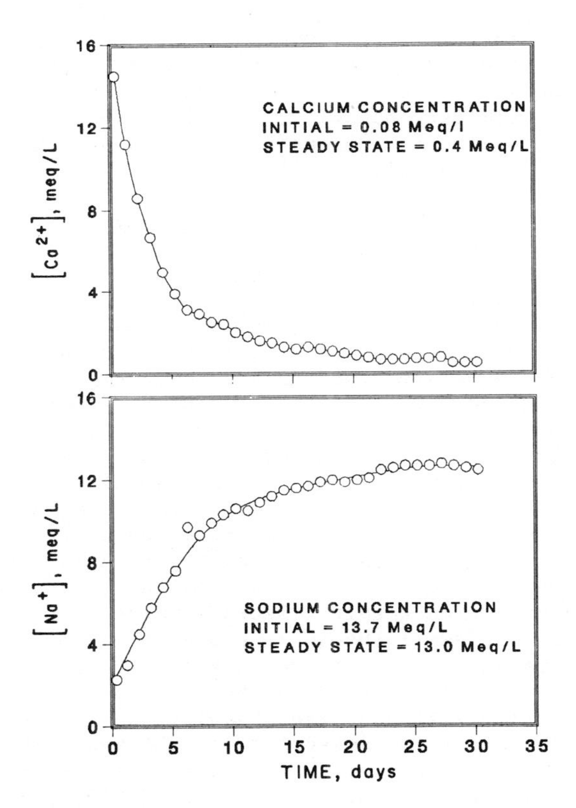

Figure 3. Attainment of steady-state concentration of sodium and calcium in an auxiliary experiment.

13.7 meq/L to its steady-state value of 13 meq/L, but, rather, it initially falls to a much lower concentration, ~2 meq/L. Gradually, over an "equilibration" period, the sodium-ion concentration rises to its steady-state value. The analog experimental data, Figure 2, show a steady-state $[Na^+]$ of only 9.6 meq/L. This difference may be a function of a slightly different surface area to volume ratio for the two experiments, or, perhaps, to a bentonite perturbation of the basalt/groundwater interaction. The calcium-ion concentration shows an opposite, yet similar, behavior. Calcium ion's initial groundwater concentration of 0.08 meq/L increases dramatically to 15 meq/L before gradually dropping to a steady-state concentration of 0.4 meq/L.

Table III shows ICP, ion chromatography, and uranium concentration data for outlet groundwaters from the first analog experiment. The data for the second and third analog experiments show the same trends as those seen in Table III. In general, alkaline-earth elements show the same behavior as does calcium; an initial very high concentration gradually falls to a steady-state concentration much higher than that of the initial groundwater. Some preliminary data for potassium ion (not shown in Table III) shows that it also follows the trend of calcium ion. The variation in anion concentrations is small in comparison to that of cationic species.

Table IV shows the major cationic and anionic constituents in the second analog experiment groundwater in (1) its initial condition, (2) its condition at steady-state on exiting the second vessel, and (3) its condition at steady-state on exiting the unaltered basalt fissure. Quite apparent from these data is the small change in groundwater composition caused by interaction of the groundwater with the bentonite and waste form, when compared to that of the effect of groundwater interaction with unaltered basalt. The major difference in groundwater after it passed through the two bentonite-containing vessels was a slightly lower pH and sodium-ion concentration. (Lower concentrations of $H_3SiO_4^-$, $H_2BO_3^-$, and CO_3^{2-} are all attributable to the lower pH value.) Passage of the groundwater through the basalt fissure greatly modified its cationic content and its pH (third column). Again, there was no perceptible change in the total SO_4^{2-}, F^-, Cl^-, silicate, or borate concentration. The charge balance is accomplished by the decrease in pH (10 to 8) that occurs in the rock fissure. This decrease in pH decreased the negative ions in solution by neutralizing much of the $H_3SiO_4^-$, $H_2BO_3^-$, and CO_3^{2-} in the groundwater. The total (carbonate + bicarbonate) also decreased by ~40% from passage through the rock core.

Thus far we have discussed only the interaction of analog groundwater with unaltered bentonite and basalt. Table V, which contains data from the 1000 y analog experiment (Exp. 4, Table I), shows the quite different, and quite limited, effect

Table III. Major Constituent Concentrations of Groundwater Samples which Exited the First Analog Experiment

Sampling Period	Collection Time, days	Chemical Constituent[1,4] B[2]	Ca[2]	Mg[2]	Na[2]	Si[2]	Sr[3]	F^-[2]	Cl^-[2]	SO_4^{2-}[2]	U[3]
	Original Groundwater	1.30	1.5	<0.02	316	53.0	<2	35	165	110	0.04
Conditioning	1-10	1.37	45.9	14.4	210	63.5	220	15	175	140	0.16
Conditioning	24-31	1.43	39.9	12.3	190	51.2	170	20	165	125	0.12
Conditioning	31-38	1.42	35.6	7.2	210	49.3	130	25	165	125	0.13
Radioactive	1-3	1.50	19.7	5.0	210	47.4	110	ND	ND	ND	ND
Radioactive	6-9	1.58	22.4	3.1	220	51.7	100	ND	ND	ND	ND
Radioactive	13-17	1.53	25.7	6.5	200	55.7	120	20	165	115	2.7
Radioactive	27-29	1.46	20.3	2.8	210	47.8	90	ND	ND	ND	ND
Radioactive	37-45	1.51	22.8	5.6	200	51.2	110	ND	ND	ND	ND
Radioactive	52-59	1.46	18.8	5.4	200	54.8	100	25	165	110	6.1
Radioactive	66-73	1.46	14.7	4.7	210	49.2	80	ND	ND	ND	ND
Radioactive	78-85	1.38	12.7	2.4	220	49.6	60	30	160	105	.6.7
Radioactive	98-105	1.38	6.4	0.5	215	50.9	44	30	160	115	ND
Radioactive	105-114	1.38	10.6	0.7	211	54.0	48	ND	ND	ND	ND
Radioactive	114-120	ND	ND	ND	ND	ND	ND	30	165	115	ND

[1] ND - no data collected.

[2] Reported in mg/L.

[3] Reported in μg/L.

[4] Total [CO_3^{2-} + HCO_3^-] concentration was 1.54 mMol/L in the original groundwater and was a nearly constant 1.15 ±.05 for outlet samples taken during the experiment.

Table IV. Major Constituents of the Second Analog Experiment Groundwater Solution in meq/L (Unaltered Basalt)[1]

Solution	pH	Na^+	Ca^{2+}	Mg^{2+}	$\sum$Cations
Initial Groundwater	9.9	13.7	0.075	nil	13.8
2nd Vessel Outlet ∿150 Days	9.6	12.3	0.018	nil	12.3
Core Outlet ∿100 Days	7.8	9.6	0.63	0.20	10.4

Cl^-	F^-	SO_4^{2-}	OH^-	$H_3SiO_4^-$	$H_2BO_3^-$	CO_3^{2-}	HCO_3^-	$\sum$Anions
4.65	1.84	2.29	0.08	1.10	0.10	0.96	1.33	12.4
4.84	1.82	2.43	0.04	0.67	0.08	0.82	1.05	11.8
4.51	1.58	2.19	nil	0.02	0.01	nil	1.33	9.6

[1] Concentrations of OH^-, $H_3SiO_4^-$, $H_2BO_3^-$, CO_3^{2-}, and HCO_3^- and pH are listed for solutions at 25°C. These would, of course, change dramatically at 90°C; where $H_2SiO_4^{2-}$ is also an important anionic species.

Table V. Major Constituents of the Fourth Analog Experiment Groundwater Solutions in meq/L (Altered Basalt)[1]

Solution	pH	Na^{+}	Ca^{2+}	K^{+}	$\sum$ Cations	Cl^{-}	F^{-}	SO_4^{2-}	OH^{-}	$H_3SiO_4^{-}$	$H_2BO_3^{-}$	CO_3^{2-}	HCO_3^{-}	$\sum$ Anions
Groundwater	9.9	13.7	0.060	0.05	13.8	6.0	1.9	2.4	0.10	1.56	0.11	0.86	0.82	13.8
2nd Vessel Outlet ∿10 Days	9.5	9.5	0.19	0.3	10.0	4.1	1.4	1.9	0.04	0.79	0.07	0.50	2.03	10.8
Core Outlet ∿90 Days	9.5	13.6	0.061	1.4	15.0	5.8	1.9	2.7	0.04	0.80	0.08	0.52	2.00	13.9

[1]Concentrations of OH^{-}, $H_3SiO_4^{-}$, $H_2BO_3^{-}$, CO_3^{2-}, and HCO_3^{-} and pH are listed for solutions at 25°C. These would, of course, change dramatically at 90°C; where $H_2SiO_4^{2-}$ is also an important anionic species.

on groundwater composition of contact by the hydrothermally-altered basalt fissure. A comparison of the first and third data rows of Table V (original analog goundwater *vs*. that exiting the rock core ~90 days into the experiment) shows only limited compositional changes in the two solutions. The steady-state compositions of major groundwater constituents in the exiting groundwater are far closer to those of the original groundwater than those noted for fresh basalt experiments. The major differences that did occur in the fourth analog experiment were substantial increases in the concentration of potassium ion, and the total of carbonate plus bicarbonate. The decreases seen for $H_3SiO_4^-$ and $H_2BO_3^-$ are due to their partial neutralization at the lower pH of the exiting groundwater.

The data for the second vessel outlet in Table V seem to show that groundwater composition is more affected by aged bentonite than by its unaltered form. The sodium ion concentration dropped dramatically as calcium ion and potassium ion both rose. Anionic species concentrations also dropped. These data, however, were collected early in the experiment (*i.e.*, after ten days) and are likely to be not representative of the steady-state condition of the system. The steady-state composition of the second vessel outlet solution will be measured, when the rock core is taken off-line at the end of the experiment.

The differences in the sums of the anionic and the cationic charges in the data presented in Tables IV and V are within the experimental error limits of the analytical methods (ICP, ion chromatography, and acid/base titration and pH measurement) used to measure the concentrations of these species.

Actinide Migration in the Rock Core

Table VI shows the concentrations of plutonium, neptunium, and uranium measured at the inlet and outlet of the unaltered and hydrothermally-altered basalt core fissures in the first five analog experiments (see Table I). Under conditions simulating a repository that was unaltered by groundwater interaction (Table I, Exp 1-3), both Np and Pu, in the concentrations developed in these analog experiments from the leaching of the waste form, were substantially retarded within the 14.6-cm basalt fissure. In fact, as can be seen from Figure 4, almost all of ^{237}Np activity was sorbed on the first one-third of the rock fissure. The data in Figure 4 have an estimated error, based on counting statistics above, of approximately 2 counts per 1000 seconds. Uranium retardation was determined to be not as complete.

By filtering groundwater samples from the first and second vessel outlets through 0.1 μm filters, it was determined that the neptunium was in a soluble form, and that most of the plutonium was associated with filterable bentonite particles

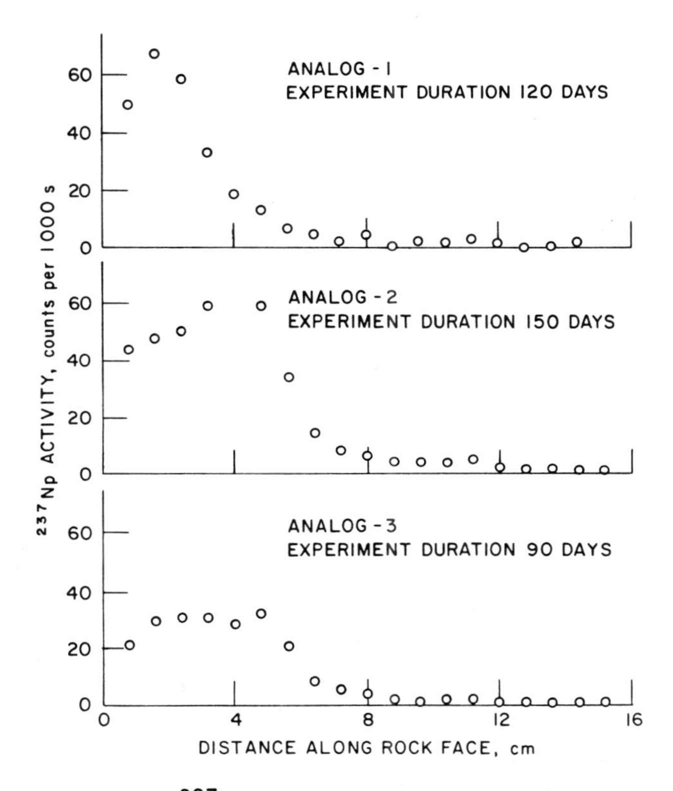

Figure 4. ^{237}**Np sorption on unaltered basalt core fissures.**

Table VI. Sorption of Actinides by Basalt Fissures

Experiment		Pu $(\frac{dpm}{mL})$	Np$(\frac{dpm}{mL})$	U$(\frac{ng}{mL})$
1	inlet	0.4	5.7	35.6
	outlet	$\leq$0.006	0.03	9.6
2	inlet	0.1	5.2	27.2
	outlet	$\leq$0.007	0.07	11.1
3	inlet	5.3	0.8	7.5
	outlet	$\leq$0.001	$\leq$0.001	0.6
4[1]	inlet	29.3	40.2	357
	outlet	13.5	24.7	396
5[1]	inlet	42.2	45.3	-
	outlet	34.4	43.6	-

[1]U, and perhaps Pu and Np, are not at steady-state concentrations in the core inlet or outlet streams. The apparent increase of U as it passed through the core is due to the differing sampling times for the two streams.

suspended in the groundwater. Filtering the eluant from the first vessel reduced the eluant's ^{239}Pu activity by >60%. Filtering the second vessel eluant made its ^{239}Pu activity level immeasurable above background.

The oxidation states of the actinides are not known, but are controlled by their initial state in the simulated waste glass and by the mechanism of their leaching in the groundwater environment. Their oxidation states in the basalt fissure presumably would be controlled by the buffering capacity of the basalt due to the Fe^{2+}/Fe^{3+} redox couple and the available ferrous ion on the rock surface (9). At the pH range of 10-8, this couple would reduce higher oxidation states to Np(IV), Pu(IV), and U(IV) (10-13). This is most likely still true when complexants, e.g., the CO_3^{2-} and F^- that are present in the analog groundwater, are in solution.

The differences in actinide element concentrations in the second vessel outlet samples for experiments with and without the presence of a gamma field (experiment 3 vs. experiments 1 and 2) are clearly evident from the data of Table VI. These differences are probably related to differing leaching characteristics of the waste with and without gamma radiation. Such effects have been reported by others (14, 15). We are planning experiments to verify and to further study this result.

The actinide retardation data for the hydrothermally-altered repository component experiments (4 and 5) in Table VI are preliminary and their interpretation is hindered by the lack of attainment of a steady-state condition for actinide migration in the systems. These data indicate that only a small fraction of the actinides are retained by the rock core. Comparing these results to those of the unaltered fissure experiments, where Pu and Np were almost completely retarded by the rock core, one could conclude that altering the rock tends to lower its ability to retard actinide migration. Another way to discuss these same data, and one that leads to a completely opposite conclusion, is in terms of the amount of activity retained by the rock core, or, the rate of actinide loss from the groundwater in terms of Δ (dpm/mL) of ^{239}Pu and ^{237}Np. These quantities are both higher for experiment 4 and 5 than for the three experiments with unaltered repository components. The loss of Pu through the core for experiments 1 and 2, for example, was 0.4 and 0.1 dpm/mL, respectively; in experiments 4 and 5 the loss of Pu through the core appears to be 20 to 40 times greater. This exercise is intended to show that there is no legitimate way to compare the behavior of altered and unaltered basalt from these data. Experiment 6, yet to begin, should clarify this situation. What is clear from comparing the data in Table VI is that actinide behavior in altered and unaltered repository situations will be quite different.

Discussion

The laboratory analog approach to identifying and measuring repository interactions has been demonstrated by a series of experiments. The data obtained in this program suggest a strong correlation between changes in groundwater composition and radionuclide retardation by basalt. Our conclusions, based on initial data are:

- Fresh, unaltered basalt fissures and surfaces will not be in equilibrium with groundwater that enters the nuclear waste repository area. (The differences between the mineralogy of the Pamona flow basalt, used in these experiments, and those of other basalt flows in the Umtanum formation are not great enough to explain the observed changes in groundwater composition.)
- A steady-state groundwater composition will be achieved in a fairly short time period (20-25 days for the laboratory analog experiments). This steady state is one where the $[Ca^{2+}]$ and $[K^{+}]$ are increased above their initial concentrations, and where the $[Na^{+}]$ and pH are decreased. The reason for these groundwater compositional changes are likely due to the hydrolysis and alteration of minerals on the basalt surface. Because Si concentration of the groundwater did not vary, these reactions act primarily as ion exchange processes and not dissolution.

- While bentonite is transported by flowing, unaltered groundwater, interaction with unaltered basalt surfaces appears to agglomerate and to bind it to the basalt.
- Unaltered basalt appears to effectively limit the migration of actinides dissolved or suspended in the groundwater.
- There was no perceptible difference in groundwater/basalt interactions or actinide migration due to laboratory analog apparatus construction materials (Hastelloy vs. Monel).
- Gamma radiation does not substantially affect groundwater compositional changes due to interactions with unaltered bentonite and basalt but appears to generate identifiable differences in actinide leaching from the waste form.
- Gamma radiation perceptably modifies leaching characteristics of the waste form.

At this time, it is not possible to identify the mechanisms that are responsible for actinide retardation in unaltered basalt fissures. There are four possible mechanisms that may be responsible for the fissure actions:

1. Decomposition of Pu, Np, and U carbonate complexes due to drop in pH and CO_3^{2-} concentration.
2. Reduction of higher oxidation states of U, Np, and Pu by ferrous ion in basalt.
3. Changing of the surface charge on bentonite due to changing groundwater composition thus effecting its adsorption and agglomeration properties.
4. Coprecipitation of actinides with divalent-cation carbonates and sulfates.

Auxiliary experiments are underway and are being planned to elucidate the mechanisms of actinide retardation in altered and unaltered repository conditions.

Future work in this program will emphasize the modeling of laboratory analog experiments with computer codes designed to predict repository behavior. This relatively simple system is a good test for these computer codes' predictive reliability.

Also, laboratory experiments will be run with rock cores containing naturally-aged fissures. This will be an important step in testing the appropriateness of laboratory aging of rock surfaces to simulate natural aging processes.

Literature Cited

1. Apps, J; Doe, T; Doty, B; Doty, S; Galbraith, R; Kearns, A; Kohrt, B; Long, J; Monroe, A; Narasimhan, T. N.; Nelson, P.; Wilson, C. R.; Witherspoon, P. A. "Geohydrological Studies for Nuclear Waste Isolation at the Hanford Reservation, Volume II, Final Report"; LBL-8764, Vol II, July 1979.
2. Ames, L. L.; "Hanford Basalt Flow Mineralogy;" PNL-2847; September 1980.
3. Steindler, M. J.; et al.; "Fuel Cycle Programs Quarterly Progress Report, October-December 1981;" ANL-82-18, May 1982; Section IV.
4. Steindler, M. J.; et al.; "Fuel Cycle Programs Quarterly Progress Report, January-March 1982;" ANL-82-34, December 1982; Section IV.
5. Steindler, M. J.; et al.; "Fuel Cycle Programs Quarterly Progress Report, April-June 1982;" ANL-82-58, December 1982; Section IV.
6. Steindler, M. J.; et al.; "Fuel Cycle Programs Quarterly Progress Report, October-December 1982;" ANL-83-19, in press; Section V.
7. Steindler, M. J.; et al., "Fuel Cycle Programs Quarterly Progress Report, July-September 1981;" ANL-81-82, May 1982; Section IV, C.
8. Bates, J. K.; Jardine, L. J.; Steindler, M. J.; Science; 1982, 218, 51.
9. Jacobs, G. K.; Apted, M. J.; EOS, Trans. Am. Geo. Phys. Union; 1981, 62, 1065.
10. Rydberg, J; "Groundwater Chemistry of a Nuclear Waste Repository in Granite Bedrock"; UCRL-53155; September 1981.
11. Duffy, C. J.; Ogard, A. E.; "Uranite Immobilization and Nuclear Waste"; LA-9199-MS, February 1982.
12. Moody, J. B.; "Radionuclide Migration/Retardation Research and Development Technology Status Report"; ONWI-321; March 1982.
13. Apted, M. J.; Meyer, J.; "Comparison of the Hydrothermal Stability of Simulated Spent Fuel and Borosilicate Glass in a Basaltic Environment"; RHO-BW-ST-38P, July 1982.
14. Nash, K. L.; Fried, S; Friedman, A. M.; Susak, N; Rickert, P; Sullivan, J. C.; Karim, D. P.; Lam, D. J., in "Scientific Basis for Nuclear Waste Management"; Topp, S. V., Ed.; North Holland Press, New York, 1982; p 661.
15. McVay, G. L.; Pederson, L. R; J. Am. Ceram. Soc.; 1981, 64, 154.

RECEIVED December 5, 1983

FIELD MEASUREMENTS OF RADIONUCLIDE MIGRATION

Role of Organics in the Subsurface Migration of Radionuclides in Groundwater

A. P. TOSTE, L. J. KIRBY, and T. R. PAHL

Pacific Northwest Laboratory, Richland, WA 99352

Research is underway at a commercial shallow-land burial site at Maxey Flats, Kentucky, to identify the chemical forms of migrating radionuclides. An experimental study area, consisting of an experimental slit trench in five sections and a number of inert atmosphere wells, was constructed adjacent to a waste-filled trench to monitor any subsurface migration of radionuclides in groundwater. Research to date indicates that certain radionuclides, ^{3}H, $^{238,239,240}Pu$, ^{90}Sr, ^{60}Co and ^{137}Cs, have migrated within the study area over relatively short distances (4-7m). Organic analyses indicate that the chelator EDTA and three "EDTA-like fragments" are the major hydrophilic organic species (ppm levels) in leachate from two waste trenches and one inert atmosphere well. The absence of other chelating agents like EDTA strongly suggests that the "EDTA-like fragments" are degradation products of EDTA, perhaps via microbial or radiolytic diagenesis. Trace levels of hydrophobic organic species (generally ppb levels), including two barbiturates, were also identified in waste leachate and water from an inert atmosphere well. On the basis of steric exclusion chromatography, it is clear that polar organic species in the groundwaters co-elute with certain radionuclides and are, therefore, presumably chelated. Pu and ^{60}Co co-elute with EDTA. An association, perhaps electrostatic, also appears to exist between ^{137}Cs and ^{90}Sr and carboxylic acids, e.g., palmitic and stearic acids. The appearance of EDTA and the barbiturates in only one inert atmosphere well argues that communication between the waste-filled area and the experimental study area is very specific, presumably via fracture flow. Radionuclide, inorganic and organic data

0097–6156/84/0246–0251$06.00/0

indicate that the groundwater flow patterns are complex, at least in the vicinity of the experimental study area.

A critical issue facing the nuclear industry today is the long-term disposal of radioactive waste. An area of particular concern is the shallow land burial of low-level waste. The concern is not only about waste being generated now or in the future but also about waste generated in the past which has already been stored under less than ideal conditions. Much remains to be understood about the subsurface migration of radionuclides in soil and groundwater. In spite of this, commercial shallow land burial of low-level radioactive waste has existed for over two decades at six sites in the U.S.: West Valley, New York; Maxey Flats, Kentucky; Sheffield, Illinois; Barnwell, South Carolina; Beatty, Nevada; and Hanford, Washington. Unforeseen problems necessitated the closing of the West Valley and Maxey Flats sites in 1975 and 1977, respectively. Problems included trench cap subsidence and water seepage into waste trenches, resulting in dissolution of buried waste by groundwater and subsurface transport of some of the radionuclides buried at both sites.

At Pacific Northwest Laboratory we have had the opportunity to study the subsurface migration of radionuclides at the Maxey Flats burial site, where the groundwater is anoxic. The Maxey Flats site is one of several commercial shallow land burial sites in the eastern United States where rainfall is comparatively high. Trench cap subsidence and infiltration of surface water into waste trenches are greater problems at wet eastern sites compared to arid western sites. At Maxey Flats, the resulting seepage of contaminated groundwater, coupled with remedial construction efforts, has resulted in some low-level contamination within the site. Atmospheric transport of radionuclides in the plume from an on-site water evaporator and spillage of radioactive solutions have also influenced the distribution of radioactivity. A new wave of remedial engineering work is currently underway to correct past oversights and minimize the infiltration of surface water into the waste trenches. One of the most significant pieces of remedial action has been the installation of a plastic cover over the waste burial area as an infiltration barrier to rainfall.

The factors governing the subsurface migration of radionuclides appear to be extremely complex. Research in our laboratory is aimed at identifying the physicochemical forms of radionuclides in groundwater and their role in the subsurface migration of radionuclides. Research indicates that the mobility of radionuclides depends strongly on their physicochemical forms. For example, hexavalent plutonium and uranium migrate through soil faster than the tetravalent species, and technetium migrates faster as pertechnetate ion than it does in the reduced form (1-2). Recent research strongly suggests that organic chelating agents may play an important role in the subsurface migration of radionuclides (3-6). Decontaminating agents such as EDTA buried with radioactive waste or biodegradation products of organic material in waste or soil may complex or chelate with radionuclides, enhancing their mobility.

In this report we describe two broad research efforts at Maxey Flats. First, we have completed a study of radionuclide migration and groundwater movement in an experimental study area we constructed adjacent to a waste-filled trench. Second, we are engaged in a research effort aimed at elucidating the role of organic species in the subsurface migration of radionuclides. We have completed a survey study aimed at mapping the organic content of various groundwater samples from waste trenches and the experimental study area. We have also begun a detailed chemical speciation study to determine which organics identified in the survey study are complexed or chelated to radionuclides.

EXPERIMENTAL METHODS

Experimental Study Area

In 1979 we constructed an experimental study area adjacent to the waste-filled area at Maxey Flats. A slit trench consisting of five sections was installed adjacent to waste trench 27 to intercept subsurface water flow from the waste-filled area (Figure 1). The distance between the experimental trench and waste trench 27 ranges from 4m at experimental trench section T1 to 7m at section T5. Groundwater was sampled during and after the construction of the experimental trench from sumps placed in each trench section. Nonradioactive tracers were added to sections of the experimental trench during its construction to monitor groundwater flow and test various design features of the experimental trench. Sodium bromide was added to trench sections 1, 2, 3 and 5 as a groundwater tracer to test whether groundwater communicates between the experimental study area and the waste-filled area. Pentafluorobenzoic acid was added to the cap of trench section 4 as a cap tracer to test whether water seeps through the experimental trench cap. A series of inert atmosphere wells were subsequently installed around the experimental trench (Figure 1). The inert atmosphere wells were filled with argon between the periodic sampling trips.

Sample Collection

The groundwater samples from the waste trenches, experimental trench sumps, and inert atmosphere wells were sampled on the dates indicated in Tables I - VI. The samples were pumped from the wells or trench sumps using peristaltic pumps and collected under argon in Teflon or glass bottles. They were degassed with argon and sealed tightly for shipment to the laboratory, where they were stored in a refrigerated room at ~4°C.

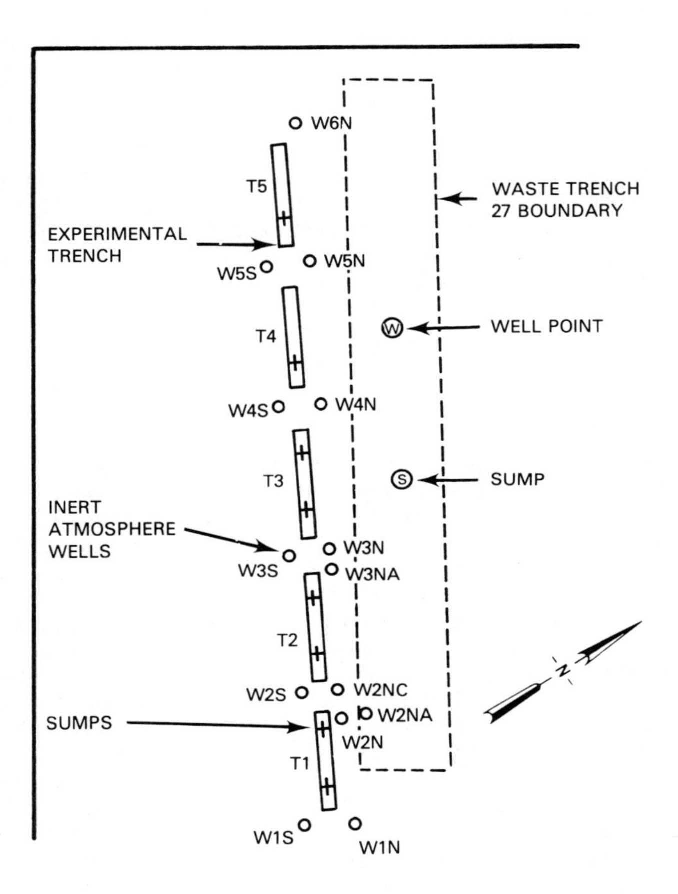

FIGURE 1. Experimental study area at the Maxey Flats shallow-land burial site. A series of experimental trench sections (T1-T5) and inert atmosphere wells (e.g. W1N) were installed adjacent to waste trench 27 to permit concurrent sampling of water from the waste trench (sump or well point), experimental trenches, and wells.

Radiochemical Analyses

A number of procedures were used to analyze the radionuclide content of the groundwater samples. Tritium analyses were performed by liquid scintillation spectrometry; ^{90}Sr was chemically separated and counted on a beta proportional counter. Gamma-emitting radionuclides were analyzed by gamma-ray spectrometry using a Ge(Li) detector. Plutonium and americium were determined using solid state detectors and alpha energy analysis following radiochemical separations. Nonradioactive species were analyzed by neutron activation analysis using a subcritical neutron multiplier, (7) and by chemical and instrumental analysis.

Organic Analyses

Sample Preparation. Organic analyses were performed on whole water samples, i. e., no chromatographic fractionation, (survey study) and on fractionated water samples following steric exclusion chromatography (detailed speciation study). For the survey study, a specific volume (25-50 mL) of each water sample was concentrated to near dryness by rotary evaporation prior to organic derivatization. The water samples set aside for chromatographic fractionation were concentrated to specific volumes ranging from 15-20 mL and the pH adjusted to that of the original sample (~pH 7.2 - pH 7.6). Each solution was filtered, loaded on a Sephadex G-15 column, and chromatographed with Milli Q-purified water. Replicate runs were made, with and without Blue Dextran as a void volume marker. The column effluent was monitored by ultraviolet absorbance at 254 nm and collected using a fraction collector. Each column fraction was divided into two portions, one for radionuclide analysis and the other for organic analysis.

Each concentrated survey sample and column fraction was extracted with chloroform to remove hydrophobic organic compounds; this extract, the hydrophobic organic fraction, was concentrated and set aside for analysis by gas chromatography (GC) and combined GC-mass spectrometry (GC-MS). The extracted water sample, containing hydrophilic organic compounds, was evaporated to dryness under nitrogen. The residue of each aqueous fraction was then methylated in a sealed vial with 1 ml of BF_3/methanol (14% w/v) at 100°C for 40 min. After cooling, 1 ml of chloroform was added, the mixture was transferred to a test tube containing 3 ml of 1M KH_2PO_4 buffer solution (pH 7) and vortexed. Part of the chloroform layer (0.5 ml), which contained the methylated hydrophilic organics, was then evaporated under nitrogen. The residue was redissolved in chloroform and analyzed by GC and by GC-MS.

GC Analysis. GC analyses were performed on a Hewlett-Packard 5880 gas chromatograph equipped with a 60 m x 0.25 mm I.D. fused silica capillary column coated with a 0.25 μm film of SE-52. From an initial value of 40°C, the column temperature was programmed at

20°C per minute for three minutes to 100°C followed by 8°C per minute for 25 minutes to 300°C, and finally maintained isothermally at 300°C for 10 minutes.

GC-MS Analysis. GC-MS analyses were performed on a Hewlett Packard 5985 GC-MS instrument in the electron-impact (70-eV) mode. The gas chromatograph on the 5985 instrument was equipped with a 60 m x 0.25 mm I.D. fused silica capillary column coated with 0.25 μm of SE-54; the column was programmed from 40°C to 300°C at 5°C/min, where it was maintained isothermally for 8 min. A splitless injection system was used to introduce the sample onto the GC-MS instrument. A mass range of 50 to 400 amu was scanned every 1.0 sec by computer (HP-2100MX equipped with the HP-7920 Large Disc Drive).

Quantitation. The organic species identified by the GC-MS analyses were quantitated by GC analysis using internal and external standardization methods. Pure compounds representative of the various compound classes identified by GC-MS were selected as standards and methylated. A specific amount of each standard was co-injected with each sample to confirm the GC-MS identifications. For quantitation purposes, each standard was injected onto the gas chromatograph prior to and following sample analyses. The response factor of each standard was calculated under analytical conditions identical to those of the sample analyses.

Materials

Standards. The standards used in the GC analyses were purchased from Aldrich Chemical Company (Milwaukee, Wisconsin), United States Pharmacopeial Convention, Inc. (Rockville, Maryland) and Sigma Chemical Company (St. Louis, Missouri).

Chromatographic Columns. The glass columns (2.5 cm x 45 cm) and Sephadex G-15 used in the detailed speciation study were purchased from Pharmacia Fine Chemicals (Piscataway, New Jersey). The SE-52 and SE-54 silica capillary columns used in the GC and GC/MS studies were purchased from J & W Scientific, Inc. (Rancho Cordova, California).

Reagents, Solvents and Glassware. The BF_3/Methanol (14% w/v) used in the methylation reaction was purchased from Regis Chemical Company (Morton Grove, Illinois). All of the solvents used in the organic analyses were redistilled-in-glass solvents purchased from Burdick and Jackson Laboratories, Inc. Deionized water, prepurified for laboratory use, was further purified on a milli-Q system (Millipore) containing two ion exchange resins and two charcoal filters. All glassware was acid cleaned in a hot solution of sulfuric acid/nitric acid (4:1 v/v).

RESULTS AND DISCUSSION

Radionuclide Migration and Groundwater Movement

Results of several radiochemical studies at the experimental study area we installed at Maxey Flats have already been described in detail elsewhere (8-10). The areal distribution of radionuclides in the surface soil throughout the site has been mapped. Groundwater flow patterns near the experimental study area have also been mapped using tritium as a groundwater tracer.

In addition to the above studies, radiochemical analyses were performed on water which flowed into the experimental trench during its construction. Initial radionuclide concentrations in the water were comparable to those in water from waste trench 27 (Table I), indicating that subsurface migration of radionuclides has occurred over short distances on site at Maxey Flats. When the experimental trench was resampled later, the radionuclide concentrations were much lower, perhaps because the trench design perturbed the migration of radionuclides. Consequently, we installed and sampled a series of inert atmosphere wells around the experimental trench.

TABLE I. Comparison of Radionuclide Concentrations in Water from Trench 27 (5-78) and Initial Experimental Trench Inflow (9-79)

Trench	Concentration, pCi/l				
	^{3}H	Total Pu	^{90}Sr	^{60}Co	^{137}Cs
Waste:	5.9+08	4.8+03	2.1+05	1.3+03	8.0+03
Experimental:					
Section T1	3.7+08	3.6+01	6.7+03	3.6+01	6.5+01
Section T2E	1.7+09	6.1+01	1.7+05	2.2+02	1.6+03
Section T2W	4.2+09	3.4+03	1.6+05	9.7+03	4.6+03
Section T3E	3.4+09	5.0+03	1.1+05	6.0+03	9.0+01

Evidence of the groundwater tracer sodium bromide was found in water from waste trench 27 and several inert atmosphere wells. Changes in bromine concentrations were observed in water from waste trench 27, the experimental trench, and inert atmosphere wells, indicating that there is some communication by groundwater flow between these locations (Table II). Bromine concentrations in water samples from waste trench 27 were quite low prior to the construction of the experimental trench. Following trench construction, the bromine concentration in water from waste trench 27 increased from 2.8 to 460 ppm, indicating that some of the sodium bromide tracer had moved into waste trench 27. Bromine concentrations in water from inert atmosphere wells W3N, W3NA, and W2NA were also high indicating strong communication, whereas wells W1N, W2N and W5N do not appear to be connected to the sodium bromide source. When radionuclide concentrations are compared (Table II), only wells W2NA and W3N appear to be closely related to waste trench 27, possibly connected to it by subsurface fractures.

TABLE II. Comparison of Radionuclide and Bromine Concentrations in Water from Waste Trench 27 (5-78 and 9-80) and Waters from Inert Atmosphere Wells (9-80)

Waste Trench 27	Radionuclide Concentration (pCi/1)				Bromine (ppm)
	^{3}H	^{238}Pu	$^{239,240}Pu$	^{90}Sr	
Sump (5-78)	5.9+08	4.1+03	6.7+02	2.1+05	2.8+00
Sump (9-80)				4.6+02	
Well Point (9-80)				2.7+00	
Experimental Wells (9-80)					
W1N	4.7+05	3.7+01	6.0+00	1.5+02	1.0+00
W2N	2.9+06	2.1+01	7.7+01	3.3+02	2.0+00
W2NA	7.7+08	2.4+02	4.8+00	1.7+04	1.2+02
W3N	1.2+09	2.4+03	4.4+01	3.4+04	8.2+02
W3NA	2.0+07	2.3+01	5.0+00	2.4+02	5.2+02
W5N	1.0+03	9.4+00	5.0+00	2.1+02	5.0+00

Organic Content of Groundwaters

Quite a variety of hydrophilic and hydrophobic organics have been identified to date in groundwaters from the Maxey Flats site, as listed in Tables III - VI. The chelating agent EDTA is the most abundant of the hydrophilic species, appearing in groundwater from three locations at ppm levels (Tables III and IV). It was identified in water from both waste trenches 27 and 19S (Table III). Water from waste trench 27 sampled on April 7, 1981, contained 1000 ppb EDTA whereas a more recent sample

TABLE III. Hydrophilic Organic Compounds [a] in Groundwater from Trenches at Maxey Flats

	Waste Trenches(ppb)[b]			Experimental Trench Sumps(ppb)[b]	
	WFT-27[c]		WFT-19S	S4	S5
	(4/7/81)	(7/27/82)	(7/22/82)	(8/18/81)	(8/18/81)
Ethylenediaminetetraacetic Acid[d] (EDTA)	1000	3830	1340		
"EDTA fragment"[d] (MW 219)	trace	trace			
"EDTA fragment"[d] (MW 244)	145	758	259		
"EDTA fragment"[d] (MW 288)		90	88		
Pentafluorobenzoic Acid[d]	trace			trace	
Lactic Acid[d]					4
Oxalic Acid[d]	28		35		4
Dimethyl Sulfate				7	
4-oxo-pentanoic Acid[d]	19		76	28	7
Succinic Acid[d]	23		29	2	4
Methyl Succinic Acid[d]	trace				
Benzoic Acid[d]			200	7	6
Nonanoic Acid[d]					7
Hexanedioic Acid[d]	26	45	71	17	8
Dimethyl Phthalate	90	26	558	52	
Citric Acid[d]	8	trace			
Nonanedioic Acid[d]	20		192	14	
Tetradecanoic Acid[d]					10
Hexadecanoic Acid[d]	28	267		16	41
Tridecanedioic Acid[d]					13
Dibutyl Phthalate	68	51		9	14
Octadecanoic Acid[d]	32	26		11	26
Dioctyl Phthalate	35			23	14
Silicone Oil Oligomers	67				
Brominated Organics					e

a Methylated, BF_3/methanol;
b No entry indicates compound is below detection levels;
c Multiple sampling dates;
d Methyl Ester
e Not quantitated

TABLE IV. Hydrophilic Organic Compounds[a] in

	W1N[c]	
	(8/17/81)	(7/26/82)
Ethylenediaminetetraacetic Acid[d] (EDTA)		
"EDTA fragment"[d] (MW 219)		
"EDTA fragment"[d] (MW 244)		
Lactic Acid[d]	2	
Oxalic Acid[d]	1	
Dimethyl Sulfate		trace
Hexanoic Acid[d]	0.6	
Propanedioic Acid[d]		
4-oxo-pentanoic Acid[d]	2	3
Butanedioic Acid[d]		
Succinic Acid[d]	0.6	1
Methyl Succinic Acid[d]		
Benzoic Acid[d]	2	1
Octanoic Acid[d]	1.4	0.8
Pentanedioic Acid[d]		
Nonanoic Acid[d]	3	2
Hexanedioic Acid[d]	1	1
Phenoxyacetic Acid[d]		
Decanoic Acid[d]	2	1.4
Dimethyl Phthalate	2	4
Octanedioic Acid[d]		
Citric Acid[d]		
Dodecanoic Acid[d]		2
Nonanedioic Acid[d]	1	2
Tetradecanoic Acid[d]	2	2.7
Hexadecanoic Acid[d]	8	7
Tridecanedioic Adid[d]		
Dibutyl Phthalate	6	10
Octadecanoic Acid[d]	6	4
Dioctyl Phthalate	5	2.7
Silicone Oil Oligomers		

a Methylated, BF_3/methanol;
b No entry indicates compound is below detection level;
c Multiple sampling dates;
d Methyl Ester

Groundwater from Experimental Wells at Maxey Flats

Inert Atmosphere Wells(ppb)[b]				
W2NA[c]			W3N	W3NA
(5/18/81)	(8/17/81)	(6/14/82)	(8/17/81)	(8/17/81)
1088	1009	793		
trace	trace	trace		
125	145	171		
	52	82		
61	67		10	3
	33		14	11
1			8	1
48	92		3	15
20	14			
	41		16	2
7			10	2
78	21	trace	5	4
			3	1
	trace		trace	
9	8		43	8
123				
115	14	13	13	3
	4			
trace	18		10	1
			2	
17	33	trace	8	1
			1	
58	14	15	8	4
15	18		14	
17	28	51	7	4
111	16	51	8	3
	13		3	2
			33	7

TABLE V. Hydrophobic Organic Compounds in Groundwater

	Waste Trenches (ppb)[a]		
	WFT-27[b]		WFT-19S
	(4/7/81)	(7/27/82)	(7/22/82)
Pentafluorobenzoic Acid	1067		
Barbital	6	1	
Pentobarbital	2	1	
Piperidinone		0.2	0.5
Nicotine			
Methyl Methacrylate	2		
Caprolactam		0.3	
1,3-dihydro-2H-indol-2-one	6		
Methyl Butenone	9		
Dichloroiodomethane	4	0.2	
Bromodichloromethane	3	0.7	0.2
Dichloromethylbutane	2		
Chloromethylbutene	10		
Toluene	23		
Benzofuran	10		
Benzothiazole	13		
Benzotriazole	3	trace	
2(3H)-benzothiazolone	835	8	
C_2-toluenesulfonamide	1	1	1
C_3-toluenesulfonamide	0.1		
2,5-dimethyl Benzene Butanoic Acid	17		
Phthalic Anhydride	17		
Undecane	3		
Dodecane	4		
Tridecane	3		
Alkanes (C_{20}-C_{31})	197		4
Alkylphenoxy Oligomers	102		

a No entry indicates compound is below detection level;
b Multiple sampling dates

from Trenches at Maxey Flats

	Experimental Trench Sumps(ppb)[a]	
	S4	S5
	(8/18/81)	(8/18/81)
trace		
	trace	
	0.2	

TABLE VI. Hydrophobic Organic Compounds in Groundwater from Experimental Wells at Maxey Flats

	Inert Atmosphere Wells(ppb)[a]						
	W1N[b]		W2NA[b]			W3N	W3NA
	(8/17/81)	(7/26/82)	(5/18/81)	(8/17/81)	(6/14/82)	(8/17/81)	(8/17/81)
Barbital			2		trace		
Pentobarbital			1		0.2		
Pyridine	97						
Nicotine			2			4	
Methyl Methacrylate	9						
Methyl Butenone			99			21	
Dichloroiodomethane			6		0.2		
Bromodichloromethane	0.3	trace	0.7		1		0.6
Dichloromethylbutane	7		38	3		21	36
Chloromethylbutene	20		107	21		57	94
Toluene			64			92	72
Benzofuran			16			42	37
C_2-toluenesulfonamide	trace						
N,N-bis(phenylmethyl)amine	3						4
Undecane						27	15
Dodecane						19	19
Tridecane						17	18
Alkanes(C_{20}-C_{31})	9						158
Alkylphenoxy Oligomers	197		963	trace		1144	735

a No entry indicates compound is below detection level;
b Multiple sampling dates

(July 27, 1982) contained 3830 ppm EDTA. The dramatic increase in EDTA concentration may be due to the installation of the plastic cover over the waste trenches. The presence of this barrier should decrease infiltration by rainwater into the waste trenches resulting in less water accumulation in the "bathtub-like" trenches, thereby possibly resulting in a concentration of any chemical species dissolved in the contaminated groundwater. Water from waste trench 19S, near the center of the burial site, contained 1340 ppb EDTA. Of the many water samples analyzed from the experimental wells and sumps, only groundwater from inert atmosphere well W2NA contained EDTA (1088 ppm, 1009 ppm and 793 ppm on three different sampling dates, see Table IV).

The appearance of EDTA in waste trench 27 and in the vicinity of experimental trench section 2 (well W2NA) but not in the vicinity of experimental trench sections 1, 3, 4 and 5 argues that there is a very specific communication between waste trench 27 and the vicinity of inert atmosphere well W2NA, perhaps via groundwater flow along subsurface fractures.

EDTA-like species of various molecular weights were also identified in the samples containing EDTA. Three distinct methylated species of molecular weights 219, 244, and 288 were identified in water from waste trenches 27 and 19S. The mass spectra of these species closely resemble those of EDTA, suggesting that they arise from degradation of EDTA, perhaps via microbial- or nuclear-mediated diagenesis. Proposed structures are listed in Figure 2. We have been unable to find man-made chelators with such molecular weights or mass spectra. The older sample from waste trench 27 lacked the MW 288 species, perhaps because this species is labile. Water from waste trench 19S lacked the MW 219 species but the levels in water from waste trench 27 were quite low as well. The EDTA-like species constitute ~16-19% of the total EDTA fraction (EDTA plus EDTA-like fragments) in trenches 27 and 19S, respectively.

The EDTA-like species also appeared in the water samples from well W2NA. These species constitute ~10-17% of the total EDTA fraction in well W2NA. The MW 288 species is conspicuously absent in these samples, suggesting that this particular species is indeed more labile or is preferentially retarded during subsurface migration. The latter hypothesis seems less likely since the structural differences between EDTA, the MW 288 species, and the other EDTA-like species are not great enough to suggest different soil sorption behavior.

The other hydrophilic organics identified in the various water samples are generally much less abundant (ppb levels), and consist mainly of carboxylic acids. A trace of the organic groundwater tracer pentafluorobenzoic acid (added to the cap of experimental trench section 4) was detected in water from sump S4 in expermental trench section 4 (sampled 8/18/81). A trace of this compound was also detected in water from waste trench 27 (sampled on 4/7/81), as a result either of subsurface transport in

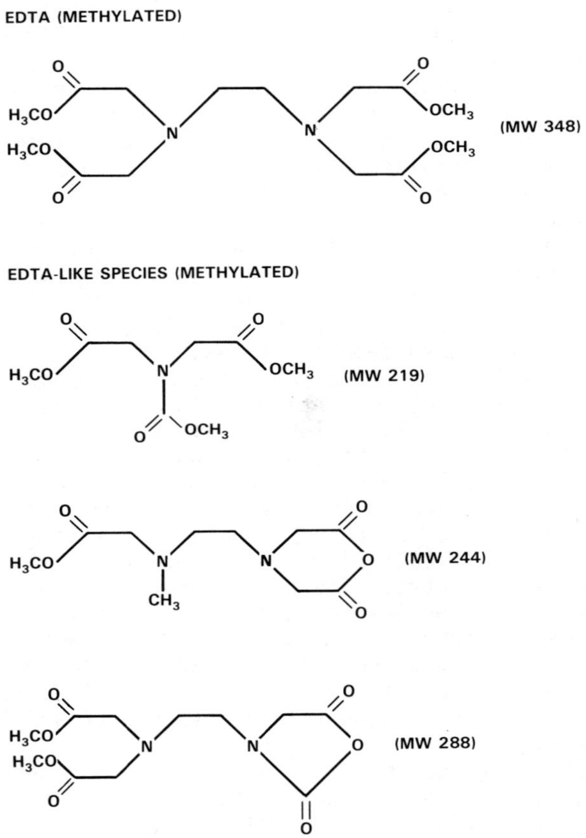

EDTA (METHYLATED)

(MW 348)

EDTA-LIKE SPECIES (METHYLATED)

(MW 219)

(MW 244)

(MW 288)

FIGURE 2. Chemical structure of EDTA (tetramethyl ester) and proposed structures of EDTA-like organic compounds identified in water from waste trenches 19S and 27 and from inert atmosphere well W2NA. The proposed structures for the MW 219, MW 244, and MW 288 organic species are based on mass fragmentation patterns from GC-MS analyses.

groundwater or surface transport in water run-off followed by infiltration through the cap of the waste trench.

A variety of hydrophilic brominated organic species were detected in water from experimental trench sump S5 (Table III). During construction of the experimental trench, sodium bromide was placed in the bottoms of trench sections 1, 2, 3 and 5 as a groundwater tracer. It seems likely, therefore, that the appearance of the brominated organic species in experimental trench section five is due to the bromination of polar organics present in the groundwater. Why these experimental species do not also appear in water from other sections of the experimental trench is more difficult to explain. However, we have observed that water inflow into section 5 is much slower than that observed for sections 1, 2 and 3. It is possible that the increased contact time between the sodium bromide tracer and stagnant groundwater in experimental trench section 5 has resulted in the bromination of organic compounds in the groundwater.

Most of the hydrophobic organics isolated from the water samples were present at ppb levels (Tables V and VI). The most interesting exception was pentafluorobenzoic acid (1067 ppb) which appeared in water from waste trench 27, confirming its presence in the vicinity of the waste trench and arguing again for communication between experimental trench section 4 and the waste trench (Table V). Many of the hydrophobic compounds are not typically found in uncontaminated water samples and some possess toxic properties. Two barbiturates, barbital and pentobarbital, have been identified in water from waste trench 27 and inert atmosphere well W2NA (Tables V and VI). The two barbiturates are undoubtedly waste-related, presumably associated with biomedical waste. Pentobarbital is often used in pharmacological studies and barbital has been commonly used as a buffering agent, for example. Interestingly, the two barbiturates were identified only in older water samples from waste trench 27 (4/7/81) and well W2NA (5/18/81). Their levels in these samples are real but quite low (ppb levels). Their absence in later water samples may be due to their depletion from the buried waste or, more likely, because their levels have fallen below detection limits. Nevertheless, their appearance in water from well W2NA, coupled with the presence of EDTA in the water, argues strongly for specific communication between waste trench 27 and well W2NA.

Organics and Radionuclide Migration

The presence of relatively high levels of strong organic chelators like EDTA in the water samples prompted a detailed chemical speciation study aimed at determining whether the organic compounds identified in the survey study are chelated or complexed to radionuclides. Water samples from waste trenches 19S and 27 and inert atmosphere wells W1N and W2NA were fractionated by steric exclusion chromatography and subsequently analyzed for their radionuclide and organic content. Each water sample exhibited a distinct chromatographic pattern (Figure 3).

The chromatograms of water from the two waste trenches are much more complex than those of water from the inert atmosphere wells. Based on the chromatograms, water from well W2NA appears to contain lower molecular weight aggregates (i.e., later-eluting UV-absorbing species) compared to the water from both waste trenches, under the chromatographic conditions used in this study.

GC analyses of the fractions collected during the chromatographic separations revealed that EDTA eluted in approximately the same region (1.5 - 1.7 hours) in three of the samples (waste trenches 27 and 19S and well W2NA). Neither EDTA nor any radionuclides were detected in water from well W1N. Alpha-emitting radionuclides and ^{60}Co co-eluted with the EDTA, strongly suggesting that EDTA is complexed with these species in the water samples.

This finding agrees well with an earlier observation that plutonium in water from waste trench 27 exists as a strong anionic complex (3,8,9). Earlier binding studies with solutions of Pu^{3+}, Pu^{4+} and EDTA also revealed that the presence of a strong chelating agent like EDTA, even at low concentrations comparable to those at Maxey Flats, may actually be more important than the oxidation state of plutonium in determining whether plutonium adsorbs or migrates in soil (10).

A band of ^{90}Sr and ^{137}Cs activity eluted from the Sephadex G-15 column between 2-3 hr. A variety of organic acids and other hydrophilic compounds co-eluted with the ^{90}Sr and ^{137}Cs. It appears, therefore, that association of radionuclides with polar organic species such as organic acids may provide a mechanism for the migration of radionuclides at the Maxey Flats commercial shallow land burial site. Detailed radionuclide analyses are under way to determine precise correlations between organic species and specific radionuclides.

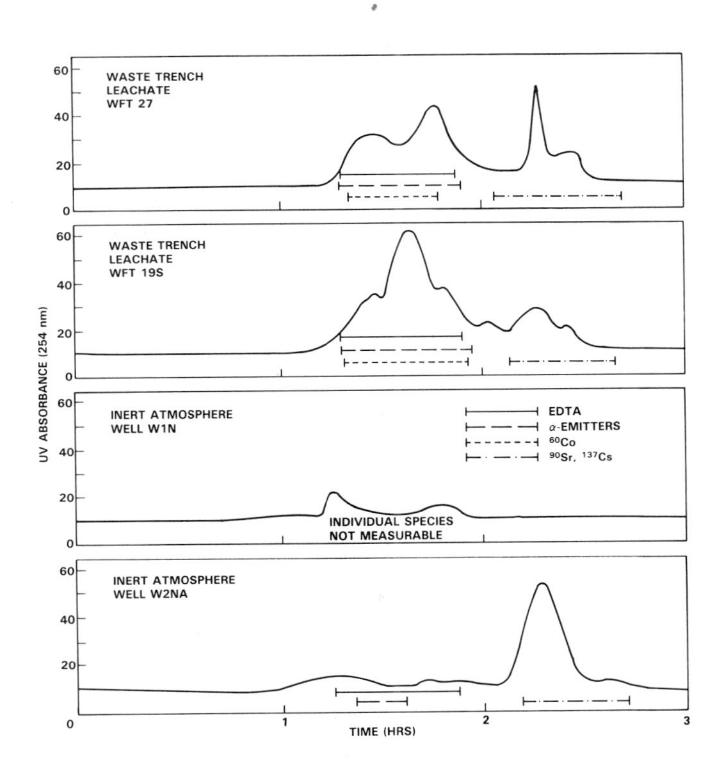

FIGURE 3. Steric exclusion chromatography of water from waste trenches and inert atmosphere wells at Maxey Flats, sampled July, 1982. Water samples from waste trenches 27 and 19S (440 ml) and wells W1N (585 ml) and W2NA (238.5 ml) were concentrated to 10 ml, pH-adjusted to their original values and chromatographed on a Sephadex G-15 column at flow rates ranging from 28-33 ml/hr. Individual column fractions were collected every 10 min and analyzed for specific organic compounds and radionuclides.

ACKNOWLEDGMENTS

The authors are indebted to Ms. Roberta Myers, Margaret McCulloch, Dorothy Harless and Lois Abbey and Mr. Quentin Dierks for their assistance in the laboratory. Dr. Roger Schirmer assisted greatly in the interpretation of GC-MS data. The research was supported by the U.S. Nuclear Regulatory Commission. Ms. Susan Kesterson and Mrs. Katherine Burton of PNL's Word Processing Team helped make this manuscript a reality. Pacific Northwest Laboratory is operated by Battelle Memorial Institute for the U.S. Department of Energy under contract DE-AC06-76RLO-1830.

LITERATURE CITED

1. Dahlman, R. C.; Bondietti, E. A.; Eyman, L. D. in "Activities in the Environment"; Friedman, A. M., Ed.; ACS SYMPOSIUM SERIES No. 35, American Chemical Society: Washington, D. C., 1976; p 147.
2. Stone, J. A. in Proceedings of the Fourth Annual Participants Information Meeting, DOE Low-Level Waste Management Program, ORNL/NFW-82/18, 1982, pp 469-476.
3. Kirby, L. J.; Toste, A. P.; Wilkerson, C. L. in "Environmental Migration of Long-Lived Radionuclides"; International Atomic Energy Agency: Vienna, Austria, 1982, pp. 63-69.
4. Means, J. L.; Alexander, C. A. Nuclear and Chemical Waste Management, 1981, 2, 183-196.
5. Rees, T. F.;Cleveland, J. M. in "Environmental Migration of Long-Lived Radionuclides"; International Atomic Energy Agency: Vienna, Austria, 1982, pp. 41-52.
6. Means, J. L.; Crerar, D. A.; Duguid, J. O. Science, 1978, 200, 1477-1480.
7. Wogman, N. A.; Rieck, H. G., Jr; Laul, J. C.; MacMurdo, K. W. Nucl. Instr. and Meth. 1977, 141, 539-547.
8. Kirby, L. J.; Toste, A. P. in Proceedings of the Fourth Annual Participants' Information Meeting, DOE Low-Level Waste Management Program, ORNL/NFW-82/18, 1982, pp. 477-503.
9. Toste, A. P.; Kirby, L. J.; Robertson, D. E.; Abel, K. H.; Perkins, R. W. IEEE Transactions on Nuclear Science, 1983, NS-30(1), pp. 580-585.
10. Kirby, L. J. Ed.; "Radionuclide Distributions and Migration Mechanisms at Shallow-Land Burial Sites: 1981 Annual Report of Research Investigations on the Distribution, Migration ad Containment of Radionuclides at Maxey Flats, Kentucky," NUREG/CR-2383, 1982.

RECEIVED October 20, 1983

16

Releases of Radium and Uranium into Ralston Creek and Reservoir, Colorado, from Uranium Mining

I. C. YANG

U.S. Geological Survey, Denver Federal Center, MS-407, Lakewood, CO 80225

K. W. EDWARDS

Colorado School of Mines, Golden, CO 80401

Dissolved U concentrations were determined in wastewater from a uranium mine, in water from Ralston Creek upstream from and downstream from the discharge point, and from Ralston Reservoir, located 2.5 mi downstream from the mine. Dissolved U concentrations, in micrograms per liter, were: 1,700 in discharge water; approximately 4 in creek water upstream from the mine; and 600-1,200 in creek water downstream from the mine; and 100 in reservoir water. Dissolved ^{266}Ra concentrations were less than 2 pCi/L in both the creek and reservoir waters. The concentration of suspended solids in the mine effluent was 6 mg/L. The ^{226}Ra content of suspended sediment in the mine effluent was 1,300 pCi/g; U content was 2,400 μg/g. Suspended sediments in the creek contained 130 to 1,000 μg/g U and 30 to 600 pCi/g ^{226}Ra. In the bottom sediment, the greatest U and ^{226}Ra concentrations (350 μg/g and 150 pCi/g respectively) were found 0.5 mi downstream from the mine. Both U and ^{226}Ra concentrations in the suspended and bottom sediments decreased toward the reservoir, with greater concentrations in the fine particle-size fraction (less than 270 mesh). Extracted U and ^{226}Ra with various leaching reagents were: 2N HCl, 62 and 84%; 0.25 M Na_2EDTA at pH 10, 18 and 67%; 1N KCl, 4 and 30%; and 1N $BaCl_2$, 4 and 10%. The fact that most of the ^{226}Ra and U were not displaced by K^+ or Ba^{+2} ions indicates that these radionuclides are not absorbed on the particle surfaces but rather are associated with minerals of U ores or are partly trapped inside the colloidal hydroxides of Fe and Mn coated on the sediment.

0097-6156/84/0246-0271$06.00/0

With the public concern over the problem of hazardous radioactive wastes in the environment, extensive attention needs to be given to determination of the presence, distribution, and concentration of radioactive materials in stream water, ground water, and sediments. The Schwartzwalder uranium mine, located approximately 8 mi northwest of Golden, Colorado, discharges about 600,000 gal/d of wastewater into Ralston Creek, which flows into Ralston Reservoir and Upper Long Lake at the east edge of the foothills (fig. 1). These reservoirs provide drinking water for approximately 100,000 residents of the western suburbs of Denver. Of particular concern are radionuclides of long-lived ^{226}Ra and U, both dissolved and in stream-borne sediment.

A study by Parsont (1) indicated ^{226}Ra concentrations of 1 to 5 pCi/L in the creek water and 20 to 100 pCi/g in the creek sediment. In his analyses, more than 95% of ^{226}Ra activity was found in the suspended sediment instead of the water. In that study, Parsont probably was precipitating the ^{226}Ra unknowingly before analysis by adding the acid and Ba carrier to the raw waters that contained significant concentrations of SO_4^{-2} ions. This would cause the ^{226}Ra to precipitate as $Ba(Ra)SO_4$, which would be retained on the filter paper. Chemical treatment of the wastewater from the mine was started in October of 1972. Lammering (2) of the U.S. Environmental Protection Agency reported that between May and September of 1972, ^{226}Ra in the creek water ranged from 2.2 to 86 pCi/L, and U in the creek water ranged from 400 to 450 µg/L with greater concentrations near the mine; ^{226}Ra in the sediment ranged from 17 to 188 pCi/g, and U in the sediment ranged from 50 to 520 µg/g.

In 1976, Hazen Research, Inc., (3) of Golden, Colorado, at the request of the mine operator, started a project called "Spill Prevention and Containment Plan for Control of Pollutant Discharge Streams at Schwartzwalder Mine". The results of their work (1977) decreased the dissolved ^{226}Ra concentration in the creek water to 0.7 pCi/L. No data were given for U.

The Colorado State Health Department has set a stream standard of 40 pCi/L for U (or 58.8 µg U/L) and 10 pCi/L for ^{226}Ra. The maximum permissible concentration of combined ^{226}Ra + ^{228}Ra in drinking water is 5 pCi/L, as set by the U.S. Environmental Protection Agency (4); and the "guidance level" for U is 10 pCi/L (or 14.7 µg/L) (5).

The present study encompasses: (1) Distributions of ^{226}Ra and U in the creek water, suspended material, and bottom sediment; and (2) examination of the leachability of ^{226}Ra and U from stream sediments, as a function of the particle size and lixiviant composition.

Sample Collection

Sixteen 20-L water samples containing suspended material, and 14 samples of bed sediments were collected from different sites in Ralston Creek and Reservoir during the winter of 1980 and the spring of 1981. The study area and sampling sites are shown in Figure 1. The stream was shallow, 1 to 3 ft deep, about 10 to 20 ft wide; turbulence was moderate. No specialized equipment was needed for sample collection.

Water and Suspended Solids. Water and suspended solids were collected simply by immersing a 1-L plastic bottle (wide mouth) under the water surface several times at different depths. Collected water was transferred to a 5-gal plastic container until the latter was filled. Samples were shipped to the laboratory and filtered immediately through a 0.45-µm membrane filter. The filtrate was acidified to pH 2 with HCl.

Bottom Sediment. Bottom sediments were scooped from the upper 3-cm of the stream bed close to where the water samples were collected. A plastic scoop was used and a 2-L wide-mouth plastic bottle was used as a sample container, with a minimum quantity of water in the sediment. In the laboratory, sediment samples were dried and sieved into 6 different particle-size fractions, (between U.S. standard size No. 12 and 20, 20 and 30, 30 and 60, 60 and 120, 120 and 270, and less than 270), except for site 4, (400 ft downstream from mine-outflow) which was separated into 10 different particle-size fractions for studies of particle-size distributions and leaching experiments.

Experimental

Water Samples. Eight hundred mL of the filtered water was analyzed for ^{226}Ra by a direct de-emanation method, a modified version of Chung (6). Sample water in a 1-L bubbler (fig. 2) was purged with He gas at a flow rate of 120 cm^3/min for about 30 to 35 min. Stopcocks 1 and 2 were then closed, and the bubblers were set aside for 12 d to allow ingrowth of Rn. On second de-emanation, He gas was purged through the sample bubbler again to expel the Rn gas, which was then trapped on activated charcoal. The activated charcoal was kept at liquid nitrogen temperature. The charcoal trap was heated to 350°C for 20 min and flushed slowly with He to carry Rn gas into the scintillation cell until ambient pressure was reached. The counting cell was placed in a light-tight chamber and allowed to age 3 h before counting for Rn and daughters. Uranium was measured by a fluorometric method (7).

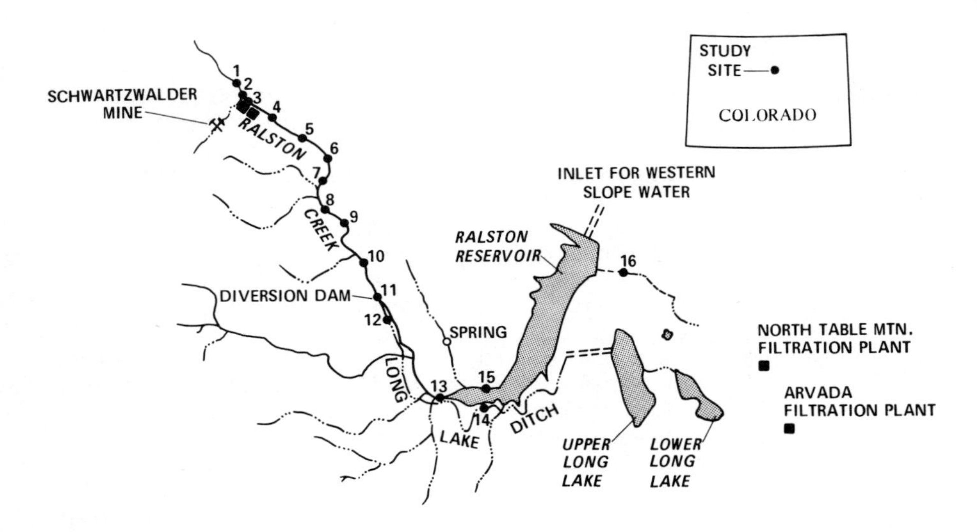

Figure 1. Map of Ralston Creek and Reservoir, near Golden, Colorado.

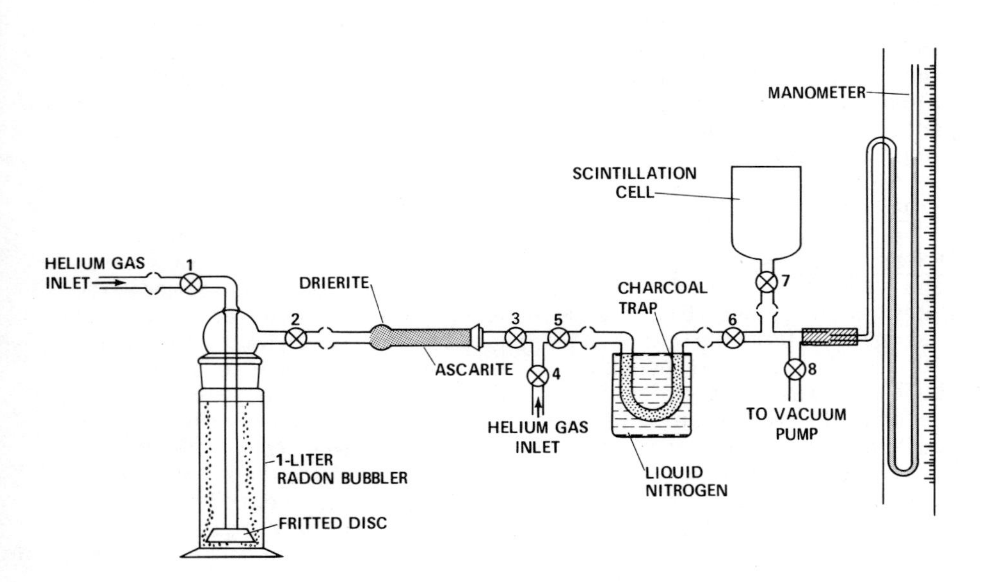

Figure 2. Radon transfer system.

<u>Suspended Sediment.</u> Dried suspended sediment on the membrane filter was ignited in a porcelain crucible covered with a watch glass simply by touching the membrane with a hot glass rod (to burn the membrane), followed by placing the crucible inside a muffle furnace at 450°C to burn off any organic material. The suspended-solids residue was then decomposed, using the same method as for bed sediments (see following section). The solution from the decomposition was diluted to 800 mL with distilled water; the pH was adjusted to 2 and analyzed for ^{226}Ra by the direct de-emanation method as described above. Ten mL from the same solution were saved for U analysis.

<u>Bed Sediments.</u> Creek sediments collected from 12 sites (sites 1, 5, 6, 7, 8, 9, 10, 11, 12, 13, 14, and 15, in figure 1), sieved into 6 fractions, were separately packaged into aluminum cans for ^{226}Ra determinations by gamma spectrometry. One-, two- and four-oz cans were used for different sample sizes in the range of 30 to 200 g, depending on the quantity of each fraction available for packaging. Analysis for ^{226}Ra was performed by allowing the sample to age for about 3 wk to allow the ^{214}Bi daughter to ingrow into near-secular equilibrium. Following ingrowth, the samples were counted on a 76 x 76 mm NaI(Tl) gamma detector coupled to a system of photomultiplier tube, preamp, amplifier, and Nuclear Data Model ND60A 2048 multichannel analyzer (Any use of brand names is for identification purposes only and does not constitute endorsement by the U.S. Geological Survey). The intensity of the 1.76 MeV ^{214}Bi peak was measured and related to the ^{226}Ra contents of the samples. Uranium was measured by a fluorometric method.

Three g of bed sediment from site 4, (through 45 mesh and retained on 60 mesh, -45+60, size-fraction), containing 51.6 pCi/g of ^{226}Ra and 83.7 µg/g of natural U, were leached with 200 mL of different leaching solutions for 3 h. Hydrochloric acid, Na_2EDTA, $BaCl_2$, and KCl at different concentrations were selected for study. The leach solution was filtered through a 0.45-µm membrane filter, and the filtrate was diluted to 800 mL (pH 1 to 2) and stored in a 1-L bubbler for ^{226}Ra analysis. A 10-mL aliquot of this solution was saved for U analysis.

Results and Discussion

<u>Creek Water.</u> Concentrations of dissolved ^{226}Ra and U in Ralston Creek and Reservoir water are plotted in figure 3. These data indicated that the ^{226}Ra concentration in creek water 400 ft upstream from the mine (background concentration) was 0.13 pCi/L, and the U concentration in the same water was 4.0 µg/L.

The mine outflow to Ralston Creek contained 0.58 to 1.03 pCi/L of ^{226}Ra and 1,700 µg/L of U; therefore, mine contribution to dissolved ^{226}Ra in the creek water was fairly small, whereas mine contribution to U was very large. Downstream from the mine, dissolved ^{226}Ra concentrations in the creek water ranged from 1 to 2 pCi/L, which is less than the acceptable drinking-water standard of 3 pCi/L; yet, concentration of U ranged from 600 to 1,200 µg/L, significantly greater than the stream standard of 58.8 µg/L. At Ralston Reservoir inlet (sample site 15), ^{226}Ra concentration was 0.46 pCi/L, and U concentration was 100 µg/L; and at the outlet (sample 16), ^{226}Ra concentration was 0.24 pCi/L, and U concentration was 34 µg/L. This decrease in both concentrations of the reservoir sample sites probably is due to dilution by the large volume of reservoir water. The major part of this water comes from western Colorado through the Moffat Tunnel and enters the reservoir's extreme northwest part, at a point not far from the reservoir outlet.

Hazen Research, Inc.(3), reported that the creek water had a ^{226}Ra concentration of 0.7 pCi/L in March 1977. In contrast, Lammering (2) reported a ^{226}Ra concentration of 81 pCi/L in a creek sample collected several hundred yards downstream from the mine in July 1972. The reason for the large value reported by Lammering is that the waste-treatment system at the Schwartzwalder Mine was not in operation until October 1972 after 2 1/2 years of research conducted primarily by the Colorado School of Mines Research Institute. Concentrations of ^{226}Ra and U 400 ft upstream from the mine (0.13 pCi/L for ^{226}Ra and 4 µg/L for U) are comparable to data reported by Lammering (2) of 0.1 to 0.2 pCi/L for ^{226}Ra and 1.5 to 6 µg/L for U.

Suspended Material. The material retained by the 0.45-µm membrane filters during filtration of the water samples collected from Ralston Creek and Reservoir was classified as suspended solids; concentrations are shown in figure 4. The ^{226}Ra content of suspended sediment in the mine effluent was 1,300 pCi/g; U content of suspended sediment was 2,400 µg/g. Suspended-material load for this sample was 6 mg/L during the sample collection. Total release of both radionuclides (dissolved and suspended) was then calculated to be 8.4 pCi/L for ^{226}Ra and 1,710 µg/L for U. Assuming an average daily discharge of 600,000 gal of mine effluent to Ralston Creek, about 19 µCi/d of ^{226}Ra and 3,900 g/d of U were being discharged to the creek. Thus, the effect on Ralston Creek by effluent from the Schwartzwalder Mine mainly is from dissolved U and suspended materials containing large ^{226}Ra and U concentration. Suspended ^{226}Ra and U concentrations were found to decrease as distance downstream from the mine increased (figure 4). The significant decrease in suspended ^{226}Ra and U concentrations just downstream from the mine may result partly

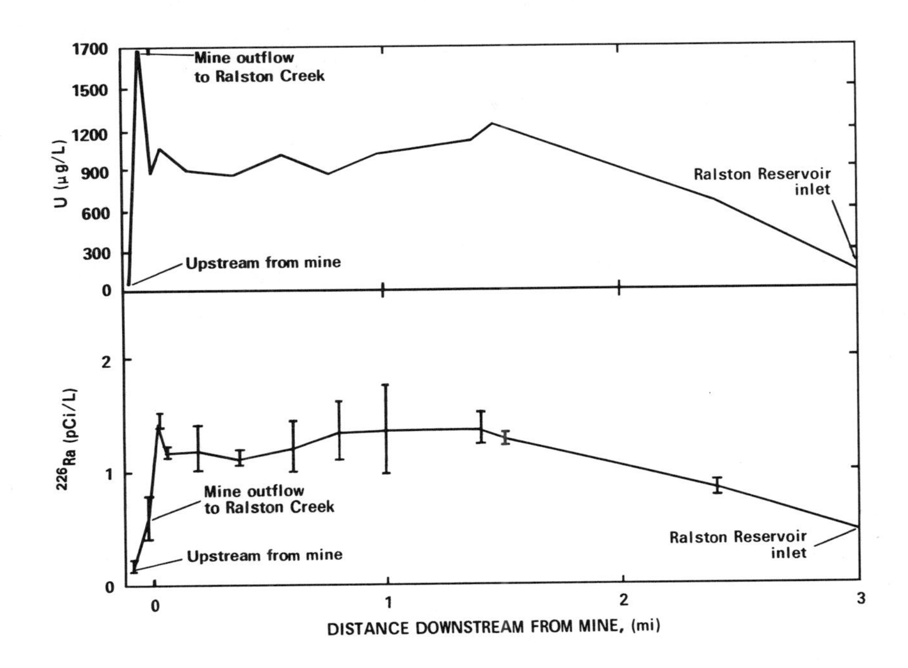

Figure 3. Dissolved ^{226}Ra and U in Ralston Creek.

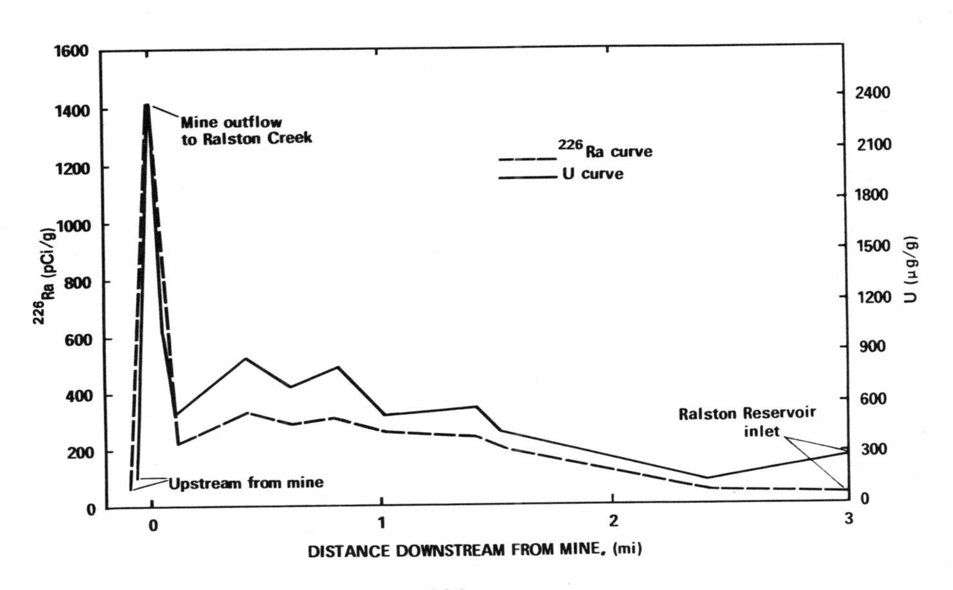

Figure 4. Suspended ^{226}Ra and U in Ralston Creek.

from dilution by suspended particles of the stream sediment and partly from settling of the suspended particles from the mine effluent into the creek bed, because of the change in stream temperatures, less turbulence, and interaction with bed sediments. The reason for the large value of U 0.2 mi downstream from the mine is not known; it is conceivable that an accidental inclusion of concentrated U-ore particles was sampled. This also is reflected in the large value of ^{226}Ra at the same site.

Most of the ^{226}Ra associated with the suspended solids in the mine effluent probably is in the form of $Ba(Ra)SO_4$, because $BaSO_4$ is used as a decontaminant for ^{226}Ra by the mine water-treatment plant. This evidence also is supported by the fact that little ^{226}Ra in these suspended materials was leached by 2N HCl, as shown in the leaching experiments. Uranium probably is in the form of fine ore particles that are not settled out in settling ponds used in $BaSO_4$ treatment.

Concentrations of ^{226}Ra and U in suspended sediment upstream from the mine (68 pCi/g for ^{226}Ra and 210 µg/g for U) were greater than expected. Four hundred ft upstream from the mine outflow may not be far enough to represent the true background concentration of suspended ^{226}Ra and U in Ralston Creek. It is possible that U ore might have been transported from the mine area to the sampling point upstream from the mine by wind and water erosion during rainfall or snowmelt.

The smallest concentrations of suspended ^{226}Ra and U were found in samples collected from the inlet and outlet of Ralston Reservoir. Previous ^{226}Ra concentrations in the suspended material for samples collected at different locations in Ralston Reservoir by Lammering (2) were in the range of 2.1 to 12 pCi/g.

Bottom Sediment. The ^{226}Ra concentration in the bottom sediment was determined by using gamma-spectrometric analyses; U concentration was determined by fluorometric analyses. The concentration of ^{226}Ra was analyzed for six-size fractions, whereas U was analyzed for three-size fractions.

Generally, concentration of both ^{226}Ra and U in the bed sediment increases as particle-size decreases, and as the distance downstream from the mine outflow decreases (figs. 5 and 6). Both ^{226}Ra and U have similar distribution patterns. It is likely that U-ore particles discharged by the mine during early days of the mine operation are present in the sediment. The U-ore particles apparently have been accumulated in certain stream reaches due to the stream condition or variable discharge of effluent from the mine during the history of the mining operation. Suspended particles of $Ba(Ra)SO_4$ and fine U ore that are not settled out in settling ponds used in $BaSO_4$ treatment probably go through discharge pipe and settled out in these reaches and contributed to the large ^{226}Ra and U contents

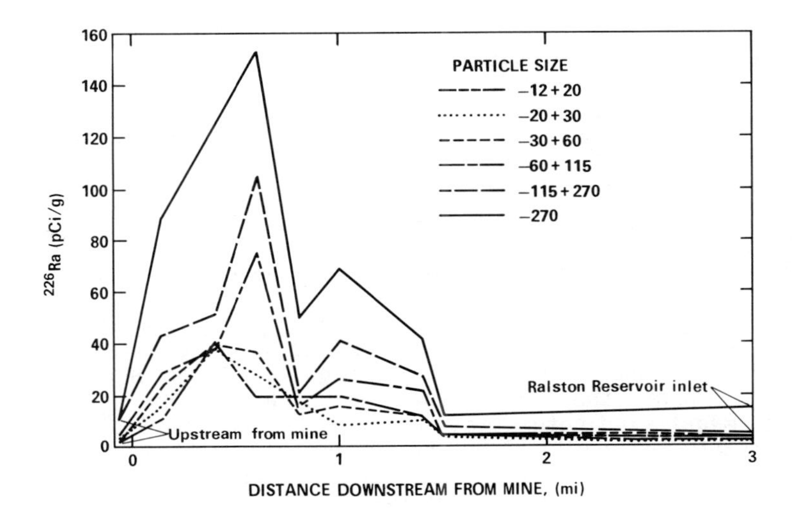

Figure 5. Distribution of ^{226}Ra in Ralston Creek as a function of particle size and distance downstream from the mine.

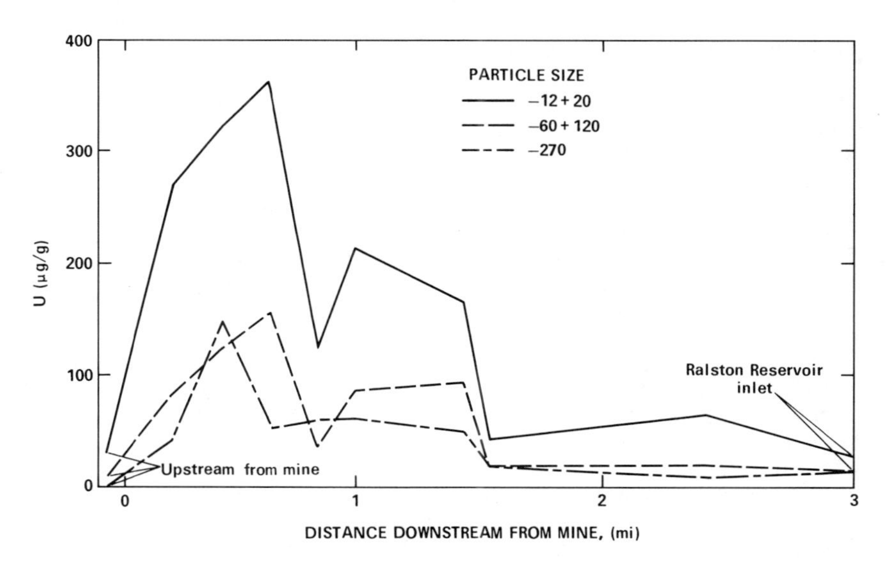

Figure 6. Distribution of U in Ralston Creek as a function of particle size and distance downstream from the mine.

of the fine particle sizes. The decrease in ^{226}Ra and U concentrations at distances greater than 1.5 mi downstream from the mine outflow probably indicates that most of the suspended material and U-ore bearing sediment settled out and dissolved radionuclides adhere to particle surfaces before these points.

Samples collected 400 ft upstream from the mine had much smaller ^{226}Ra and U contents than the corresponding fractions immediately downstream from the mine. This is consistent with the ^{226}Ra and U upstream from the mine being from the native U mineralization rather than coming from treated waste.

Distribution of ^{226}Ra and U in Different Particle Sizes. These analyses were performed on the bed sediment of sample site 4. Results are shown in table I and fig. 7. In general, the finer the particle, the larger the concentrations of ^{226}Ra and U. This might indicate that mineralogy and surface area are important factors in the retention of both nuclides. The cation-exchange capacity of creek sediment for the size fraction passing through 120 mesh and retained on 270 mesh, (-120+270), was 5.6 meq/100 g and that for the size fraction -120+170 was 8.5 meq/100 g of dry sample. These values are rather small and do not show increased exchange capacity for smaller particle size. Masuda and Yamamoto (8) also found that adsorption of the uranyl ion was not dependent on the cation-exchange capacity of the soil.

Leaching Studies (9-11). The data shown in table II indicate that, generally, the leachable quantity of ^{226}Ra and U from Ralston Creek bed sediment is dependent on concentration of the leaching solution. Either 1N or 2N HCl will liberate 84% of ^{226}Ra and 50 to 60% of U with 1 h of shaking; whereas, using 0.0001N HCl, the quantity liberated was 0.04% for ^{226}Ra and 2% for U. Increasing the shaking time of the leach to more than 1 h did not affect the quantity of ^{226}Ra liberated, but increased the quantity of uranium from 60% to 80%. Under the same conditions, when KCl, $BaCl_2$, and Na_2EDTA solutions were used as lixiviants, it was found that the quantities of U leached were significantly less than the quantities leached by HCl; ^{226}Ra leached was significantly less when KCl and $BaCl_2$ were used compared with HCl, but only about 20% less than HCl, in the case of Na_2EDTA. It is interesting to note here that Na_2EDTA has a greater leaching capacity for ^{226}Ra than U. Thus, the leaching power of the lixiviant used in the tests for Ralston Creek bed sediment decreases in the order: HCl> Na_2EDTA> KCl> $BaCl_2$.

The fact that most of the ^{226}Ra and U was not displaced by Ba^{+2} or K^+ ions indicates that these radionuclides are not significantly adsorbed on the sediment-particle surfaces by cation-exchange mechanism, but rather are associated with minerals of U ores, with $Ba(Ra)SO_4$ from the wastewater

Table I. Distribution and leaching efficiency of ^{226}Ra and U in different particle sizes of Ralston Creek and Reservoir sediments

Particle size[a] (mesh)	weight %	Unleached		Leached[b]			
		$^{226}Ra^{c}$ (pCi/g)	U^{c} (μg/g)	$^{226}Ra^{c}$ (pCi/g)	%	U^{c} (μg/g)	%
1-Upstream from mine (400 ft)							
-140 + 270	6	9.23[d]	23 ± 2	1.61 ± 0.98	17	---	--
-270	2	7.25[d]	31 ± 3	3.13 ± 1.89	42	---	--
2-Downstream from mine (0.2 mi)							
- 12 + 20	28	---	---	---	---	---	--
- 20 + 30	15	47.3 ± 5.1	130 ± 10	30.7 ± 2.5	65	40 ± 3	31
- 30 + 45	21	43.2 ± 2.6	82 ± 7	34.7 ± 4.6	80	42 ± 3	51
- 45 + 60	13	51.6 ± 1.6	83 ± 7	43.5 ± 3.0	84	57 ± 4	68
- 60 + 80	10	61.5 ± 2.0	101 ± 8	49.6 ± 3.3	81	58 ± 4	58
- 80 + 100	3	75.5 ± 8.4	132 ± 10	62.5 ± 3.0	83	63 ± 5	48
-100 + 120	1	80.4 ± 1.9	169 ± 12	69.3 ± 4.4	86	88 ± 7	52
-120 + 140	3	82.3 ± 6.3	156 ± 13	71.7 ± 12.0	87	109 ± 8	70
-140 + 170	1	112.3 ± 2.3	176 ± 15	80.2 ± 8.2	71	143 ± 10	81
-170 + 270	2	121.9 ± 4.9	328 ± 30	90.7 ± 1.2	74	213 ± 20	65
-270	3	175.2 ± 7.2	618 ± 50	114.2 ± 8.8	65	303 ± 30	49
	100						
3-Mine Outflow (suspended)	---	1,290.0 ± 220.0	---	208.9 ± 16.1	16	---	--
4-Reservoir[e]	---	3.29[d]	---	2.03 ± 1.2	62	---	--

a) U.S. standard size sieve no.
b) Leaching conditions are 2N HCl, 1 h shaking at room temperature.
c) Errors terms are expressed at the 90% confidence level.
d) ^{226}Ra by gamma; all other ^{226}Ra by direct de-emanation; U values by fluorometric.
e) Unfractionated (total) sediment.

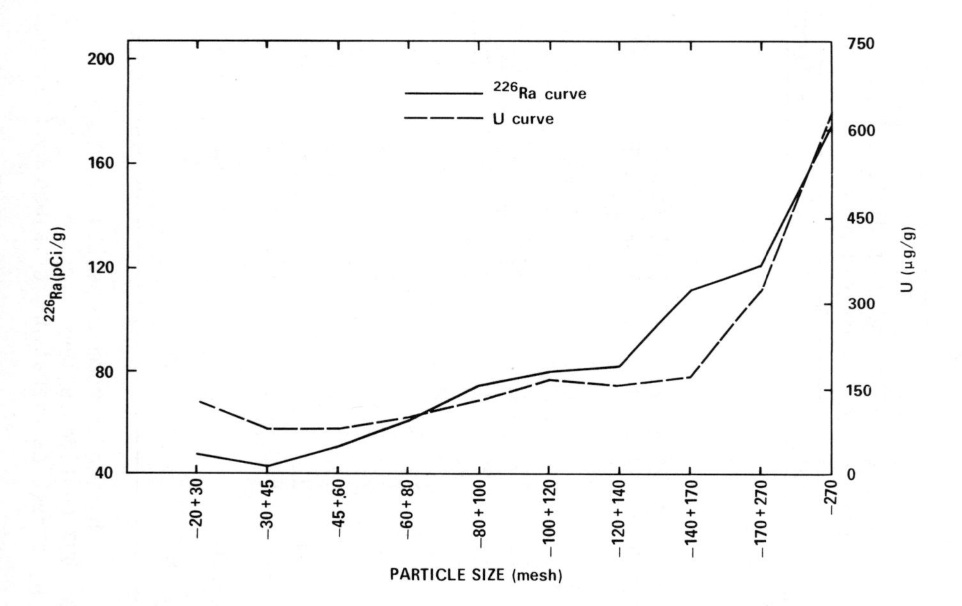

Figure 7. Concentration of ^{226}Ra and U as a function of particle size.

Table II. Leachability of ^{226}Ra and U from Ralston Creek Bed Sediment. (-45+60 mesh fraction)[a]

	Leached ^{226}Ra		Leached U	
Leaching Reagent[b]	pCi/g[c]	%[d]	µg/g	%[d]
HCl, 2N	44 ± 6	84	52 ± 5	62
HCl, 2N (2 h)	43 ± 1	83	67 ± 6	81
HCl, 1N	43 ± 5	84	47 ± 5	57
HCl, 0.0001N	.22 ± .04	.04	1.9 ± .2	2
KCl, 1N	15.4 ± .5	30	3.6 ± .3	4
KCl, 0.1N	4.8 ± .3	9	1.1 ± .1	1
KCl, 0.01N	1.5 ± .4	3	1.6 ± .2	2
$BaCl_2$, 1N	5.2 ± .6	10	3.7 ± .4	4
$BaCl_2$, 0.1N	3.0 ± 1.0	6	1.2 ± .1	6
$BaCl_2$, 0.01N	3.6 ± .5	7	1.6 ± .2	2
Na_2EDTA, 0.1M (pH=10)	34 ± 4	66	9.5 ± .9	11
Na_2EDTA, 0.1M (2 h) (ph=10)	33 ± 8	65	14 ± 1	16
Na_2EDTA, 0.25M (pH=10)	35 ± 9	67	15 ± 1	18
Na_2EDTA, 0.1M (pH=4.5)	17 ± 3	33	12 ± 1	15

a) Sample collected from Site 4 (0.2 mi downstream from mine).
b) Leaching done at room temperature for 1 h (unless otherwise indicated) with shaking, using 200 mL of lixiviant and 3 g of sediment.
c) Error terms are based on reproducibility of replicate measurements and expressed at the 90%confidence level.
d) Percentage (%) extracted based on 51.6 pCi ^{226}Ra/g, and 83.7 µg U/g, found on total decomposition of this sediment fraction.

treatment, or are partly trapped inside the colloidal hydroxides of Fe and Mn coated on the sediment.

From the above leaching studies, it is clear that the best lixiviant and leaching conditions for extracting creek sediments are 2N HCl and 1 h shaking time at room temperature for ^{226}Ra, and 2 h or longer shaking time for U. This leaching reagent and 1 h shaking time at room temperature were adopted to the study of the leachability of ^{226}Ra and U as a function of the particle size. Results are shown in table I.

The quantity of ^{226}Ra extracted from the fine particles (less than 270 mesh) in the sample from 400 ft upstream from the mine outflow is significantly greater than the quantity extracted from the coarser particles (-140 + 270 mesh). In contrast, the percentage of ^{226}Ra leached from the creek sediment in size fractions -140+ 170 and -170+270 is about 75% which is slightly greater than the percentage from the finer particles (less than -270 mesh) of 65%. Apparent greater leachability of ^{226}Ra in the fine-particle size relative to the coarse particles in the background (upstream from mine) probably is due to inclusion of acid-decomposable, fine-grained U ore that was transported from the mine area, (as discussed in the section on suspended material). Less leachability of fine particles relative to the coarse particles in the creek sediment may be due to the existence of $^{226}RaSO_4$ derived from the mine discharge water that only barely can be leached by the HCl. This is evidenced by the fact that only 16% of ^{226}Ra was extracted from the suspended material in the mine effluent. For the seven particle size classifications, smaller than 20 mesh and larger than 140 mesh in the creek sediment, the quantities of ^{226}Ra extracted were about 84% and showed little variation.

From U-leaching results, it can be seen (from table I) that the percentage U leached from size-fraction -20 + 30 is 31%, increasing to 81% for size-fraction -140+170, and then decreasing to 49% for size-fraction less than 270 mesh. The increase in leachability of U with decreasing particle size to -140+170 probably is related to increased surface areas of the decomposable U ore. As noted in the leaching study in table II, a longer shaking time of 2 h significantly increased the percentage of U leached. The reason for the decrease in percentage leached toward finer particle-size of -270 mesh in U leaching efficiency is not known. Further study by X-ray diffraction on sediment minerals may help to answer these questions.

Conclusions

Dissolved ^{226}Ra concentrations in the mine effluent, Ralston Creek, and Ralston Reservoir were all less than the limit of 3 pCi/L set by drinking-water regulations, whereas dissolved U in Ralston Creek ranges from 600 to 1,200 µg/L, and is 1,700 µg/L in the mine outflow. The U.S. Environmental Protection Agency is planning to set a "guidance level" of 10 pCi/L (14.7 µg/L) of U in drinking water (5). The mine operator also is studying ways to remove U from the water. Suspended ^{226}Ra and U concentrations were very high, as much as 600 pCi/g for ^{226}Ra and 1,000 µg/g for U in the creek, and 20 pCi/g for ^{226}Ra and 200 µg/g for U in the reservoir. Bottom sediment from the creek bed indicates that concentrations of both radionuclides increase as particle-size decreases, and as distance downstream from the mine decreases. The greatest concentrations of ^{226}Ra and U in the sediment are several tens to hundreds of times greater than the background (upstream from the mine) concentrations. It is believed that the greater than natural radioactivity in the sediment of Ralston Creek is partly due to U-ore particles discharging directly into the creek during the early days of the mine operation and partly due to suspended solids of $Ba(Ra)SO_4$ and fine U-ore from present-day discharges.

Leaching studies indicate that these radionuclides associated with sediment are not in the form of adsorbed ions, but in the form of discrete minerals, or are partly trapped inside the colloidal hydroxides of Fe and Mn coated on the sediment. Further studies by other leaching agents, such as hydroxylamine hydrochloride, and ammonium oxalate, and by X-ray diffraction and scanning-electron microscopy on sediment minerals will reveal the geochemical status of these radionuclides in the sediment.

Literature Cited

1. Parsont, M. A. Ph.D. Thesis, Department of Radiology and Radiation Biology, Colorado State University, Fort Collins, Colorado, 1967.
2. Lammering, M. "Impact of Schwartzwalder Mine on the Water Quality of Ralston Creek, Ralston Reservoir, and Upper Long Lake", Technical Investigation Branch, Surveillance and Analysis Division, U.S. Environmental Protection Agency, Region VIII, SA/TIB-25, 1973, 16 p.
3. Carrasco, M. "Spill Prevention and Containment Plan for Control of Pollutant Discharge Streams at Schwartzwalder Mine", Hazen Research, Inc., HRI Project 4135, 1977, 26 p.
4. "Drinking Water Regulations, Radionuclides, Part II", U.S. Environmental Protection Agency, July, 1976.
5. Lappenbush, W. L. Health Physics, 1980, 39, 1018, P/54.
6. Chung, Y. Ph.D. Thesis, University of California, San Diego, 1971.
7. Thatcher, L. L.; Janzer, V. J.; Edwards, K. W. "Methods for Determination of Radioactive Substances in Water and Fluvial Sediments"; U.S. Geological Survey, TWRI, Book 5, 1977, Chapter A5, 95 p.
8. Masuda, K.; Yamamoto, T. J. Radiat, Res. 1971, 12, 94.
9. Shearer, S. D. Jr.; Lee, G. F. Health Physics, 1964, 10, 217-227.
10. Havlik, B.; Nycova, B.; Grafova, J. Health Physics, 1968 14, 423-430.

11. Landa, E. "Isolation of Uranium Mill Tailings and Their Component Radionuclides from the Biosphere -Some Earth Science Perspectives." U.S. Geological Survey Circular, 1980, 814, 32 p.
12. Szalay, S. Geochim. Cosmochim. Acta, 1964, 28, 1605.

RECEIVED October 28, 1983

17

Uranium Mobility in the Natural Environment

Evidence from Sedimentary Roll-Front Deposits

W. J. DEUTSCH and R. J. SERNE

Pacific Northwest Laboratory, Richland, WA 99352

Roll-front deposits consist of naturally occurring ore-grade uranium in selected sandstone aquifers throughout the world. The geochemical environment of these roll-front deposits is analogous to the environment of a radioactive waste repository containing redox-sensitive elements during its post-thermal period. The ore deposits are formed by a combination of dissolution, complexation, sorption/precipitation, and mineral formation processes. The uranium, leached from the soil by percolating rainwater, complexes with dissolved carbonate and moves in the oxidizing ground water at very low concentration (parts per billion--ppb) levels. The uranium is extracted from the leaching solution by the chemical processes, over long periods of time, at the interfaces between oxidized and reduced sediments. The Eh of the ground water associated with the reduced sediments (Eh = -100 mv to +100 mv) is higher than the Eh expected for most waste repository environments (Eh = -100 mv to -300 mv); this suggests that uranium solids will not be very soluble in the repositories. Data from in-situ leach mining and restoration of roll-front uranium deposits also provide information on the potential mobility of the waste if oxidizing ground water should enter the repository. Uranium solids probably will be initially very soluble in carbonate ground water; however, as reducing conditions are re-established through water/rock interactions, the uranium will reprecipitate and the amount of uranium in solution will again equilibrate with the reduced uranium minerals.

0097–6156/84/0246–0287$06.00/0

During the past ten years, in situ leach mining of uranium deposits has become a commercially successful method of recovering uranium from roll-front deposits that are either too deep or of too low a grade to mine by conventional techniques. The roll-front deposits are formed through a secondary enrichment process, and the mechanism of roll-front emplacement provides us with information on the mobility of uranium in the natural environment. At Pacific Northwest Laboratory (PNL) we are currently conducting research, sponsored by the Nuclear Regulatory Commission (NRC) into methods of minimizing contamination from in-situ leach mining. As part of this work we are studying the geochemistry of uranium in an aquifer environment, and the effect of leach mining on uranium mobility. Also, we are studying the chemical interactions between the leaching solution and the sediments surrounding the leached ore zone. In this paper we discuss the natural occurrence of uranium in an aquifer environment and the effect of in situ leach mining on uranium mobility. We also present the results of our laboratory studies on the interaction of uranium-rich solutions and sediments containing reducing minerals.

The behavior of uranium stored as nuclear waste in a geologic repository will be partially controlled by the geochemical environment of the repository. The effect of ground-water leaching on the waste can be simulated by estimating a ground-water composition and then using an equilibrium thermodynamic computer model to simulate the interaction of the waste with the solution. We present the results of such a modeling study and discuss the effects of solution pH, Eh, and temperature on the expected concentration of uranium in solution.

Uranium in Ground Water Associated with Roll-Front Deposits

Uranium occurs as a trace element in the major rock forming minerals (quartz, feldspar, and mica) and tends to be concentrated in accessory minerals, such as allanite, apatite, monazite, sphene, and zircon. The uranium concentration in most rock types is low. Rogers and Adams (1) have compiled available data on the abundance of uranium in igneous and sedimentary rocks. They show that the average uranium concentration in silicic rocks (granites, rhyolites, and tuffs) is approximately 5 parts per million (ppm), whereas in more mafic igneous rocks the concentration is less than 1 ppm. Uranium concentrations differ with rock type because uranium is segregated into rock types characteristic of the later stages of petrologic evolution. In common sandstones, uranium averages about 1 ppm, whereas in shales the concentration average is on the order of 3 or 4 ppm.

The uranium that is found in roll-front deposits is generally believed to be derived from the dissolution and leaching of host minerals by soil water and ground water (2,3). Typical source rocks for the uranium are granites, tuffs, and tuffaceous sandstones that have relatively high concentrations of uranium in their minerals. Cowart and Osmond (4) state that the uranium found in the south Texas uranium ore zones in the Oakville Sandstone and the Catahoula Tuff came from the leaching of ash (tuffaceous) material associated with the original host sandstones. The source of uranium in the roll-front deposits of Wyoming is not known; however, tuffaceous and granitic material is known to occur upgradient of the deposits in most Wyoming aquifer systems and may have been the host for the original uranium.

When soil water and ground water contact the uranium source material, the minerals will dissolve and leach, thereby releasing uranium into solution. In the oxidizing, carbonate-bearing waters characteristic of ground waters near recharge zones, the uranium will be mobile as U(VI) carbonate complexes [$UO_2(CO_3)_2^{2-}$ and $UO_2(CO_3)_3^{4-}$]. The uranium is transported in oxidizing ground water but is removed from solution by chemical processes (e.g., oxidation-reduction, adsorption, chemical precipitation) that take place at the interface between oxidizing and reducing zones in the aquifer. Figure 1 shows the spatial relation between oxidizing and reducing zones in an aquifer and the presence of a roll-front deposit.

Typical uranium concentrations in ground water in aquifers containing roll-front deposits are listed in Table I. Although the concentrations of uranium found by investigators at different sites diverge widely, as do the values for each site studied, we can frame some generalities from the measurements reported in Table I. The uranium concentration of ground water sampled from the ore zone is typically higher than the concentration in waters sampled up and down the hydrologic gradient from the uranium deposit. The low content of uranium in the upgradient ground water reflects the fact that uranium is a trace constituent in the source rocks. Uranium is removed from ground water at the redox interface because of oxidation-reduction processes that change U(VI) to U(IV) and because of mineral precipitation associated with the low solubility of U(IV) minerals, principally uraninite and coffinite. The chemical system is dynamic and the roll-front deposit can migrate as a result of the ingress of ground water from the oxidized side of the redox interface. Thus, uranium may be dissolved on one side of the deposit and precipitated on the other. This effect would lead to locally high concentrations of uranium in the ground water at the ore zone, which is what is found at the sites listed in Table I. Downgradient of the roll front, reducing conditions exist and the uranium concentration in these waters is generally the lowest of the three regions represented in Table I.

The concentration of uranium in the sediments of roll-front deposits is typically in the 1000 to 2000 ppm range (8). The uranium occurs as coatings on grains and as interstitial material; the predominant uranium mineral is uraninite. Coffinite is often

Table I. Concentration of Uranium in Parts per Billion in Ground Water Associated with Roll-Front Uranium Deposits

Ground-Water Zone	Location				
	Pawnee, Texas	Oakville, Texas	South Texas	Bruni, Texas	Red Desert, Wyoming
Upgradient from Roll Front	0.1 to 61	0.1 to 350	10 to 50	30 to 250	20 to 300
Ore Zone	0.7 to 6.8	0.2 to 32	37 to 317	40 to 760	5 to 2000
Downgradient from Roll Front	0.02 to 0.4	<1	3 to 25	53 to 170	0.14 to 1.9
References	(4)	(5)	(6)	(7)	(4)

reported as an accessory mineral in roll-front deposits. Equilibrium calculations often show a relation between the saturation of the ground water with respect to uraninite (and sometimes coffinite) and the presence of ore-grade material (5,6). Upgradient from the ore zone the oxidizing ground water is undersaturated with respect to uraninite, within and downgradient of the ore zone the ground water computes to be at equilibrium or slightly oversaturated.

Whether ground water is associated with source rocks containing less than 10 ppm uranium, as is the case in the recharge zones of the roll-front aquifers, or if the ground water is from an ore zone containing thousands of ppm of readily accessible uranium material, the concentration of uranium dissolved in ground water rarely exceeds 1 ppm and is often at the low ppb level. Uranium is generally a trace constituent in ground water, either because of its low concentration in the source material or because its concentration in the ground water is limited by relatively insoluble U(IV) minerals. The solubility of uranium minerals is governed by the local environmental condition of the aquifer, and, as is shown in the following sections, a much higher concentration of uranium in the ground water is possible if oxidizing conditions become established where U(IV) minerals are present.

In Situ Leach Mining of Uranium

Roll-front uranium deposits in confined aquifer systems are amenable to extraction by in situ leach techniques. This method of mining was first tested in Wyoming approximately twenty years ago, and sites in Wyoming, Texas, New Mexico, and Colorado are now being mined commercially or are involved in pilot-scale testing (9).

In situ leach mining of uranium involves locating the ore zone, drilling a number of injection and recovery wells within the ore zone, pumping lixiviant (leaching solution) through the ore zone, and extracting the dissolved uranium from the pregnant lixiviant at a surface facility (Figure 2). The lixiviant used at most facilities is ground water that has been fortified with oxygen and carbon dioxide. Oxygen in the lixiviant oxidizes U(IV) in the ore-zone uranium minerals (uraninite and coffinite) to U(VI). Carbon dioxide is added to the lixiviant because it increases the carbonate concentration of the ground water and forms stable dissolved carbonate complexes with U(VI). The carbonate complexation further increases the concentration of uranium that can exist in solution in equilibrium with uranium minerals.

Although pyrite is the predominant mineral containing species with reduced valence states in the roll-front deposits, uraninite appears to be preferentially dissolved by the oxidizing lixiviant (10). Depending on the amount of uraninite present and on competition by other reduced minerals for the available oxygen dissolved in the lixiviant, uranium can reach concentrations in the range of 100 to 200 ppm in the lixiviant. Although the amount of oxygen and carbonate in the lixiviant used for in situ mining is well above that of typical confined aquifers, the potential for mobilization of uranium by the natural ingress of oxidizing, carbonate-bearing waters is shown by the effect that the lixiviant has on the stability of the reduced uranium minerals. Furthermore, the precipitation of an oxidized uranium mineral (such as schoepite or carnotite), if it does occur in the leach mining system, does not significantly limit the concentration of uranium in solution. Consequently, the uranium present in the roll-front deposit is stable and relatively immobile only if reducing conditions are maintained in the aquifer. If conditions become oxidizing the reduced uranium minerals can be expected to dissolve rapidly and uranium will be mobile.

Interaction of Uranium-Rich Solution With Sediments Containing Reduced Minerals

The localized enrichment of uranium in solution in an in situ leach field could be a source of contamination in the aquifer if the pregnant lixiviant migrates from the mining zone. During mining a series of monitoring wells (shown in Figure 2) are sampled to test for unwanted movement of the lixiviant out of the leach field. After mining, the ore zone is restored to a predetermined chemical condition in accordance with regulatory guidelines. The aquifer is restored through induced-restoration techniques and through the natural solution/sediment interactions. We

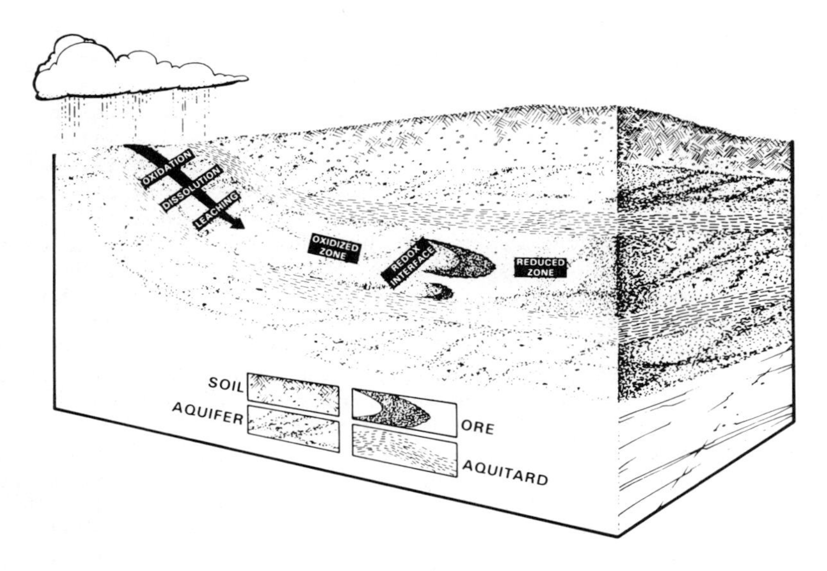

Figure 1. Schematic representation of a roll-front uranium ore deposit.

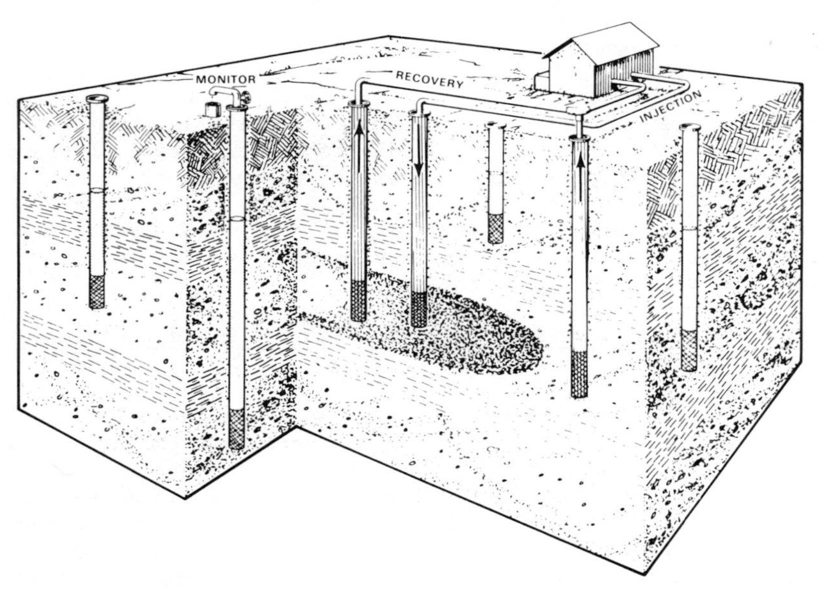

Figure 2. Injection, recovery, and monitoring well pattern for an in situ uranium mine.

have studied the interaction between pregnant lixiviant and sediments containing reduced minerals and have evaluated the mobility of uranium in such an environment.

We obtained samples of pregnant lixiviant containing 52 ppm uranium from an operating in situ leach facility in south Texas. Also, the reduced sediments downgradient from the ore zone were sampled. The dominant mineral in these sediments that contains components in a reduced valence state is pyrite, which makes up a small percentage of the sediment (7). Marcasite is also present, but at a relativey low concentration.

To simulate interaction of the lixiviant and the sediment in the environment of an aquifer, we built a flow-through column test apparatus (Figure 3) that pumped lixiviant through the sediment. We monitored the pH and Eh of the column effluents by means of in-line sampling cells equipped with glass and platinum electrodes, respectively. Effluent samples were collected with a fraction collector. Three separate tests were conducted with columns of identical diameter but different lengths: 11, 22, and 44 cm. The different column lengths allowed us to investigate the effect on solution chemistry of residence time in the column and surface area of minerals contacted by the solution. At the flow rates used in our experiment, the residence times were approximately 1, 2, and 4 days, respectively, for the three columns. The mineral surface area contacted by the solutions is proportional to the column lengths.

Pregnant lixiviant was pumped through the 11- and 44-cm columns for a month and through the 22-cm column for 2 weeks. The uranium concentration in the effluents from these columns was measured by laser fluorimetry and is shown in Figure 4. In this figure we divided the cumulative volume eluted by the pore volume of the respective column to generate the x-coordinate of the plot. This normalizes the results on a pore volume basis. Figure 4 shows that the concentration of uranium in the effluent solution did not reach the influent concentration (52 ppm) at any time during the experiment for any of the columns. Following an initial peak in uranium concentrations for the two shorter columns, the amount of uranium stabilized in the 5 to 10 ppm range as the 2nd to 5th pore volumes were eluted. After 5 pore volumes of effluent were collected from the shortest column, the uranium concentration in the effluent dropped to the 10 to 50 ppb range. In the longest column the uranium concentration in the effluent rises more slowly, but also appears to stabilize in the 5 to 10 ppm range.

We observed from the column data that uranium in solution is not very mobile when the solution contacts the sediments used in the experiment. We expected that the oxidized uranium [U(VI)] in the pregnant lixiviant would be reduced and immobilized by solution/sediment interactions, and this is what happened in the experiments after two to three pore volumes were eluted. The actual removal of uranium from solution may occur by adsorption onto mineral surfaces, which produces localized high concentra-

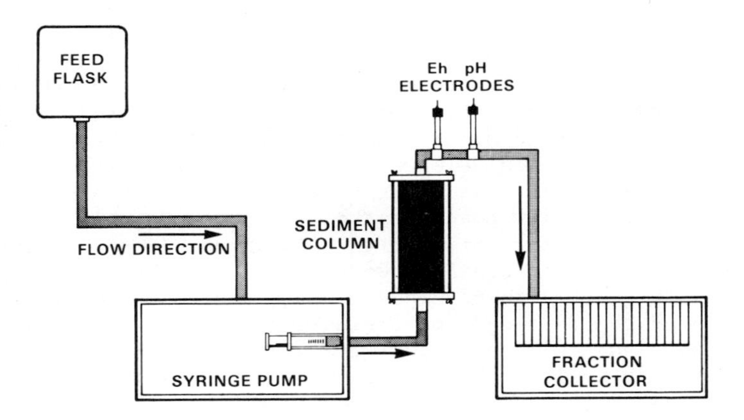

Figure 3. Test apparatus used in column experiments of solution/sediment interactions.

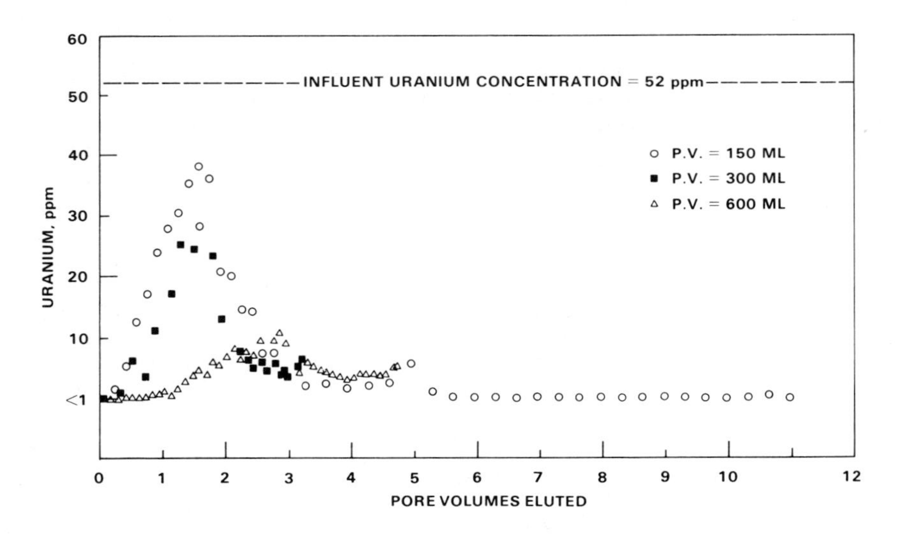

Figure 4. Uranium concentration in effluents from column experiments.

tions of uranium on the substrate and fosters the formation of a U(IV) solid. The initial peak in uranium concentrations for the two shorter columns requires an additional explanation and two hypotheses have been proposed. If the reduction of U(VI) to U(IV) is controlled by bacteria, then the early peak in uranium concentration may be due to the amount of time necessary for the bacteria to acclimate to the lixiviant. An initial lag time is common in bacterial systems. Bacterial mediation may also affect the rate at which the Eh of a solution equilibrates with a new environment. The redox potential data for the columns showed a decrease from +200 mv to -300 mv after 1 to 1.5 days of contact time, independent of residence time in the columns.

A second possible explanation for the uranium peak is that the surfaces of the reducing minerals were initially at least partially blocked from reacting with the lixiviant by a coating on the grains. As lixiviant flowed through the columns, this coating dissolved, exposing the reactive mineral surfaces. The initial effluents from the columns consisted principally of residual pore water retained in the sediment. The initial effluents had sulfate concentrations twice those of later effluents and the high sulfate solution samples were in equilibrium with gypsum ($CaSO_4$), which may have been coating some of the sediment. The lixiviant is well undersaturated with respect to gypsum and rapidly dissolves this mineral on contact. A gypsum coating, if present, probably covered only a certain percentage of the grains, which means that a longer column would have proportionally more grains exposed to the lixiviant as the solution passed through the column. This would explain why the shortest column had the highest uranium concentration peak and the longest column did not show a definite peak. As the initial lixiviant flowed through the longer column it would contact more reactive (uncoated) surfaces than would be the case for the shorter columns; consequently, the reactions that lower the uranium concentration in solution would be enhanced in the longer column. As additional lixiviant flows through the columns the coating is dissolved and a point is reached (after two to three pore volumes of solution are eluted) when enough reactive surfaces are exposed in each column to make their responses to uranium transit in solution identical.

Implications of the Natural Mobility of Uranium on the Storage of Nuclear Waste

Information concerning the effect of in situ mining on the potential concentration of uranium in solution and concerning the experimentally determined effect of solution/sediment interactions on uranium mobility in the aquifer environment can be used to estimate the response of a geologic nuclear waste repository to the ingress of oxidizing water.

The placement of nuclear waste in deep, geologic repositories is currently being studied as a permanent disposal method. The waste may be either in the form of spent fuel, which is predominantly UO_2, or it may be a reprocessed material, which will contain a small percentage uranium. Because uranium will be present in the waste as either a major or minor component, the stability of uranium compounds and the mobility of uranium in the repository environment should be evaluated in the performance assessment of a repository. The most probable scenario that could lead to loss of integrity of the repository is a breach in which ground water leaches or dissolves the waste. We can use a geochemical model of the system to investigate the effect of ground-water chemistry on waste form stability and potential uranium concentrations in the solution.

The computer code with which we modeled the system is MINTEQ. This code was developed at Pacific Northwest Laboratory (11) and basically consists of a combination of the best features of two other geochemical codes: MINEQL (12) and WATEQ3 (13). In this study we used the ion speciation, solubility, and mass transfer subroutines from MINTEQ to calculate the amount of uranium that could be expected in ground water contacting uraninite. The thermodynamic data on uranium solids and solution species in MINTEQ were obtained principally from the compilation of Langmuir (14). These data are being continually revised based on new experimental work (15) and future modeling studies will be enhanced by using an updated data set. In this study, thermodynamic equilibrium was assumed between the solution and the solid and kinetic effects were not considered.

The expected compositions of ground water that might enter geologic repositories in selected rock types are listed in Raines et al. (16). The data for basalt, tuff, and granite ground waters are listed in our Table II, along with the composite water composition used in our modeling study. The dominant cation in all waters is sodium, followed in abundance in most cases by calcium, potassium, and magnesium. Chloride is the dominant anion in basalt and granite ground waters, whereas carbonate dominates in tuff. Silica concentrations are high in the basalt and tuff ground water compared to granite ground water, probably as a result of the presence of soluble glass in the tuff and basalt and less soluble quartz in the granite. The primary components of the ground water that we expect to affect uranium concentration are the carbonate concentration, pH, Eh, and temperature. We have fixed the carbonate concentration at a value of 50 mg/l and varied pH, Eh, and temperature in our modeling study.

The effect of solution pH on the concentration of uranium is shown in Figure 5. To produce the results shown in this figure, the temperature of the model solution was held constant at 25°C and uraninite was equilibrated with the solution at various redox potentials (Ehs). Figure 5 shows that the amount of uranium that

Table II. Composition of Reference and Model Ground Waters (All Concentration Units in ppm)

Constituent	Basalt	Tuff	Granite	Model
Calcium	1.3	14	59	30
Magnesium	0.04	2.1	0.5	1
Sodium	250	51	125	60
Potassium*	1.9	4.9	0.4	3
Chloride	148	7.5	283	100
Sulfate	108	22	19	30
Carbon, as CO_3	46	118	16	50
Fluoride	37	2.2	3.7	5
Iron	-	0.04	0.02	*
Phosphate	-	0.12	-	0.12
Aluminum	-	0.03	-	0.03
Lithium	-	0.05	-	*
Nitrate	-	5.6	-	*
Strontium	-	0.05	-	*
Barium	-	0.003	-	*
Silica, as SiO_2	92	61	11	60
pH	9.7	7.1	9	7-10
Eh, mv	-500	mildly reducing	+170	-400 to +200
Temperature, °C	46	-	7.5	5-50
References	(16,17,18)	(16,19,20)	(16,21)	

\- Not reported
* Not considered

can be maintained in a solution in equilibrium with uraninite is highly dependent on the pH and Eh of the solution. For each unit increase in pH in the low Eh range of the figure, the concentration of uranium in solution increases by an order of magnitude. The reason for the increase in concentration with pH change is the increasing stability of U(IV) hydroxide and U(VI) carbonate complexes at the higher pH's. These results are highly dependent on the U(IV) speciation scheme chosen. We have used Langmuir's compilation of uranium data (14), which assumes a regular progression of stepwise stability constants from $U(OH)^{3+}$ to $U(OH)_5^-$, however there is no experimental evidence as to which U(IV) hydrolysis specie is dominant in the range of this modeling study. Using the estimated thermodynamic data, $U(OH)_5^-$ was the dominant specie, however if $U(OH)_4^\circ$ is dominant there would be no pH dependence. For the higher Eh range shown in Figure 5, the effect on uranium concentration in solution of changing pH at constant Eh is even greater. This occurs because U(VI) becomes dominant over U(IV) as Eh increases and U(VI) forms very stable complexes with carbonate. The redox potential has little effect on uranium concentration at the lower Eh values up to an Eh of -200 mv to -100 mv. At the low Eh range shown in Figure 5, uranium is principally in the +4 valence state, and the amount of uraninite that would dissolve at a constant pH does not vary appreciably until the more oxidized forms of uranium become important in the solution.

Because uraninite is composed of uranium in the lower valence state, as Eh is raised and the higher valence state of uranium begins to dominante over U(IV), uraninite will not be stable. This effect is shown on Figure 5 by the steep increase in the slope of the lines. At Eh's of 0 mv to 100 mv, uranium will dissolve to concentrations of tenths of a mg/l to hundreds of mg/l, depending on the exact pH and Eh. Raising the Eh by adding oxygen to ground water is the primary means of leaching uranium by the in situ technique; results of the modeling support the conclusion that high concentrations of uranium are possible in oxidizing solutions in equilibrium with uraninite.

The solubility of uraninite in a waste repository will also be affected by the temperature of the system at the time that ground water interacts with the waste form. Normal ground-water temperatures at depths for a waste repository range from 5 to 50°C. Using the thermodynamic data of Langmuir (14) we have modeled how this temperature change might affect uraninite solubility and the results are shown in Figure 6. Temperature variations over the range considered do not have as drastic an effect on uraninite solubility as pH changes do. However, throughout most of the Eh range considered, increasing the temperature from 5 to 50°C would raise the amount of uranium possible in solution by an order of magnitude. Above an Eh of approximately 100 mv, which would not be considered highly oxidizing at a pH of 7, temperature appears to have a much greater effect on the dissolution of uraninite. This is caused by the increasing importance of U(VI) at higher Eh values and the fact that U(VI) carbonate complexes are more stable at higher temperatures.

The modeling results shown in Figures 5 and 6 allows us to estimate the response of uraninite to dissolution by ground waters of a given composition and set pH, Eh, and temperature. The thermodynamic data in the computer code, upon which the estimates are based, are subject to revision and the concentrations given in the figures will certainly change. Major revisions in the data may even change the trends displayed in the figures. Using the available data it appears that increasing pH, Eh, or temperature will increase the solubility of uraninite, and at higher Eh values changes in temperature and pH will have an even larger effect on the total amount of uranium that could be expected in a solution in equilibrium with uraninite.

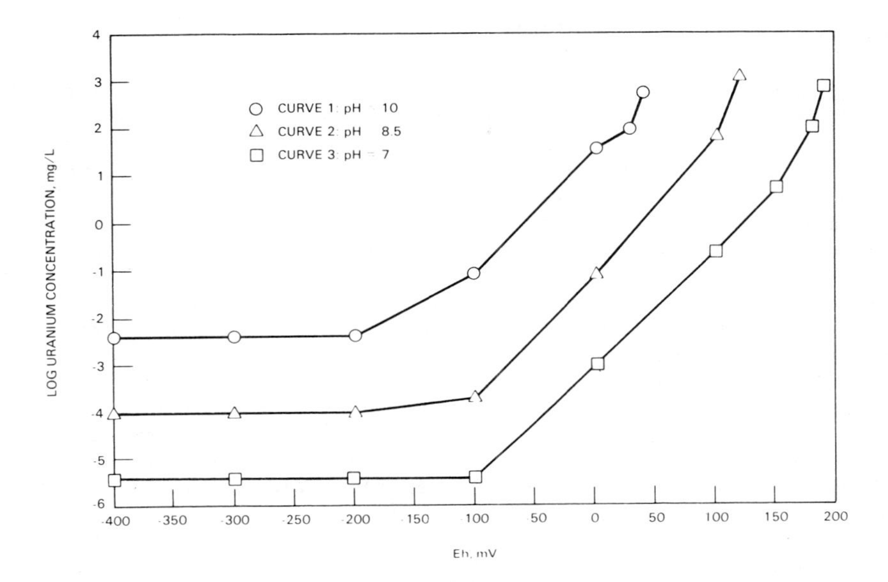

Figure 5. Computed uranium concentration in solutions in equilibrium with uraninite. Temperature equals 25°C.

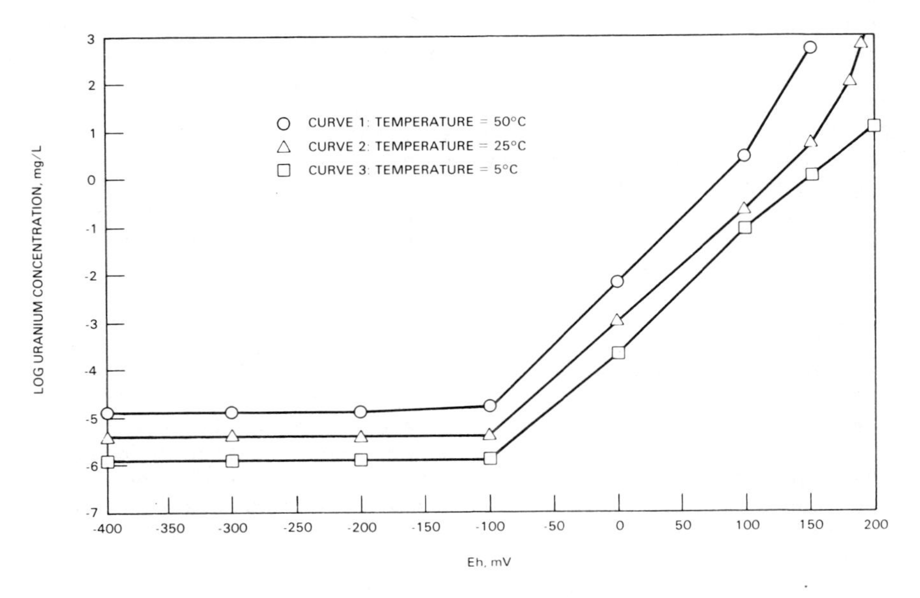

Figure 6. Computed uranium concentration in solutions in equilibrium with uraninite. pH equals 7.

Summary

Uranium is a trace constituent in most ground waters. In typical aquifers with oxidizing ground water, uranium's concentration in solution appears to be limited by its abundance in source rocks. The uranium in ground water is removed from solution and deposited at a redox interface between sediments with reducing minerals and sediments without. In the resultant roll-front deposit, uranium is concentrated in the sediment, and its ground-water concentration remains low because of the low solubility of the uranium minerals that compose the deposit. Consequently, it is possible to have localized high concentrations of uranium in the earth's crust that are stable.

If the environmental condition of the ore zone changes as a result of either natural or manmade causes, the uranium can become very mobile. This is evidenced by the in situ mining process in which relatively minor modifications are made to the ground-water chemistry to produce a solution that can rapidly dissolve uranium and maintain uranium solution concentrations of hundreds of parts per million. We have shown in our laboratory experiments that when this uranium-rich solution contacts aquifer sediments containing minerals capable of reducing uranium from (VI) to (IV), large portions of the uranium are rapidly removed from solution and immobilized.

Using a geochemical model of a ground-water/uraninite system, we analyzed the effect of changing environmental conditions on the stability of the uraninite and estimated the amount of uranium that might be expected in solution. We found that the pH, Eh, and temperature of the ground water will each significantly affect the dissolution of the uraninite. This type of simulation can be useful in assessing the environmental compatibility of various proposed deep geologic repository sites to the waste form. Although a given waste form may dissolve rapidly under certain breach conditions, uranium may not be very mobile in a deep aquifer where reducing conditions are re-established after the ground water moves away from the repository.

Acknowledgments

This work was sponsored by the Nuclear Regulatory Commission, Office of Nuclear Research. It was conducted at the Pacific Northwest Laboratory, which is operated by Battelle Memorial Institute for the Department of Energy under contract number DE-AC06-76RLO 1830. Additional data and information are given in the report "Aquifer Restoration at In-Situ Leach Uranium Mines - Laboratory Evidence for Natural Restoration Processes," NUREG/CR-3136, PNL-4604, 1983. We appreciate the efforts of Wayne Martin who did much of the laboratory work.

Literature Cited

1. Rogers, J. J. W.; and Adams, J. A. S. "Handbook of Geochemistry"; Wedepohl, K. H., Ed.; Springer-Verlag: New York, 1974; Chap. 92.
2. Harshman, E. N. "Geology and Uranium Deposits, Shirley Basin Area, Wyoming"; Professional Paper 745, U. S. Geological Survey: Washington, D.C., 1972.
3. Galloway, W. E.; and Kaiser, W. R. "Catahoula Formation of the Texas Coastal Plain: Origin, Geochemical Evolution, and Characteristics of Uranium Deposits"; Report of Investigations No. 100, Bureau of Economic Geology; The University of Texas at Austin: Austin, Texas, 1980.
4. Cowart, J. B.; and Osmond, J. K. "Uranium Isotopes in Ground Water as a Prospecting Technique"; Report No. GJBX-119(80); U. S. Department of Energy: Grand Junction Office, Colorado, 1980.
5. Chatham, J. R.; Wanty, R. B.; and Langmuir D. "Groundwater Prospecting for Sandstone-Type Uranium Deposits: The Merits of Mineral-Solution Equilibria Versus Single Element Tracer Methods"; Report No. GJBX-129(81); U. S. Department of Energy: Grand Junction Office, Colorado, 1981.
6. Runnells, J. J.; and Lindberg, R. D. "Hydrogeochemical Exploration for Uranium Ore Deposits: Use of the Computer Model WATEQFC"; J. Geochem. Exp. 1981, 15, 37-50.
7. Deutsch, W. J.; Serne, R. J.; Bell, N. E.; and Martin, W. J. "Aquifer Restoration at In-Situ Leach Uranium Mines: Evidence for Natural Restoration Processes"; Report No. NUREG/CR-3136; U. S. Nuclear Regulatory Commission: Washington, D. C., 1983.
8. De Voto, R. H., "Uranium Geology and Exploration"; Colorado School of Mines: Golden, Colorado, 1978.
9. Larson, W. C. "Uranium In Situ Leach Mining in the United States"; Information Circular 8777; U. S. Dept. of the Interior, Bureau of Mines: Washington, D. C., 1978.
10. Goddard, J. B.; and Brosnahan, D. R. "Rate of Consumption of Dissolved Oxygen During Ammonium Carbonate In Situ Leaching of Uranium"; Mining Engineering 1982 34 (11), 1589-1596.
11. Felmy, A. R.; Girvin, D. C.; and Jenne, E. A. "MINTEQ - A Computer Program for Calculating Aqueous Geochemical Equilibria"; Battelle, Pacific Northwest Laboratories: Richland, Washington (in preparation).
12. Westall, J. C.; Zachary, J. L.; and Morel, F. M. M. "MINEQL A Computer Program for the Calculation of Chemical Equilibrium Composition of Aqueous Systems"; Technical Note No. 18; Dept. of Civil Engineering, M. I. T.: Cambridge, Massachusetts, 1976.
13. Ball, J. W.; Jenne, E. A.; and Cantrell, M. W. "WATEQ3: A Geochemical Model with Uranium Added"; Open File Report No. 81-1183; U. S. Geological Survey: Denver, Colorado, 1981.

14. Langmuir, D. "Uranium Solution Mineral Equilibria at Low Temperatures with Applications to Sedimentary Ore Deposits", Geochim. Cosmochim. Acta. 1978, 42, 547-569.
15. Johnson, L. H.; Shoesmith, D. W.; Lunansky, G. E.; Bailey, M. G.; and Temaine, P. R. "Mechanisms of Leaching and Dissolution of UO_2 Fuel," Nucl. Tech. 1982, 56, 238-253.
16. Raines, G. E.; Rickertsen, L.D.; Claiborne; H. C.; McElroy, J. L.; and Lynch, R. W. "Development of Reference Conditions for Geologic Repositories for Nuclear Waste in the USA", in Proceedings of the Third International Symposium on the Scientific Basis for Nuclear Waste Management, v. 3, Moore, J. G., Ed.; Plenum Press: New York, New York, 1981.
17. Gephart, R. E.; et al. "Hydrologic Studies within the Columbia Plateau, Washington: An Integration of Current Knowledge"; Report No. RHO-BWI-ST-5; Rockwell Hanford Operations: Richland, Washington, 1979.
18. Apps, J.; et al. "Geohydrological Studies for Nuclear Waste Isolation at the Hanford Reservation, v. 2", Report No. LBL-8764; Lawrence Berkeley Laboratory, Univ. California: Berkeley, California, 1979.
19. Winograd, I. J. "Hydrogeology of Ash-Flow Tuff: A Preliminary Statement"; Water Resources Res. 1971, 7(4), 994-1006.
20. Wolfsberg, K.; et al. "Sorption-Desorption Studies on Tuff I. Initial Studies with Samples from the J-13 Drill Site, Jackass Flats, Nevada"; Report LA-7480-MS; Los Alamos Scientific Laboratory: Los Alamos, New Mexico, 1979.
21. Fritz, P.; Barker, J. F.; and Gale, J. E. "Geochemistry and Isotope Hydrology of Groundwater in the Stripa Granite - Results and Preliminary Interpretation"; Technical Information Report No. 12. National Technical Information Service: Springfield, Virginia, 1979.

RECEIVED October 14, 1983

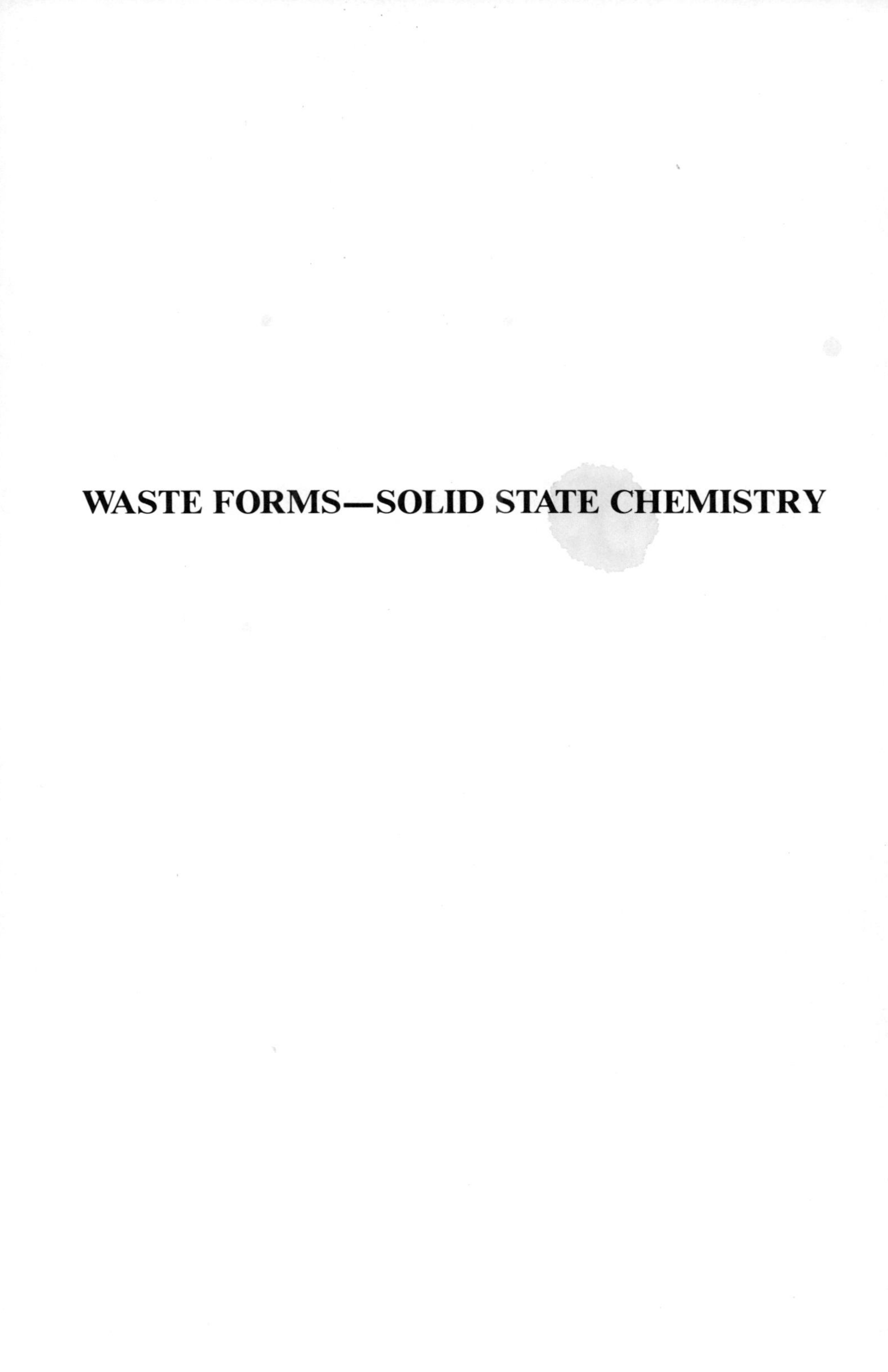

WASTE FORMS—SOLID STATE CHEMISTRY

18

Crystal Chemistry of ABO_4 Compounds

A. T. ALDRED

Argonne National Laboratory, Materials Science and Technology Division,
Argonne, IL 60439

To evaluate the factors affecting the structural stability of some crystalline materials that are potential hosts for radioactive wastes, the crystal structures of a series of $A^{3+}P^{5+}_{1-x}V^{5+}_{x}O_4$ compounds, where A is lanthanum or a member of the rare-earth series, were determined. The end-member phosphates (APO_4) have the monoclinic Monazite structure ($P2_1/n$) for A = La, Ce-Gd, and the tetragonal Zircon structure ($I4_1/amd$) for A = Tb - Lu. The corresponding vanadates have the Monazite structure only for $LaVO_4$, and the Zircon structure for A = Ce - Lu. When the end members are isostructural, e.g., $LaPO_4/LaVO_4$, Monazite, $YbPO_4/YbVO_4$, Zircon, complete solid-solution behavior is observed, and a plot of the unit cell volume against x shows that Végard's Law is followed. When the end members are not isostructural, a systematic change in the solubility range in both structures is found as A is varied, and the data have been systematized in terms of a simple, potentially predictive, structure-field map. The pervasive polymorphism of these ABO_4 compounds, involving both reconstructive and displacive transformations and metastable structures produced by different sample preparation methods, indicates that the crystal structural stability of substituted compounds needs to be carefully evaluated as a function of temperature to assess the structural integrity of waste-form materials.

The nature of bonding and its relationship to structure in actinide-containing materials is a subject of continuing interest. From the perspective of nuclear waste isolation, it is important to understand all the factors affecting the long-term stability of actinide ions in potential host materials, be

0097-6156/84/0246-0305$06.00/0

they amorphous or crystalline. The present work forms part of a systematic study to evaluate factors affecting structural stability, and therefore integrity, of some crystalline materials that are either potential hosts, or may yield insight into possible hosts, for nuclear waste isolation. Data from this work may also be used to ascertain structural changes that may occur when sufficient actinide or fission-product lanthanide ions in the host material have decayed. The particular group of ABO_4 compounds was chosen because the four major structure types that occur, namely; Monazite (monoclinic, $P2_1/n$, prototype $CePO_4$), Zircon (tetragonal, $I4_1/amd$, prototype $ZrSiO_4$), Scheelite (tetragonal, $I4_1/a$, prototype $CaWO_4$), and Fergusonite (monoclinic, I2/c, prototype $YNbO_4$), show a number of structural similarities. They are characterized by strong tetrahedral bonding of oxygen atoms around the B atoms, with the tetrahedra well isolated from one another, and eight-fold (nine-fold for monazite) coordination around the A atoms. The structures will accommodate a range of valences at both A and B sites as long as overall charge balance is maintained. There is extensive polymorphism in compounds having these structures, involving both reconstructive and displacive transformations. The pathways from one structure to another include temperature, pressure, preparation techniques, and small changes in atom size (Figure 1). The existence of polymorphism suggests that the free-energy differences among the structures are small, and therefore, indicates an enhanced possibility that multicomponent substitutions, of the type necessary in a nuclear waste host, may promote phase transformations and structural changes over the long term.

The concept of a structure-field map (1) has proven useful in systematizing the occurrence of different structures among a range of fixed-stoichiometry compounds of the $A_xB_yO_z$ type studied here. A binary phase diagram is constructed in which the axes represent the crystal radii (2) of the A and B ions, r_A and r_B, for the appropriate near-neighbor configuration.

The published (3) structure-field map (SFM) for the $A^{3+}B^{5+}O_4$ compounds does show regular regions of stability for the structures studied here, and it is evident that the rare-earth series provides a fine grid size in terms of variations of r_A. However, the lack of appropriately-sized B^{5+} ions produces wide gaps in the plot. In an attempt to remedy this, we have prepared and characterized a series of substituted compounds of the form $A^{3+}(B_{1-x}B'_x)^{5+}O_4$. If we presume that a compound of this form has a mean B-ion radius ${}^4\bar{r}_B = (1 - x)\ {}^4r_B + x\,{}^4r_{B'}$, then we can produce a more detailed and precisely defined SFM.

To evaluate the concept of a mean ionic radius, we proceed to analyse literature data with respect to A site substitution in $A^{3+}B^{5+}O_4$ compounds. Schwartz (4) has reported that compounds prepared in which A^{3+} is replaced by 0.5 Th^{4+} + 0.5 A^{2+} (A = Ca, Sr, Ba, Cd, Pb) and B^{5+} is P, As, or V have either the Monazite or Zircon structures. If we take a mean radius for this composite A ion $^{8}\bar{r}_A$ (the superscript 8 denotes the near-neighbor configuration) based on the tabulated radii for the tetravalent and divalent ions in eight-fold coordination (2), then Schwartz's results can be added to the SFM. This is shown in the left hand side of Figure 2, and the data do indeed fit the original systematics, apart from $Th_{0.5}Pb_{0.5}VO_4$ and $Th_{0.5}Sr_{0.5}VO_4$ which have the Zircon structure rather than the anticipated Monazite structure. Figure 2 also includes the results of Fonteneau, et al. (5) on the corresponding niobate compounds (B^{5+} = Nb) containing either Th^{4+} or U^{4+} which have the Fergusonite structure, and of Davis et al. (6) on $Ca_{0.5}U_{0.5}PO_4$ which has the Monazite structure. To push the analysis one stage further, the unit cell volumes V of the compounds studied by Schwartz have been calculated, and are plotted as $V^{1/3}$ vs $^{8}r_A$ in the right hand side (rhs) of Figure 2. Again, the data of Fonteneau et al. (5), who produced a similar, but more restricted, plot, are included. The solid lines represent the results of linear least squares of $V^{1/3}$ vs $^{8}r_A$ for the APO_4 (Monazite), APO_4 (Zircon), $AAsO_4$ (Zircon), AVO_4 (Zircon) and $ANbO_4$ (Fergusonite) compounds where A is a rare earth (Ce-Lu). The values for all the series except $AAsO_4$ were determined in this Laboratory (7). The results for $AAsO_4$ were obtained from National Bureau of Standards reports (8). The linear relationship between unit cell volume and the cube of the ionic radius among a series of isostructural compounds has been emphasized by Shannon and Prewitt (2) as a powerful means of systematizing crystallographic results. The data of Schwartz and Fonteneau et al. (rhs Figure 2) are consistent with the unsubstituted $A^{3+}B^{5+}O_4$ results and thus support the concept of a mean radius $\bar{r}_A$ and by analogy $\bar{r}_B$ as a predictor, in combination with the appropriate SFM, of the occurrence of particular structure types.

Experimental

Samples were prepared from starting materials at least 99.9% pure in the form of the normal sesqui- and pentoxides except for $(NH_4)_2HPO_4$. Lanthanum sesquioxide was obtained by fresh decomposition of lanthanum oxalate at 1000°C. Conventional solid-state sintering techniques were used, and care was taken to mix and grind samples under an inert atmosphere (argon) and heat treat them in pure dry oxygen. Repeated firing (in the temperature range 1100-1400°C, as appropriate), grinding and mixing cycles were used until the x-ray diffractometer scans showed no change and no evidence of the starting materials. The sensitivity level of the diffractometer traces, based on

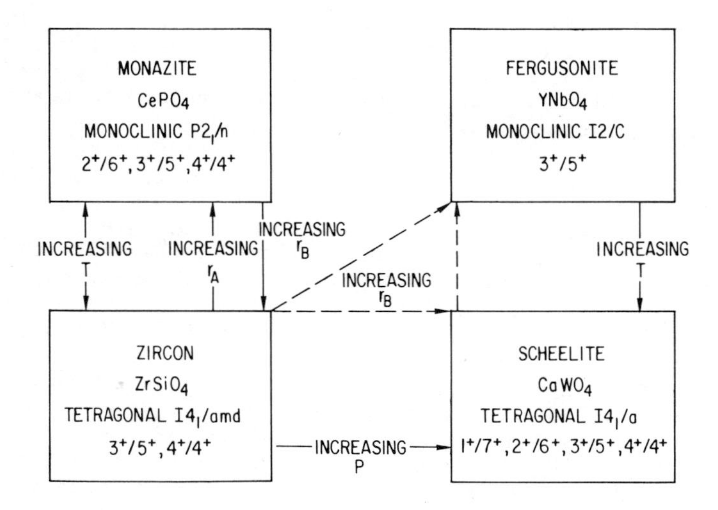

Figure 1. Possible pathways between crystal structures ocurring among $A^{3+}B^{5+}O_4$ compounds.

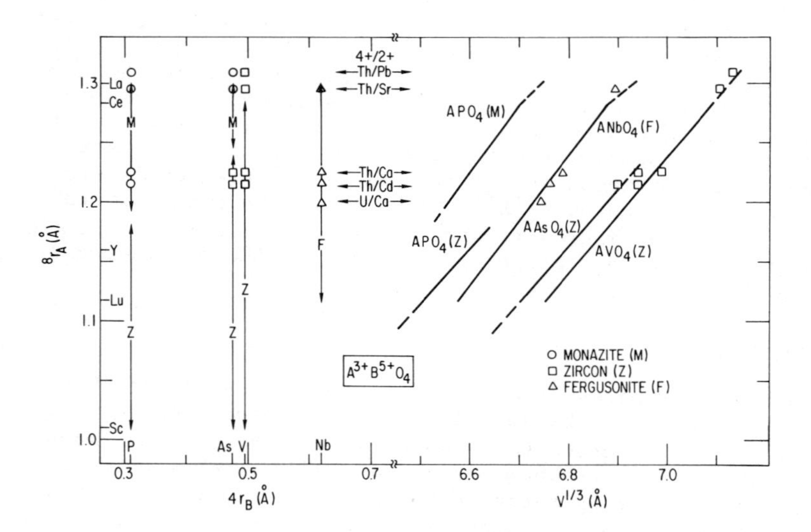

Figure 2. Left hand side: Partial Structure field map for $A^{3+}B^{5+}O_4$ compounds. Right hand side: Corresponding variation of the cube root of the unit cell volume with the Size of the A ion $^{8}r_A$.

relative intensities, is ~1%. Additionally, line widths were required to be comparable to the instrumental resolution and the line widths of standard samples (CeO_2, SiO_2). Samples that meet this criterion should be homogeneous and the substituted ions should therefore, be distributed randomly and uniformly throughout the material. Several samples were examined by scanning electron microscopy, primarily to check compositions by means of the EDX technique. Good agreement was found between nominal and measured compositions, and so nominal compositions are used in the rest of the paper.

Phase identification was done on the basis of both d-spacing and the peak height intensity of all the x-ray lines. These values were compared with values obtained for the end-member (unsubstituted) compounds and also calculated by means of the Lazy-Pulverix computer program (9). Precision lattice parameters were obtained by the Debye-Scherrer method with a 114.6 mm dia. camera and filtered Cr K_{α} radiation; standard least-squares methods were used.

Results

The phase-identification studies of the present work are summarized in Figure 3, which consists of a detailed SFM for the $A^{3+}(P_{1-x}V_x)^{5+}O_4$ series where A is either La or a rare earth. The salient features of this plot are as follows: When the two end-member compounds are isostructural, e.g., $LaPO_4/LaVO_4$ (monazite) or $HoPO_4/HoVO_4$ (zircon), complete solid-solution behavior is observed. For those $AP_{1-x}V_xO_4$ series where the end members are not isostructural, the width of the two-phase field is a systematic function of ^8r_A. At large ^8r_A (Ce, Pr) the two-phase field is narrow and dominated by a monazite solid-solution field. For intermediate ^8r_A (Nd, Sm), the two-phase field is broad with little solubility in either the monazite or zircon structures. At small ^8r_A (Gd), the two-phase field is again narrow with extensive solid solubility in the zircon phase. The relative amounts of both phases in any two-phase sample was estimated on the basis of relative x-ray intensities over the whole pattern. The composition of the phase boundaries in any $AP_{1-x}V_xO_4$ series were then determined by the lever rule. The solid line representing the boundary of the Zircon and Monazite two-phase field in Figure 3 was drawn on this basis. Some results are also included in Figure 3 for the systems $Gd_{1-x}Tb_xPO_4$, and $La_{1-x}Ce_xVO_4$ where the respective end members have the Monazite and Zircon structures. Based on room temperature x-ray results, there is, essentially, no solubility of Tb in the Monazite structure of $GdPO_4$ and no solubility of Gd in the Zircon structure of $TbPO_4$. On the other hand, at least 50% of the La ions in $LaVO_4$ may be replaced by Ce with retention of the Monazite structure and at least 20% of the Ce ions in $CeVO_4$ may be replaced by La ions with retention of the Zircon structure. The two-phase field is thus very narrow.

Some representative lattice constant data will now be presented in graphical form. Figure 4 shows the composition dependence of the four lattice constants of the monoclinic monazite structure for $LaP_{1-x}V_xO_4$ (circles), where complete solid solubility is observed, and for $CeP_{1-x}V_xO_4$ where there is substantial but not complete solid solubility. In the case of $LaP_{1-x}V_xO_4$, all the lattice parameters vary linearly with x for $0 \leqslant x \leqslant 1$ indicating that Végards Law is observed. For $CeP_{1-x}V_xO_4$, the lattice parameters again vary linearly for $0 \leqslant x \leqslant 0.7$ and are parallel to the La-series results. The lattice constants for the x = 0.8 sample, which consists of a major Monazite and a minor Zircon phase, represent values at the phase boundary, determined in the manner described earlier as x = 0.78 and shown by the vertical dashed lines in Figure 4. Transposition of these lattice parameters to x = 0.78 shows that they also fit on the straight line describing the results for lower x. Extrapolation of the results for $CeP_{1-x}V_xO_4$ in Figure 4 to x = 1 yields the following lattice constants for hypothetical (non-equilibrium) $CeVO_4$ with the Monazite structure: a = 7.00 Å, b = 7.24 Å, c = 6.69 Å, β = 105.1°. Yoshimura and Sata (10) reported the existence of a monoclinic form of $CeVO_4$ prepared by oxidation of $CeVO_3$ at temperatures below 400°C. The structure was metastable and decomposed irreversibly (and exothermically) on heating above 400°C. The reported lattice parameters: a = 6.98 Å, b = 7.22 Å, c = 6.76 Å, β = 105.0°, are in good agreement with the extrapolated values listed above.

Figure 5 shows the composition dependence of the lattice constants for $HoP_{1-x}V_xO_4$ and $GdP_{1-x}V_xO_4$ compounds with the tetragonal Zircon structure. Again, the linear variation shows that Végard's Law is obeyed for the Ho-based system where there is complete solubility in the Zircon structure. The linear variation in the Gd-based system parallels that of the Ho system over the solubility range in the Zircon structure. The composition of the phase boundary, determined in the manner described earlier, is at x = 0.30. Extrapolation of the data for Gd $P_{1-x}V_xO_4$ to x = 0.0 yields parameters of a = 6.093 (3) Å and c = 6.970 (3) Å for hypothetical $GdPO_4$ with the Zircon structure.

Discussion

The lattice parameter results for all the $AP_{1-x}V_xO_4$ series studied here are brought together in Figure 6 where $V^{1/3}$ is plotted against concentration (and ${}^4r_B^-$). The linear variation in the complete solid-solution series $LaP_{1-x}V_xO_4$ (Monazite), $HoP_{1-x}V_xO_4$ (Zircon), and $YbP_{1-x}V_xO_4$ (Zircon) again shows that

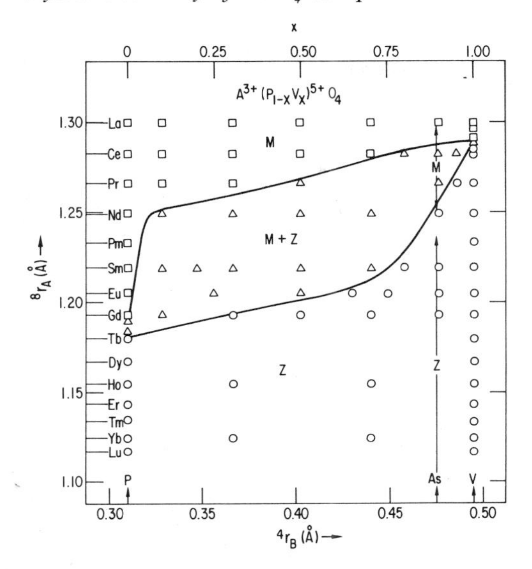

Figure 3. Structure field map for $A^{3+}B^{5+}O_4$ compounds.

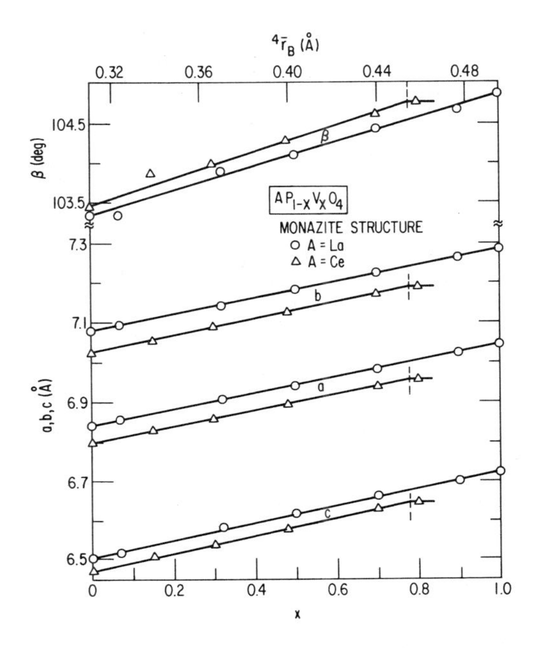

Figure 4. Variation of the monoclinic lattice constants with composition for compounds in the $LaP_{1-x}V_xO_4$ and $CeP_{1-x}V_xO_4$ systems.

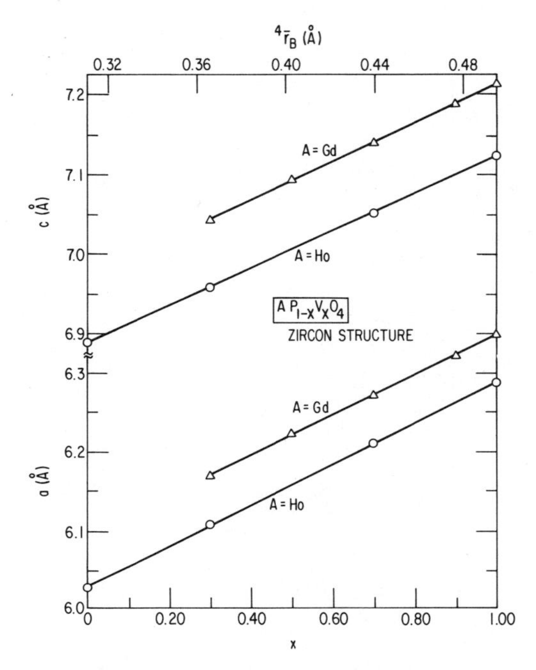

Figure 5. Variation of the tetragonal lattice constants with composition for compounds in the $HoP_{1-x}V_xO_4$ and $GdP_{1-x}V_xO_4$ systems.

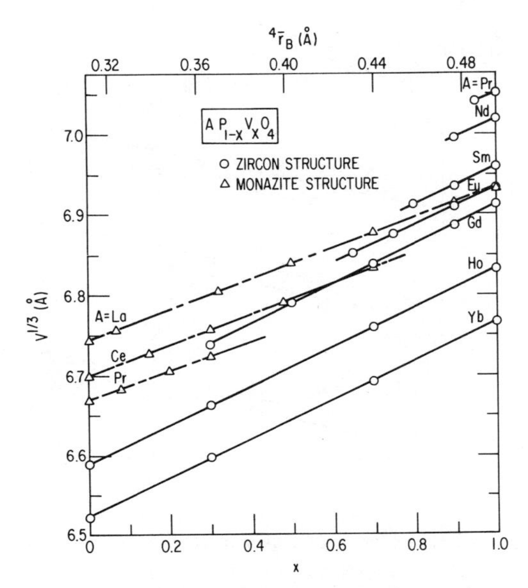

Figure 6. Variation of the cube root of the unit cell volume with composition for the $AP_{1-x}V_xO_4$ series.

Végard's Law is obeyed and that the concept of a mean radius of the B ion is a valid one. It thus provides additional justification for the use of the structure field map as shown in Figure 3 to predict structural existence and stability in a partially-substituted compound.

Even in the cases of partial solid solubility, the composition dependence of $V^{1/3}$ is parallel within a given crystal structure as demonstrated in Table I, which gives the results of linear least-square analysis of the data of Figure 6. The final column of Table I shows the relative change in $V^{1/3}$ with composition determined in this analysis normalized by the relative change in the radius of the B ion $\Delta\,{}^4\bar{r}_B = {}^4r_V - {}^4r_P = 0.185$ Å. For the Monazite structure, this ratio is essentially 1, i.e. the lattice is expanding/contracting, on average, exactly enough to compensate for the difference in size of the two B ions. On the other hand, this ratio is significantly larger (1.3) for the Zircon structure. This dichotomy is paralleled in the response of $V^{1/3}$ to changes in the size of the A ion in the APO_4 and AVO_4 series (7) and reflects a difference between the close-packed low-symmetry Monazite structure and the more open higher-symmetry Zircon structure. In cases where polymorphism between these two structures (Figure 2) is known to occur, e.g. $ThSiO_4$ (11), there is a difference of ~8% in the unit cell volumes of the two structures. This observation may have significance to the problem of accommodation of a range of ions of different sizes into a nuclear waste isolation host. It is likely that transuranic ions will be incorporated into such materials in the tetravalent form. Given the tabulated radii of the tetravalent transuranic ions (2), which are smaller than U^{4+}, substitution of these ions by appropriate valence compensation with a divalent ion would yield a composite A ion small enough to possibly stabilize the Zircon structure as opposed to the Monazite structure in the APO_4 series (Figure 1). A reconstructive transformation involving a significant volume change such as this could have a severe influence on structural integrity.

In summary, we have demonstrated that the concept of a structure-field map is a useful one in systematizing the occurrence of crystal structures in a series of iso-stoichiometric compounds. In addition, the concept of a weighted mean radius of an ion at a particular site in a substituted compound has been found to be a valid one. The use of a SFM to predict structural stability and provide warnings about possible polymorphism (and so structural integrity) in a complex multicomponent substituted system could be a useful tool in assessing potential hosts for nuclear waste isolation.

Table 1. Results of Linear Least-square fits to $V^{1/3} = B + Cx$ for $AP_{1-x}V_xO_4$ Compounds

A	Structure	X	C	$C/\Delta^{4-}r_B$
La	Monazite	0-1.0	0.191(1)	1.03(1)
Ce	"	0-0.78	0.193(2)	1.04(1)
Pr	"	0-0.44	0.181(4)	0.98(2)
Pr	Zircon	0.95-1.0	∿0.2	∿1.1
Nd	"	0.88-1.0	∿0.24	∿1.3
Sm	"	0.765-1.0	0.249(5)	1.35(3)
Eu	"	0.61-1.0	0.234(6)	1.27(3)
Gd	"	0.30-1.0	0.247(2)	1.33(1)
Ho	"	0-1.0	0.244(1)	1.32
Yb	"	0-1.0	0.244(1)	1.32

Acknowledgments

It is a pleasure to thank J. W. Downey for performing the Debye-Scherrer x-ray measurements. This work was supported by the U. S. Department of Energy.

Literature Cited

1. Muller, O.; Roy, R., Crystal Chemistry of Non-Metallic Materials, Vol. 4, "The Major Ternary Structural Families", Springer-Verlag, New York, Heidelberg, Berlin, 1974.
2. R. D. Shannon; C. T. Prewitt Acta. Cryst. 1969, B25, 925.
3. Muller, O.; Roy, R., *op. cit.*, Figure 18 p. 143.
4. Schwarz, H., Z. Anorg. Allgem. Chem. 1965, 334, 175, 261.
5. Fonteneau, G.; L'Helgoualch, H.; Lucas J., Mat. Res. Bull. 1977, 12, 25.
6. Davis. D. D.; Vance, E. R.; McCarthy, G. J., in "Scientific Bases for Nuclear Waste Management", Vol. 3, Moore; J. G., ED., Plenum, New York 1981, p. 197.
7. Aldred, A. T., to be published.
8. NBS Monograph 25, 1964, Section 3, 30-34, 54-56, 1966, Section 4, 17,33,38, 1967, Section 5, 36.
9. Yvon, K.; Jeitschko, W.; Parthe, E. J. Appl. Cryst. 1977, 10, 73.
10. Yoshimura, M.; Sata, T. Bull. Chem. Soc. Japan, 1969, 42, 3195.
11. Taylor, M.; Ewing, R. C. Acta. Cryst. 1978, B34, 1074.

RECEIVED October 24, 1983

19

Transformation Characteristics of $LaV_xNb_{1-x}O_4$ Compounds

M. V. NEVITT and A. T. ALDRED

Argonne National Laboratory, Materials Science and Technology Division,
Argonne, IL 60439

X-ray diffractometry measurements were made as a function of temperature on a series of polycrystalline $LaV_xNb_{1-x}O_4$ compounds ($0 < x \lessapprox 0.3$) to determine the effect of V substitutions on the temperature of the monoclinic/tetragonal transformation. The purpose was to provide basic information relating to crystal lattice stability in ABO_4 compounds that are either candidates or are appropriate models for candidate materials for hosting nuclear waste ions. Partial substitution of V^{5+} on the Nb^{5+} site significantly lowers the tetragonal scheelite ($I4_1/a$) to monoclinic fergusonite ($I2/c$) transformation, from 770 K in $LaNbO_4$ to approximately 215 K for $LaV_{0.25}Nb_{0.75}O_4$ (the solubility limit is close to $x = 0.35$). The transformation is displacive, of second order, involving two coupled order parameters. Heat capacity measurements on $LaV_{0.25}Nb_{0.75}O_4$ showed that the specific heat anomaly at the transformation point is extremely small. It is concluded that the two polymorphic forms of $LaV_xNb_{1-x}O_4$ have very nearly the same free energies over a substantial range of temperature below the transformation.

The need for high chemical and physical stability in the substances that will hold nuclear waste ions in an immobilized state has stimulated extensive materials research. In support of the candidacy of the cation-linked tetrahedral compounds having the monazite, scheelite or zircon crystal structure, studies have sought to determine the extent to which various elements occurring in the waste can occupy stable substitutional positions on the host lattice sites.(1)

These ABO_4 compounds, although they have high thermodynamic stability, have a pronounced tendency toward polymorphism, the rationale for which is incompletely formulated (2, 3). There is

0097-6156/84/0246-0315$06.00/0

only a modest amount of information, principally on $BiVO_4$ and to a lesser extent on $LaNbO_4$, concerning the transformation mechanism and origin (4-7).

The present work, the initial phase of a basic transformation and lattice dynamics study of relevant oxides, focuses on compounds in the system $LaV_xNb_{1-x}O_4$ ($0 < x \lesssim 0.3$). This choice is based on the following factors:

(1) There is considerable existing information on the host compound, $LaNbO_4$; (2) The substitution of V for Nb provides a simple (and hopefully interpretable) cation replacement on the tetrahedral site; (3) Preliminary studies in this Laboratory have shown that the V substitution has a significant effect on transformation temperature. The transformation from the high temperature tetragonal scheelite ($I4_1/a$) structure to the low temperature monoclinic fergusonite ($I2/c$) structure [770 K in pure $LaNbO_4$ (2)] involves two coupled order parameters (a-c) and (β-90) which describe the deformation of the base of the scheelite unit cell from a square to a parallelogram.

Materials Preparation

Polycrystalline samples having densities about 80% of their theoretical values were prepared from 99.9+% pure powders of La_2O_3, V_2O_5 and Nb_2O_5 by repetitive sintering, grinding, mixing and resintering of pellets at 1700 K in oxygen gas at one atm pressure. The La_2O_3 was obtained by decomposition of lanthanum oxalate, and care was maintained to perform all the preparation steps as far as possible under an inert atmosphere. In some cases this procedure was followed by a hot isostatic pressing (HIP) treatment of the sintered and crushed pellets at the same temperature in hermetically sealed capsules in argon gas at 2000 atm pressure. The HIP treatment produced samples with densities within 3% of theoretical values, and with an average grain diameter of approximately 20 μm. As will be discussed, the transformation characteristics of the compounds were influenced by the sample preparation process. Compositions of the conventionally-sintered samples were checked by scanning electron microscopy; the compositions of the HIP-treated samples are nominal.

Experimental Methods

X-ray diffractometry was performed, by methods described in a companion paper (8), between 80 and 600 K at ambient pressure on thin solid wafers cut from the prepared samples. Mounted with high temperature cement on relatively massive metal backing plates, these wafers could be maintained at temperatures constant to within ± 3 K, as measured by either a platinum thermometer or a copper-constantan thermocouple. The widths of the (004) and (008) lines, unsplit by the transformation, provided assurance of the chemical homogeneity of the samples; the full width at half maximum (FWHM) of these reflections did not significantly exceed the FWHM of a standard material and the instrumental resolution, thus indicating that the experimental observations could be associated with discrete, uniform V-ion substitutions [See also (8)]. Lattice parameters were determined by a least-squares procedure based on d-spacings derived from ten to twenty reflections.

The specific heat at constant pressure, c_p, of the HIP-treated sample with nominal composition $LaV_{0.25}Nb_{0.75}O_4$ was measured over the temperature range 4-400 K by the heat pulse method in a calorimeter that incorporates a feedback system to regulate the temperature of concentric radiation shields surrounding the sample (9). The c_p values are accurate to within 1%, as determined by calibration runs using a polycrystalline copper sample and a sapphire single crystal sample.

Results

The upper part of Figure 1 shows the collapse of the monoclinic *a* and *c* lattice parameters to become the tetragonal *a* parameter of $LaV_{0.25}Nb_{0.75}O_4$ as temperature is increased. These data are typical of HIP-treated samples, and the transformation, which occurs at 215 ± 1 K for this composition (see below), is displacive and of higher order than first. Similar results have been reported for $LaNbO_4$ (10). If the transformation can be described in a mean-field model, we would anticipate that the order parameters (*a-c*) and (β-90) would vary as $(T_t-T)^\alpha$ with $\alpha = 0.50$ over some temperature range near T_t. Thus a plot of the square of the order parameters versus temperature would yield straight lines extraplating to zero at T_t. Figure 2, showing typical plots of this type, indicates that the transformation point is strongly dependent on the level of V substitution. Careful, repetitive measurements made with rising and falling temperature, and after varying times at constant temperature, show that the transformation has no observable hysteresis and no measurable induction period. The upper part of Figure 3 shows the transformation temperature as a function of composition. By analogy with the upper part of Figure 1, the lower portion of Figure 3 shows the collapse of the room-temperature monoclinic lattice parameters *a* and *c* as a function of composition in the $LaV_xNb_{1-x}O_4$ system. The transformation should occur at ambient temperature and pressure at the approximate composition, $LaV_{0.22}Nb_{0.78}O_4$.

The samples prepared without the HIP treatment showed substantially the same transformation behavior as the HIP-treated material, with one notable exception: The monoclinic and tetragonal forms were seen to coexist over at least a 100 K temperature range in all vanadium-substituted samples that did

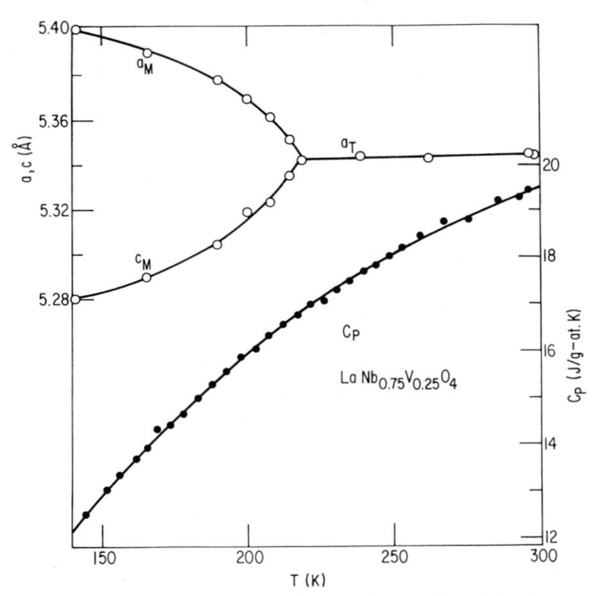

Figure 1. Lattice parameters and specific heat of $LaV_{0.25}Nb_{0.75}O_4$ as a function of temperature.

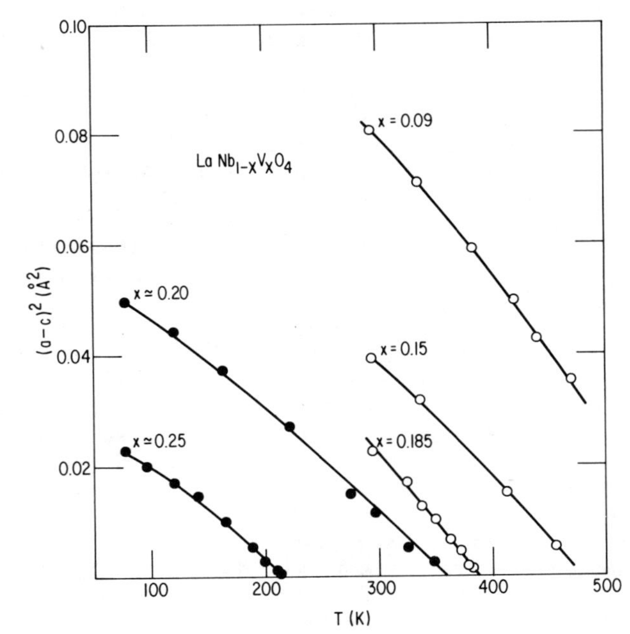

Figure 2. Variation of the square of order parameter, a-c, with temperature for $LaV_xNb_{1-x}O_4$ compounds. Solid circles represent data from HIP-treated samples.

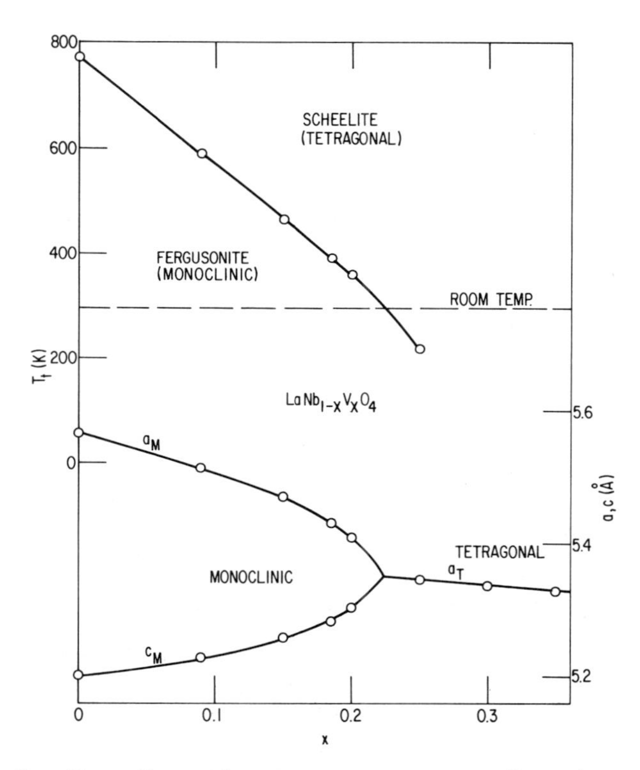

Figure 3. Transformation temperature as a function of composition for $LaV_xNb_{1-x}O_4$ compounds (upper) and variation of room temperature lattice parameters with composition for $LaV_xNb_{1-x}O_4$ compounds.

not receive a post-sinter HIP treatment and also in HIP-treated samples that were subsequently reground and examined in the powdered condition (particle size $\lesssim$ 64 μm). Faint tetragonal reflections could be observed at low temperatures, and these reflections increased in intensity with increasing temperature, merging indistinguishably with the coalescing monoclinic reflections at the transformation point. Both the intensities and the 2θ positions of these tetragonal reflections varied smoothly and reproducibly on cycling through the transformation with no evidence of hysteresis. Additionally, no time-dependent effects on the relative intensities or 2θ positions of the two phases were observed. As there is no evidence, X-ray or otherwise, of chemical inhomogeneity in any of the samples, it is tempting to associate this anomaly with a constrained phase transformation, which has been connected to the metastable retention of tetragonal ZrO_2 at room temperature (11 12). It is postulated that when ZrO_2 grains are sufficiently small, the combined effects of surface energy and specific volume favor the retention of the high temperature tetragonal structure over the monoclinic form. The applicability of this concept to the $LaV_xNb_{1-x}O_4$ transformation is under study in this Laboratory.

Figure 1 shows the temperature dependence of c_p for $LaV_{0.25}Nb_{0.75}O_4$ over a temperature range bracketing the transformation point. There is no discernable latent heat at the transformation point, thus confirming the X-ray observations that the order of the transformation is higher than first. It is also evident that the specific heat anomaly, which accompanies a higher order transformation, is exceedingly small, not greater than 0.05 J/g at. K at 215 K. The absence of a discernable specific heat anomaly at the transformation point, taken together with the observed coexistence of the two phases, indicates that there is little difference between the free energies of the two polymorphic forms over a substantial range of temperature below the transformation.

Discussion

The work reported here shows that the transformation temperature in the $LaV_xNb_{1-x}O_4$ system is significantly affected by the simple, partial substitution of one 5+ ion for another at the center of the tetrahedron. This effect is connected closely to the dynamical characteristics of the crystal lattice. A study involving inelastic neutron scattering, heat capacity and Raman scattering is under way in this Laboratory to clarify these aspects of the problem.

With reference to the candidacy of generically-related crystalline materials to host and immobilize nuclear waste ions, these results prompt several comments regarding stability. Displacive transformations, as observed here, represent relatively small distortional changes in structure that are not likely to produce physically disruptive effects such as would accompany first order, reconstructive transformations involving nucleation and growth of another phase. This is a positive consideration as regards long term physical integrity of the material. On the other hand, the occurrence of a composition-sensitive, and possibly strain-sensitive, displacive transformation, linking polymorphic forms with very small differences in structure and free energy, may reflect conditions favorable to the occurrence of the metamict state. Since these conditions will probably occur in the waste storage environment, where there will be diverse and time-variant ion substitutions, the connection to a tendency toward metamict behavior should be clarified.

Acknowledgements

The authors wish to thank their colleagues, particularly Drs. S.-K. Chan, J. Faber, G. S. Knapp, and D. J. Lam, and Mr. J. W. Downey for valuable advice, scientific help and technical support.

Literature Cited

1. Schutz, B. E.; Freeborn, W. P.; Pepin, J.; White, W. B., in "The Scientific Basis for Nuclear Waste Management"; Topp, S.V., Ed.; North-Holland: New York, 1982; Vol. 6, pp. 155-61.
2. Gingerich, K. A.; Bair, H. E. Advanc. X-ray Anal. 1964, 7, 22-30.
3. Stubican, V. S. J. Am. Ceram. Soc. 1964, 47, 55-8.
4. Pinczuk, A.; Welber, B.; Dacol, F. H. Solid State Comm. 1979, 29, 515-18.
5. David, W. I. F.; Glazer, A. M.; Hewat, A. W. in "Phase Transitions"; Gordon and Breach: London, 1979; Vol. 1, pp. 155-69.
6. Wood, I. G.; Welber, B.; David, W. I. F.; Glazer, A. M. J. Appl. Cryst. 1980, 13,224-9.
7. Wada, M.; Nakayama, Y.; Sawada, A.; Tsunekawa, S.; Ishibashi, Y. J. Phys. Soc. Japan 1979, 47, 1575-80.
8. Aldred, A. T., 1983, this Symposium.
9. Trainor, R. J.; Knapp, G. S.; Brodsky, M. B.; Pokorny, G. J.; Snyder, R. B. Rev. Sci. Instrum. 1975, 46, 95-98.
10. Brixner, L. H.; Whitney, J. F.; Zumsteg, F. C.; Jones, G. A. Mat. Res. Bul. 1977, 12, 17-24.
11. Garvie, R. C.; Hannink, R. H.; Pascoe, R. T. Nature 1975, 258, 703-4.
12. Lange, F. F. J. Mat. Sci. 1982, 17 225-34 (Part 1) and 235-39 (Part 2).

RECEIVED October 20, 1983

20

Stability of Tetravalent Actinides in Perovskites

CLAYTON W. WILLIAMS, LESTER R. MORSS, and IN-KYU CHOI

Argonne National Laboratory, Chemistry Division, Argonne, IL 60439

This paper reports the first determination of the enthalpy of formation of a complex actinide(IV) oxide: ΔH_f° ($BaUO_3$,s, 298 K) = -1690 ± 10 kJ mol^{-1}. The preparation and properties of this and other actinide(IV) complex oxides are described and are compared with other perovskites $BaMO_3$. The relative stabilities of tetravalent and hexavalent uranium in various environments are compared in terms of the oxidation-reduction behavior of uranium in geological nuclear waste storage media; in perovskite, uranium(IV) is very unstable in comparison with uranium(VI).

In simple oxides, the actinides are most stable in the +4 oxidation state; the dioxides, AnO_2, are known for all elements thorium through californium. Although the properties of ThO_2, UO_2, and PuO_2 are especially important in nuclear technology, complex actinide oxides (oxides with one or more metal ions in addition to an actinide) are also important since they may be found as fission products in nuclear fuels and they are models for possible matrices in which nuclear wastes will be stored. It is surprising that so much effort has been devoted to complex actinide oxides that contain actinides in the +6 rather than the +4 oxidation state (1).

Because the preparative conditions of complex actinide oxides indicate that they favor higher actinide oxidation states, we have selected a model system in which both +4 and +6 actinides are easily prepared in similar oxide coordination. In the dioxides each actinide is surrounded by 8 equidistant oxygens at the corners of a cube, but in complex oxides the actinide ion is usually surrounded by 6 oxygens at the apices

0097-6156/84/0246-0323$06.00/0

of a regular or distorted octahedron (2). Our model system is perovskite, the mineral $CaTiO_3$, in which structure type many complex actinide(IV), (V), (VI), and even (VII) oxides crystallize (1,2) (Figure 1).

We have previously determined the enthalpies of formation of several perovskite +4 and +6 oxides (3,4). Our objective in this study was to determine the enthalpy of formation of $BaUO_3$ and to evaluate the relative stability of uranium(IV) and (VI) in comparable complex oxides, especially in comparison with binary oxides, halides, and aqueous ions.

From a practical point of view, these compounds are models of crystalline matrices for nuclear waste disposal. One such storage material is SYNROC, a synthetic mineral whose major constituents are the complex oxides hollandite, zirconolite, and perovskite. We have chosen perovskite as a model structural family because of its efficient packing and its accommodation of a wide range of cations, both in size and oxidation state (5).

Previous Work. The best-known set of actinide(IV) perovskites is $BaMO_3$ (M=Th through Cf) (1,6). The corresponding lanthanide(IV) oxides $BaCeO_3$, $BaPrO_3$, and $BaTbO_3$ are also known and have been well characterized (7). A few compounds $SrMO_3$ (M=Pa, U, Np, Pu) are also known (1). Recently, $EuUO_3$ and $EuNpO_3$ were prepared and characterized as $(Eu^{2+})(U^{4+})(O^{2-})_3$ (8,9).

By far the greatest body of literature exists for $BaUO_3$ (10,11). Very thorough studies of the BaO-UO_2 system revealed a pseudo-cubic $BaUO_3$ phase with a_o = 4.387 Å (12), or a_o = 4.410 Å (13). In general, it was found that $BaUO_3$ could only be prepared by heating BaO with UO_2 at 1200-1900°C in inert gas or hydrogen. Partial oxidation and loss of Ba have been noted, as has the fact that $BaUO_3$ can take up more than 1 mol of BaO in solid solution (10,13).

While our work was in progress, a new report on $BaUO_{3+x}$ appeared (14). Hydrogen reduction of $BaUO_4$ was incomplete even at 1100°C, yielding $BaUO_{3.23}$ or $BaUO_{3.3}$, a_o = 4.40 Å. Reaction of Ba_3UO_6 with UO_2 in hydrogen at 1150-1200°C yielded $BaUO_{2.83}$ or $Ba_{0.98}UO_3$, a_o = 4.39 Å. These pseudocubic x-ray lattice parameters were resolved by powder neutron diffraction into the orthohombic space group Pnma with refined vacancy structures of $Ba_{0.911}U_{0.909}O_3$ and $Ba_{0.976}U_{1.001}O_3$.

The two other most important actinide(IV) complex oxides are $BaThO_3$ and $BaPuO_3$. The former has been prepared many times (15); however, there are disputes concerning its properties. Fava et al. (16) prepared $BaThO_3$ from stoichiometric amounts of BaO and ThO_2 and observed an ideal perovskite, whereas Nakamura (17) found that an excess of BaO is necessary and identified a distorted perovskite structure. It is also not obvious why BaO is not taken up in solid solution with $BaThO_3$ (15); $BaUO_3$ is

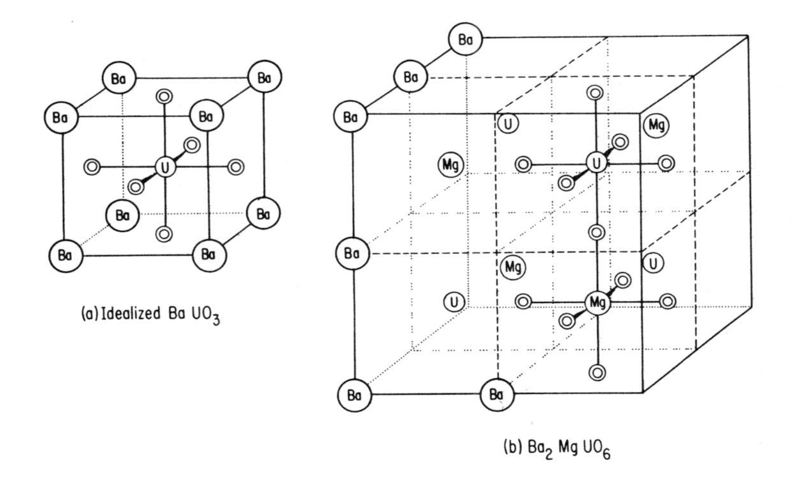

Figure 1. Perovskite structures: (a) simple perovskite, shown as idealized cubic $BaUO_3$. (b) ordered perovskite Ba_2MgUO_6 ($BaMg_{0.5}U_{0.5}O_3$).

reported to form a solid solution with up to 2 moles of BaO (10). Some doubt remains concerning the Ba/Th ratio in these samples, since no analyses were reported and loss of Ba is possible by volatilization during heating or during methanol extraction. $BaPuO_3$ was thoroughly studied by Keller (18) who found it necessary to use a $BaO:PuO_2$ ratio of 3:1 to react all of the PuO_2; excess BaO was then extracted with methanol.

Experimental

We prepared $BaUO_3$ twice, each time beginning with high-purity BaO (CERAC, assay 99.3% by acidimetric titration, found free of $Ba(OH)_2$ or $BaCO_3$ by x-ray powder diffraction) and $UO_{2.00}$ (reduced from high-purity U_3O_8 in H_2 at 1000°C; composition verified gravimetrically by ignition to U_3O_8 at 800°C). The stoichiometric amount of each oxide was weighed in a dry box (<5 ppm H_2O and O_2), mixed thoroughly in an agate mortar, and heated to 1000°C for 24 hours in 99.998% H_2. After repeating the grinding in the dry box and heating in H_2, each sample was handled only in the dry box.

The first sample was analyzed as $Ba_{0.99}UO_{3.20}$ (referred to subsequently as $BaUO_{3.20}$) by separating the cations by ion exchange followed by gravimetric analysis of Ba as $BaSO_4$ and analysis of U using weight-buret modified Davies and Gray titration (19). The oxygen content was determined gravimetrically by ignition of a sample to $BaUO_4$ at 1000°C. The second sample was analyzed similarly as $BaUO_{3.06}$. X-ray powder diffraction Debye-Scherrer films were indexed as simple cubic with $a_o = 4.4155 \pm 0.0005$ Å for $Ba_{0.99}UO_{3.20}$ and $a_o = 4.4007 \pm 0.0020$ Å for $BaUO_{3.06}$. Several weak non-cubic lines were found on the $Ba_{0.99}UO_{3.20}$ film.

Attempted preparation of $BaThO_3$ and $BaPuO_3$ by similar procedures was not as successful. As recommended by Scholder et al. (20), $BaThO_3$ was prepared from a 10% excess of $BaCO_3$ mixed with ThO_2 and fired in nitrogen at 900°C. Although the product dissolved completely in 1M HCl(aq) and was extracted with absolute methanol in a dry box to remove excess BaO, its analysis still revealed excess BaO. Although the x-ray powder diffraction film (cubic, $a_o = 4.4960 \pm 0.0010$ Å) showed no BaO, $BaCO_3$, or ThO_2 lines, the sample was not considered suitable for calorimetry.

The preparation of $BaPuO_3$ paralleled that of $BaUO_3$, beginning with $^{242}PuO_2$ ignited at 800°C, a sufficiently low temperature that it would not be refractory. Although the x-ray powder diffraction film (cubic $a_o = 4.3839 \pm 0.0004$ Å) showed no PuO_2 lines, the sample dissolved slowly in 1M HCl(aq) or $HClO_4$(aq). Radiometric (alpha assay) and spectrophotometric analyses indicated only 90% of theoretical plutonium in solution after 30 minutes in dilute acid, with some insoluble residue (perhaps amorphous PuO_2) remaining after several days. These results were consistent with those of Keller (18).

Part of the $BaPuO_3$ was oxidized in flowing O_2 in a Mettler TGA-2 apparatus. Weight gain was complete at 700°C but corresponded only to $BaPuO_{3.5}$. The product showed an x-ray diffraction pattern of a mixture of $BaPuO_3$, PuO_2, and Ba_3PuO_6.

Solution calorimetry of $BaUO_{3+x}$ was performed in an 880 cm^3 isoperibol calorimeter described by Nocera *et al.* (21). Samples of $BaUO_{3+x}$ were weighed on a Cahn GRAM electrobalance in the dry box and reacted sequentially in the calorimeter in 1*M* HCl(aq).

Results

Enthalpies of solution were measured separately for $BaUO_{3.20}$ and $BaUO_{3.06}$ samples. The data are shown in Table I. The thermochemical cycles that lead to the enthalpies of formation of these nonstoichiometric compounds are shown in Table II.

In view of the difficulty in approaching the stoichiometric composition of $BaUO_3$, it is necessary to estimate its enthalpy of formation by extrapolation. It is well known that the partial molal thermodynamic properties of solid solutions vary smoothly as a function of nonstoichiometric parameters (22) so that, by integration, the integral thermodynamic properties of corresponding nonstoichiometric compounds also vary smoothly. It is not as obvious that the enthalpies of formation of a series of compounds such as AB_x and ABC_x also vary smoothly as a function of x, even when the compounds pass through structure changes. Such a series of compounds, which is relevant to this study, is UO_{2+x}, shown at the top of Figure 2. Similar smooth variation of $\Delta H_f^\circ(BaUO_{3+x})$ is seen in the lower curve of Figure 2; two of the data points are from Table II and that for $BaUO_4$ ($\Delta H_f^\circ = -1997.1 \pm 2.1$ kJ mol^{-1}) is from O'Hare *et al.* (23). By extrapolating the lower curve of Figure 2 to $x = 0$, we estimate $\Delta H_f^\circ(BaUO_{3.00}) = -1690 \pm 10$ kJ mol^{-1}.

Discussion

It is interesting to note the differences in slopes of the two curves of Figure 2. The slope of the $BaUO_{3+x}$ data is approximately twice that of the UO_{2+x} data; this fact reflects the enhanced stability of hexavalent compared to that of tetravalent uranium in complex oxides versus simple oxides. Ackermann and Chandrasekharaiah (24) calculated similar data for UO_{2-x}, which plot on a steeper curve than the $\Delta H_f^\circ(UO_{2+x})$ because UO_{2-x} is a cation-vacancy U(IV) compound rather than a mixed-valence U(IV)-U(VI) oxide.

Table I. Enthalpy of Solution of $BaUO_{3+x}$ in 880 cm^3 of 1M HCl(aq) at 298.15 K

Composition	Mass/mg	$\frac{\Delta H^a}{J}$	$\frac{\Delta H_{soln}}{kJ\ mol^{-1}}$
$BaUO_{3.20}$	92.46	-51.62	-238.2
$BaUO_{3.20}$	73.97	-41.37	-238.6
$BaUO_{3.06}$	74.32	-48.70	-274.3
$BaUO_{3.06}$	69.10	-45.45	-278.5
$BaUO_{3.06}$	83.48	-55.39	-280.9
$BaUO_{3.06}$	68.04	-44.36	-276.0
	Ave. ΔH_{soln} ($BaUO_{3.20}$)		-238.4 ± 5.0^b
	Ave. ΔH_{soln} ($BaUO_{3.06}$)		-278.8 ± 2.9^b

a. Corrected for bulb breaking and evaporation of solvent into bulb gas.
b. 95% confidence.

Table II. Thermochemical Cycles for $\Delta H_f^\circ(BaUO_{3+x})$

Reaction	ΔH_f°(kJ mol^{-1})
$BaUO_{3+x}(c) + (6-2x)HCl(1\underline{M}) = BaCl_2$ (in 1$\underline{M}$ HCl) + $(1-x)UCl_4$ (in 1$\underline{M}$ HCl) + xUO_2Cl_2 (in 1$\underline{M}$ HCl) + $(3-x)$ H_2O (in 1$\underline{M}$ HCl)	$\Delta H(soln, BaUO_{3+x})$[a]
$Ba(c) + Cl_2(g) = BaCl_2$ (in 1$\underline{M}$ HCl)	$\Delta H_1 = -864.1 \pm 1.7$[b]
$(1-x)[U(c) + 2Cl_2(g) = UCl_4$ (in 1$\underline{M}$ HCl)]	$\Delta H_2 = (1-x)(-1247.3 \pm 2.5)$[c]
$x[U(c) + O_2(g) + Cl_2(g) = UO_2Cl_2$ (in 1$\underline{M}$ HCl)]	$\Delta H_3 = x\ (-1345.2 \pm 1.3)$[c]
$(3-x)[H_2(g) + 1/2\ O_2(g) = H_2O$ (in 1$\underline{M}$ HCl)]	$\Delta H_4 = (3-x)(-285.85 \pm 0.04)$[d]
$(6-2x)[1/2\ H_2(g) + 1/2\ Cl_2(g) = HCl\ (1\underline{M})]$	$\Delta H_5 = (6-2x)(-164.36 \pm 0.04)$[d]

For $BaUO_{3.06}$:

$$\Delta H_f^\circ = -(-278.8 \pm 2.9) + (-864.1 \pm 1.7) + 0.94(-1247.3 \pm 2.5) + 0.06(-1345.2 \pm 1.3) + 2.94(-285.85 \pm 0.04) - 5.88(-164.36 \pm 0.04) = (-1712.4 \pm 4.1)\ \text{kJ mol}^{-1}$$

For $BaUO_{3.20}$:

$$\Delta H_f^\circ = -(-238.4 \pm 5.0) + (-864.1 \pm 1.7) + 0.8(-1247.3 \pm 2.5) + 0.2(-1345.2 \pm 1.3) + 2.8(-285.85 \pm 0.04) - 5.6\ (-164.36 + 0.04) = (-1772.5 \pm 5.7)\ \text{kJ mol}^{-1}$$

a. This research, Table I.
b. Morss, L. R., Williams, C. W. J. Chem. Thermodynamics 1983, 15, 279-285.
c. Parker, V. B. NBSIR 80-2029, U. S. Government Printing Office, Washington, D.C., 1980.
d. Parker, V. B., Wagman, D. D., Garvin, D., NBSIR 75-968, U. S. Government Printing Office, Washington, D.C. 1976 (Datum for H_2O includes a small correction for partial molal enthalpy in 1$\underline{M}$ HCl).

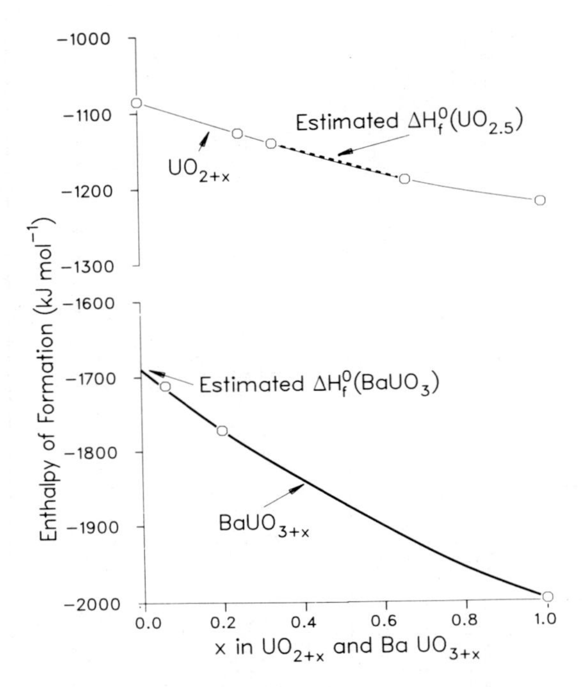

Figure 2. Enthalpies of formation of UO_{2+x} and $BaUO_{3+x}$.

In Table III are shown ΔG° data for several U(IV-U(VI) reactions, each of which is normalized by including 1/2 $O_2(g)$ as the oxidant. The first reaction is useful as a reference, paralleling the two extremes on the oxygen ΔH curve of Figure 2. The next three reactions are oxyhalides and are slightly more favorable. The fifth reaction is that of standard-state aqueous ions and includes acid-base effects [weak acid $U^{4+}(aq)$ on left and strong acid $H^{+}(aq)$ on right]. Nevertheless, it is more favorable for the oxidation U(IV)-U(VI) than the reactions above, presumably because of the formation of the strongly covalently bonded UO_2^{2+} ion. Parallel stabilizations are found for NpO_2^{2+}, PuO_2^{2+}, and AmO_2^{2+}, where the M(VI) dioxo cations are relatively stable. The last two reactions show the stability of U(VI) in complex oxides. Each of these reactions takes U(IV) in octahedral (sixfold) coordination to U(VI) in sixfold coordination; the very favorable free energy changes illustrate how strongly hexavalent actinides are stabilized in complex oxides.

Table III. ΔG°_{298K} for U(IV)-U(VI) Reactions[a]

Reaction	ΔG°(298K)/(kJ mol^{-1})
$UO_2(c) + 1/2\ O_2(g) \rightarrow UO_3(c)$	-114
$UOF_2(c) + 1/2\ O_2(g) = UO_2F_2(c)$	-123
$UOCl_2(c) + 1/2\ O_2(g) = UO_2Cl_2(c)$	-150
$UOBr_2(c) + 1/2\ O_2(g) = UO\ Br_2(c)$	-137
$H_2O(\ell) + U^{4+}(aq) + 1/2\ O_2(g) = UO_2^{2+}(aq) + 2H^{+}(aq)$	-185
$BaUO_3(c) + 1/2\ O_2(g) = BaUO_4(c)$	-278
$BaUO_3(c) + 2\ BaO(s) + 1/2\ O_2(g) = Ba_3UO_6(c)$	-400

(a) Thermodynamic data from Wagman, D. D. et al., "Selected Values of Chemical Thermodynamic Properties," NBS Technical Notes 270-3, 270-6, and 270-8, U. S. Government Printing Office, Washington, D.C., 1981. Entropies of $BaUO_3$ and Ba_3UO_6 estimated as sums of corresponding binary oxide enthalpies.

It is easy to show that $BaUO_3$ is a very strong reductant; using estimated S°($BaUO_3$), we calculate ΔG°(298 K) = -50 kJ for the reaction

$$BaUO_3(c) + H_2O(g) = BaUO_4(c) + H_2(g) \qquad (1)$$

so that $BaUO_3$ is a significantly stronger reductant than hydrogen. This calculation confirms the experimental observation (14) that hydrogen will not reduce $BaUO_4$ to $BaUO_3$.

A useful parameter by which the stability of perovskites can be judged is the Goldschmidt tolerance factor t (25). In a study of the thermochemistry of the lanthanide(IV) perovskites

Ba(Ce,Pr,Tb)O_3, Morss and Mensi (7) showed that the tolerance factor correlates roughly with ΔH(complex), i.e., ΔH for the reaction

$$BaO(c) + MO_2(c) = BaMO_3(c) \quad (2)$$

Table IV is an expanded table of the sort used by Morss and Mensi, showing that $BaUO_3$ behaves, as expected, like $BaCeO_3$ with respect to reaction (2). Because of the importance of other actinide(IV) ions in sixfold coordination, we have included estimates of ΔH(complex) for $BaThO_3$ and $BaPuO_3$. $BaThO_3$ is clearly a compound of marginal stability; $BaPuO_3$ is much more stable but difficult to synthesize free of PuO_2 unless excess BaO is used. The estimates in Table IV are useful because it may be very difficult to prepare $BaThO_3$ and $BaPuO_3$ in sufficiently high purity for thermochemical measurements.

TABLE IV. Perovskites: Structural and Thermodynamic Parameters

			$MO_2(s) + BaO(s) = BaMO_3(s)$	
Compound	IR(B^{4+}) Å	t	Δ(Molar Vol.) cm^3/mol	ΔH(complex) kJ/mol
$BaTiO_3$	0.605	0.97	-5.8	-163
$BaMoO_3$	0.650	0.95	-5.4	-92[a]
$BaHfO_3$	0.71	0.92	-3.5	-134
$BaZrO_3$	0.72	0.92	-2.4	-126
$BaTbO_3$	0.76	0.90	+0.5	-88
$BaPrO_3$	0.85	0.864	+0.5	-147
$BaPuO_3$	0.86	0.860	+0.5	(-80)[b]
$BaCeO_3$	0.87	0.856	+1.7	-52
$BaUO_3$	0.89	0.849	+0.7	-57
$BaThO_3$	0.94	0.831	+2.3	(-20)[b]

a. Unusually small ΔH(complex) since MoO_2 is stabilized by Mo-Mo bonds.
b. Estimated.

Finally, we wish to point out that our results are significant in the context of geochemical behavior of uranium in oxide hosts. For example, Schreiber *et al.* (26) found that U(IV), U(V), and U(VI) coexist in Ca-Mg-Al-silicate glasses, with high-calcium glass favoring the higher oxidation states. Pepin *et al.* (27) argue that U(VI) and interstitial oxygen are favored in fluorite solid solutions such as $(U,Th)O_{2+x}$ with unit cells larger than that of UO_2. It is well known that UO_2^{2+} is much more readily leached than U^{4+}; Wang and Katayama (28) showed that the dissolution of UO_2 and spent nuclear fuel in water, bicarbonate, and brine is initated by oxidation of U^{4+} in the solids to $UO_2^{2+}(aq)$. Our study shows that perovskite is a weak stabilizing agent for U(IV) and a very strong stabilizing agent for U(VI); we have presented quantitative data to show how much more stable is U(VI) than U(IV) in perovskite. We do not wish to imply that if $BaUO_3$ or a perovskite containing U(IV) is present in a geochemical environment it is converted to $BaUO_4$ by a reaction such as reaction (1). Oxidation of U(IV) by dissolved O_2 or a weaker aqueous oxidant would probably be followed by complexation or dissolution of the resulting UO_2^{2+}.

Most kinetic studies (e.g., 29) and a thermodynamic study (30) of nuclear waste host dissolution focus on heterogeneous reactions with major concern for temperature, pH, and complexation. We believe that strongly reducing conditions are necessary to inhibit the undesirable U(IV)-U(VI) oxidation in nuclear waste matrices. Therefore, nuclear waste matrices should incorporate reductants or oxidation-reduction Eh buffers to maintain very low oxygen partial pressure, and leach studies should be conducted under oxidation-reduction conditions that nearly match repository conditions.

Acknowledgements

This research was performed under the auspices of the Office of Basic Energy Sciences, Division of Chemical Sciences, U. S. Department of Energy under contract number W-31-109-ENG-38.

Literature Cited

1. Keller, C., in "MTP Review of Science, Inorg. Chem.", Series 1, Vol 5; Bagnall, K. W., Ed.; Butterworths: London, 1972; pp. 47-85.
2. Morss, L. R., in "Actinides in Perspective"; Edelstein, N., Ed.; Pergamon: New York, 1982; pp. 381-407.
3. Morss, L. R.; Fuger, J.; Jenkins, H. D. B. *J. Chem. Thermodyn.* 1982, *14*, 377-384.
4. Morss, L. R.; Choi, I.-K.; Gens, R.; Fuger, J. *J. Chem. Thermodyn.* 1983, *15*, in press.
5. Ringwood, T. *Amer. Scientist* 1982, *70*, 201-207.
6. Haire, R. G. *Proc. 10th Journee des Actinides* 1980, p. *19*.
7. Morss, L. R.; Mensi, N., in "The Rare Earths in Modern Science and Technology"; McCarthy, G. J., Silber, H. B., Rhyne, J. J., Eds.; Plenum Press: New York, 1982, pp. 279-282, and references therein.
8. Greedan, J.; McCarthy, G. R. *Inorg. Chem.* 1975, *14*, 772-775.
9. Berndt, U.; Tanamas, R.; Keller, C. *J. Solid State Chem.* 1976, *17*, 113-120.

10. Gmelin Handbuch der Anorg. Chem, "Uran", Erg.-Bd. C3, 1975, pp. 120-121.
11. "High Temperature Reactions of Uranium Dioxide with Various Metal Oxides", U.S. Natl. Bur. Stand. Circular 568, 1956, p. 13.
12. Trzebiatowski, W.; Jablonski, A. *Nukleonika* 1960, *5*, 499-507.
13. Voronov, N. M.; Sofronova, R. M., in "Physical Chemistry of Alloys and Refractory Compounds of Thorium and Uranium"; O. S. Ivanov, Ed., Akad. Nauk USSR (English translation, Jerusalem) 1972; pp. 204-214.
14. Barrett, S. A.; Jacobson, A. J.; Tofield, B. C.; Fender, B. E. F. *Acta Crystallogr.* 1982, *B38*, 2775-2781.
15. Gmelin Handbuch der Anorg. Chem., "Thorium" Erg.-Bd. C2, 1976, pp. 14-15.
16. Fava, J.; LeFlem, G.; Devalette, M.; Rabardel, L.; Coutures, J.-P.; Foex, M.; Hagenmuller, P. *Rev. Internat. Hautes Temp. et Refract.* 1971, *8*, 305-310.
17. Nakamura, T., *Chem. Lett.* 1974, 429-434.
18. Keller, C. *Nukleonik* 1962, *4*, 271-277.
19. U. S. Atomic Energy Commission Report NBL-289, New Brunswick Laboratory.
20. Scholder, R.; Räde, D.; Schwarz, H. Z. *Anorg. Allg. Chem.* 1968, *362* 149-168.
21. Nocera, D. G.; Morss, L. R.; Fahey, J. A. *J. Inorg. Nucl. Chem.* 1980, *42*, 55-59.
22. Sørensen, O. T. in "Nonstoichiometric Oxides", Sørensen, O. T., Ed.; Academic: New York, 1981; Chap. 1.
23. O'Hare, P. A. G.; Flotow, H. E.; Hoekstra, H. R. *J. Chem. Thermodynamics* 1980, *12*, 1003-1008.
24. Ackermann, R. J.; Chandrasekharaiah, M. S. "Thermodynamics of Nuclear Materials," Proc. Symp. 1974, Vol. II, IAEA: Vienna, 1975, pp. 3-26.
25. Goodenough, J. B.; Longo, J. M., in "Landolt-Börnstein, New Series", Group III, Vol. 4a, Springer-Verlag: Berlin, 1970; pp. 126-314.
26. Schreiber, H. D.; Balazs, G. B.; Williams, B. J.; Andrews, S. M., in "Scientific Basis of Nuclear Waste Management," Vol. 3, Plenum, N.Y.: 1981, pp. 109-114.
27. Pepin, J. G.; Vance, E. R.; McCarthy, G. J.; Rusin, J. M., in *Proc. Symp. High Temp. Materials Chem.*; Cubicciotti, D. D., Hildebrand, D. L., Eds., Electrochem. Soc.: 1982, pp. 176-185.
28. Wang, R.; Katayama, Y. B., *Nucl. and Chem. Waste Management*, 1982, *3*, 83-90.
29. Ringwood, A. E.; Oversby, V. M.; Kesson, S. E.; Sinclair, W.; Ware, N.; Hibberson, W.; Major, A., *Nucl. and Chem. Waste Management* 1981, *2*, 287-305.
30. Nesbitt, H. W.; Bancroft, G. M.; Fyfe, W. S.; Karkhanis, S. N.; Nishijima, A.; Shin, S. *Nature*, 1981, *289*, 358-362.

RECEIVED October 20, 1983

21

Chemical and Physical Consequences of α and β^- Decay in the Solid State

J. P. YOUNG and R. G. HAIRE

Oak Ridge National Laboratory, Transuranium Research Laboratory, Oak Ridge, TN 37830

J. R. PETERSON

University of Tennessee, Department of Chemistry, Knoxville, TN 37996-1600

D. D. ENSOR

Tennessee Technological University, Department of Chemistry, Cookeville, TN 38501

Interesting chemical and structural phenomena can occur when radioactive materials are stored in the solid state. Extensive studies have been made of both the chemical and physical status of progeny species that result from the α or β^- decay of actinide ions in several different compounds. The samples have been both initially pure actinide compounds--halides, oxides, etc.--and actinides incorporated into other non-radioactive host materials, for example lanthanide halides. In general, the oxidation state of the actinide progeny is controlled by the oxidation state of its parent (a result of heredity). The structure of the progeny compound seems to be controlled by its host (a result of environment). These conclusions are drawn from solid state absorption spectral studies, and where possible, from x-ray diffraction studies of multi-microgram sized samples.

The geochemical behavior of disposed radioactive wastes is controlled by a number of different physical and chemical phenomena. Some changes are caused by thermal processes which influence chemistry of both the stored material and its host. These thermal effects can also influence the physical characteristics of the stored material. Laboratory and field studies are underway in many scientific and engineering facilities to understand these effects.

The transmutation of one elemental species into another by radioactive decay is a phenomenom that will certainly influence the behavior of stored radioactive waste, but transmutation has received little attention in this context. Although the cause can

0097-6156/84/0246-0335$06.00/0

be considered a physical effect, it can provoke both chemical and structural changes. The important question is, in what way will this transmutation affect the stored material and its host? Too often transmutation is considered simply as the physical change of one element to another. The real process is, however, the change of one chemical species into another chemical species. What will be the resultant oxidation state? How will the daughter or progeny species achieve this oxidation state? What structure will the resultant species exhibit? The answers to these questions will be required to demonstrate geochemical stability of all possible storage forms and to identify those forms most suitable for storage. Note that these are questions to be answered for chemical species in the solid state. Numerous studies have been carried out in which the answers to questions such as these are based on the results of chemical analyses of solids dissolved in some solvent system prior to analysis. A survey of such studies has been published (1). The results of these studies were probably influenced by the dissolution treatment and are probably not representative of effects that actually occurred in the solid state. Our studies to date do not answer all the questions posed above. A start has been made, however, with some very interesting investigations that deal with the effect of heredity and environment on the chemical and physical consequences of α and β^- decay in the solid state.

A suitable radioactive decay series in which both α and β^- emission occur within a reasonable time is found in the heavier actinides where:

$$^{253}\text{Es} \xrightarrow[20.5\text{d}]{\alpha} {}^{249}\text{Bk} \xrightarrow[314\text{d}]{\beta^-} {}^{249}\text{Cf} \left(\xrightarrow[351\text{y}]{\alpha} {}^{245}\text{Cm} \right)$$

The Cf-249 is considered to be stable over the time of our experiments. The growth and decay of the various isotopes in this series, starting with pure Es-253, are shown in Figure 1. Each member of this series has been prepared in various compounds, mainly halides and oxides, which have been characterized by absorption spectrophotometry and X-ray powder diffraction where possible. The microchemical techniques for carrying out these studies have been previously reported (2). The growth of the respective progeny species in either the berkelium or einsteinium compounds has also been followed by the above techniques. Such studies have been carried out with initially "pure" compounds, where the parent species was at as high a concentration as possible. This was achieved by minimizing the time between isolation of the particular actinide and its conversion to a given compound for study. The conclusions derived from the studies of such bulk-phase compounds have been tested by changing the chemical composition at various stages of time after the initial isotope separation and by doping Es-253 into various non-radioactive hosts, such as lanthanide or alkaline earth halides. The results of these studies and our conclusions drawn from them will be presented in this paper.

Results and Discussion

The various bulk-phase compounds that have been prepared and studied over periods of time are summarized in Table I.

Table I. Actinide Compounds Prepared for Time Studies

Compound*	Min. Time of Study, Yrs.	Ref.	Compound*	Min. Time of Study, Yrs.	Ref.
Es_2O_3	2	Unp†	$BkBr_3$§	3	7
EsF_3	2	3	BkI_3	-	Unp
$EsCl_3$	3	4	Cf_2O_3§	-	Unp
$EsBr_3$	3	4	$CfCl_3$§	-	Unp
EsI_3	3	4	$CfBr_3$	-	8
$EsCl_2$	3	4,5	CfI_3	-	9
$EsBr_2$	3	4,5	$CfCl_2$	-	10
EsI_2	3	4	$CfBr_2$	-	8
BkF_3§	2	6	CfI_2	-	9
$BkCl_3$§	3	Unp	$CmBr_3$	5	Unp

*Isotopes used: Es-253, Bk-249, Cf-249, Cm-244
†Unpublished
§ 2 or more structures

The references given in Table I are those describing the preparation of a given compound; the reference may or may not contain information on the behavior of the compound with time. Note that the compounds have been synthesized in different oxidation states and different crystal structures where possible. Not shown in the table are einsteinium, berkelium, and californium phosphates which have also been prepared and are being studied at present (11).

Bulk-Phase Compounds. Some of our results in the studies of the bulk-phase compounds have been published (3-7). These studies have shown that oxidation state is preserved for these actinides in either α or β^- decay. Trivalent einsteinium will transmute to trivalent berkelium which transmutes to trivalent californium. It has also been observed that divalent einsteinium yields divalent californium. It is interesting to note in this latter case that it has not yet been possible to synthesize divalent berkelium in the bulk phase. Berkelium(II) has not been observed in our aged einsteinium(II) compounds either, but it would be logical to assume it has been produced there. Our inability to observe Bk(II) could be related to weak absorption intensities and/or interference by absorption bands of einsteinium(II) or californium(II). The results of our studies of bulk-phase halide compounds of ^{249}Bk have also proven that crystal structure is preserved in this β^- decay (7). The spectrum of an aged orthorhombic $BkBr_3$ sample containing the orthorhombic form of $CfBr_3$ is shown in Figure 2. This form of $CfBr_3$ has not been synthesized directly but was prepared by nature through the transmutation of orthorhombic $BkBr_3$ over a 3-year period. The absorption spectrum and X-ray diffraction pattern of this material demonstrated that it did indeed contain orthorhombic $CfBr_3$ (7). Heating this compound to 350°C converted it to the thermodynamically stable monoclinic form of $CfBr_3$, the form that can be synthesized directly. With bulk-phase compounds our results suggest that local order is also preserved in α decay. Conclusive proof of this could not be obtained, however, since useful X-ray powder diffraction data are difficult to obtain from Es compounds (12). The belief that structure is maintained, at least in the local order or coordination sense, is a subjective conclusion (4); it should be remembered that long range order in highly radioactive compounds, such as ^{253}Es halides, is apparently destroyed by self-irradiation damage. Local order is not destroyed; if it were, it would not be possible to obtain reproducible absorption spectra from these compounds. Extensive data are not available for oxides or other types of oxyanion compounds, but nothing observed using such compounds, in the studies carried out so far, is in disagreement with the conclusions described above. Es_2O_3 presents an interesting case in that it ultimately generates an unidentifiable form of Cf_2O_3; this work continues (11).

Material Balance. It has been pointed out that in our studies only trivalent progeny grow into a trivalent parent compound, and only divalent progeny grow into a divalent parent compound. No other species are seen. What of the quantitative nature of these observations? Is there a material balance of the observable species? With the irregular-shaped samples that result from the synthesis, it is not possible to obtain absorption coefficients for even the reference compounds (2), let alone the compounds of mixed species. What was done, however, was to obtain the relative absorbance ratios of several daughter Bk(III) and granddaughter Cf(III) peaks to parent Es(III) peaks in the spectra of einsteinium trihalides as a function of time. These ratios were compared to the relative absorbance ratios of the same peaks in the spectrum of a halide "freshly" synthesized from a sample of Es_2O_3 that had aged an identical period of time. It was assumed in this study that all actinide parent and progeny species, no matter what their form in a sample of oxide, would be quantitatively converted into the respective trihalides and would therefore serve as standards to which the stored halide could be compared. Since only peak ratios were compared, no absolute absorbance measurements needed to be made, and variations in the peak ratios between the

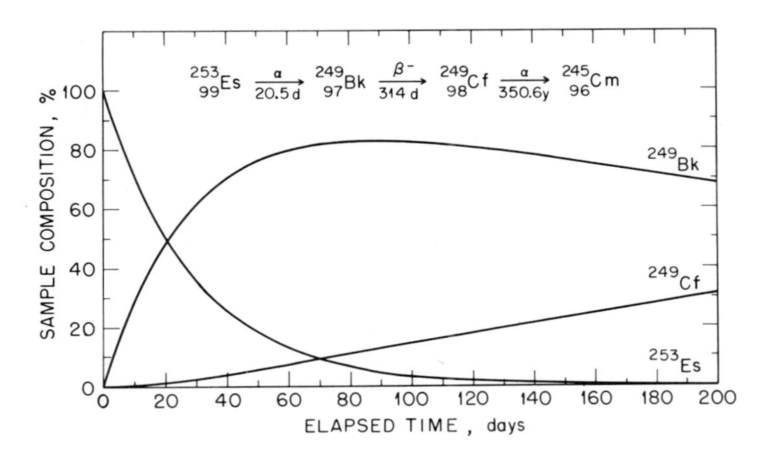

Figure 1. Growth and decay curves for progeny originating from ^{253}Es. Reproduced with permission from Ref. 2, copyright 1978, Elsevier Sequoia, S.A.

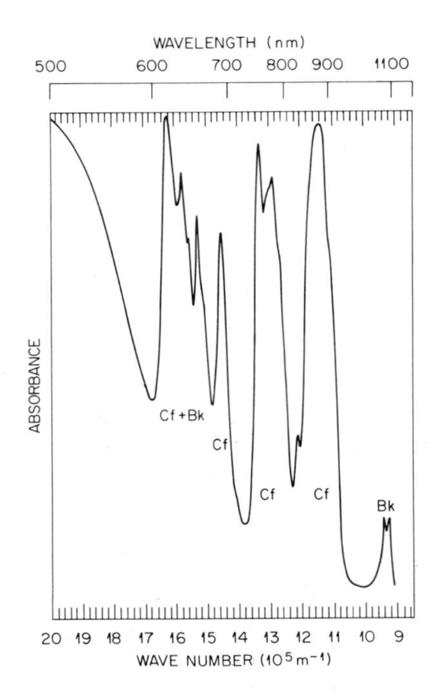

Figure 2. Spectra of aged orthorhombic $BkBr_3$ showing absorption peaks for orthorhombic $CfBr_3$. Reproduced with permission from Ref. 7, copyright 1980, American Chemical Society.

stored halides and the halides made from stored oxides should reflect differences in amounts of the actinides present in the two types of samples. One can compare, for example, the spectrum of 91-day-old $EsBr_3$ plus progeny with that of a sample of 91-day-old Es_2O_3 plus progeny converted into a tribromide. The result of such a study is shown in Figure 3. Although the absorbance peak heights in the two spectra are different, the ratios of peak heights of the various pairs are comparable. The data are given in Table II. Data for similar experiments with a series of $EsCl_3$ samples are listed in Table III.

Table II. Ratio of Progeny and Parent Absorption Peaks* in 91-Day-Old $EsBr_3$ and in 91-day-old Es_2O_3 Converted to Bromide

Sample	$EsBr_3$	Es_2O_3 ⟶	$EsBr_3$
Age (days)	91	91	0
Peak Ratio			
Cf(a)/Cf(b)	1.1		0.70
Bk/Es	0.41		0.65
Bk/Cf(a)	0.48		0.88
Bk/Cf(b)	0.53		0.62
Es/Cf(a)	1.2		1.4
Es/Cf(b)	1.3		0.96

*Peaks used in analysis: Cf(a), 875 nm; Cf(b), 640 nm; Bk, 660 nm; and Es, 790 nm.

Although in this type of comparison there is still a problem of using single beam absorption data (2) and a problem with absorption peak overlap, both of which lower the precision of analysis which might otherwise be possible, reasonably comparable ratios were obtained in both examples. This study was carried out with samples of various ages up to 130 days, and the results suggest that the same amounts of berkelium and californium were present in the stored halides as were present in the stored oxides converted to halides. This indicates that all progeny species may well be accounted for. In an auxiliary analysis of these data presently underway, the growth and decay curves for the various species involved are in reasonable agreement with that expected based on the half-lives of the respective species (13).

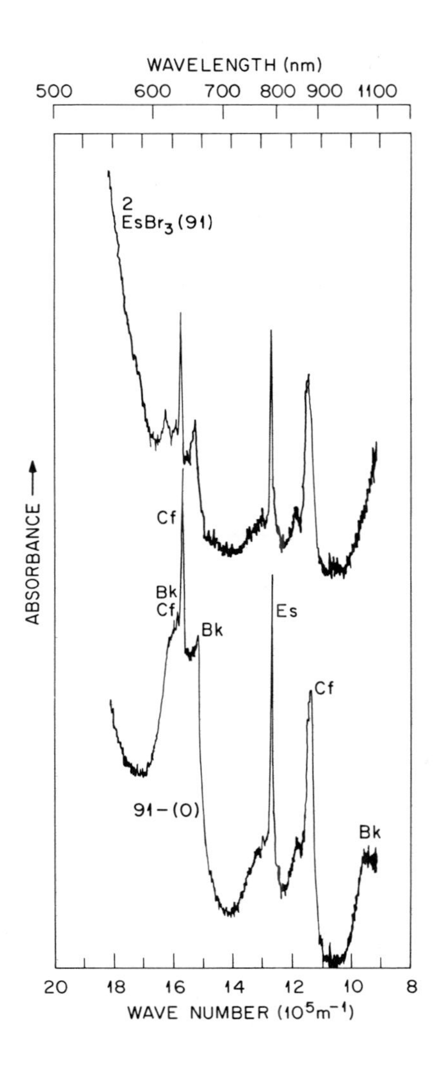

Figure 3. Spectrum of 91-day-old $EsBr_3$ compared to that of 91-day-old Es_2O_3 converted to tribromide.

Table III. Ratio of Progeny and Parent Absorption Peaks* in 81-Day-Old $EsCl_3$ and in Aged Es_2O_3 Converted to Chloride

Sample Age (days)	$EsCl_3$ 81	$Es_2O_3 \rightarrow Es_2Cl_3$ 43 24	$Es_2O_3 \rightarrow EsCl_3$ 81 0
Peak Ratio			
Cf(a)/Cf(b)	1.0	.99	1.4
Bk/Es	1.5	1.2	2.0
Bk/Cf(a)	4.2	3	4
Bk/Cf(b)	4.2	3	5.6
Es/Cf(a)	2.8	2.5	2.0
Es/Cf(b)	2.8	2.5	2.8

*Peaks used in analysis: Cf(a), 875 nm; Cf(b), 740 nm; Es, 790 nm; and Bk, 650 nm.

Solid State Dilutions. Thermalization of recoiling species formed as a result of α and β^- decay processes in a solid material can be considered a process of charge-compensated ion implantation. In bulk-phase einsteinium halides it has been established that progeny species have the same oxidation state as the parent and postulated that the local order of the progeny is the same as that of the parent einsteinium compound. These features may be caused by two different mechanisms. The maintenance of oxidation state through the decay processes may arise from the necessity for charge compensation (4) in the solid material, an effect of heredity. The maintenance of local order may arise because of the nature of radioactive decay, i.e., the replacement of ions, one by one, in an established and stable, at least on the time frame of a single radioactive event, matrix so that each implanted ion is influenced by its environment. To test the above hypotheses, solid state mixtures of einsteinium species ($\leq$ 10 mole % einsteinium) with various inorganic salts were prepared in which the einsteinium exhibited an oxidation state different from that of the host cation. Dilute einsteinium-lanthanide halide mixtures (approximately 10 mole % einsteinium) were prepared in which the host halide exhibited various structures that were not necessarily the same as the structure believed to be characteristic of the pure phase of the respective einsteinium halide. The former sample series would allow evaluation of the effects of heredity; the latter series, the effects of environment. Although both berkelium and californium spectra can be observed in aging bulk-phase einsteinium(III) halides, only californium spectra could be observed in the einsteinium-lanthanide halide mixtures due to the low absorptivity of berkelium.

In all samples prepared in which the oxidation state of the einsteinium differed from that of the host cation, over a time period of several years, only californium(III) has been identified as arising from the samples that initially contained einsteinium (III), and only californium(II) has been observed in samples that originally contained einsteinium(II). Even when the parent was in a host of differing oxidation state, the progeny species maintained the oxidation state of the parent.

The crystal structures of the various lanthanide halides into which einsteinium trihalides were incorporated are summarized in Table IV. Both bromides and chlorides were prepared. The crystal

Table IV. Crystal Structures of Lanthanide Halides Used as Hosts for Einsteinium Halide Decay Studies

Lanthanide	Trichloride	Tribromide
Lanthanum	Hexagonal	Hexagonal
Samarium	-	Orthorhombic
Gadolinium	Hexagonal (dimorphic)	Rhombohedral
Terbium	Orthorhombic	Rhombohedral
Ytterbium	-	Rhombohedral
Lutetium	Monoclinic	Rhombohedral
Actinide		
Einsteinium	Orthorhombic	Monoclinic
Californium	Orthorhombic or Hexagonal	Monoclinic

structures of the pure bulk-phase einsteinium and californium trihalides are also given in the table. The spectra of several of the bromide samples obtained 6 months after preparation of the initial mixture are shown in Figure 4. There are three different crystal forms of lanthanide(III) bromides represented in this series, and there are three different spectra for californium corresponding to these three different structures. We had expected the spectrum of californium growing into $SmBr_3$ to correspond to that of the orthorhombic form (the form of $CfBr_3$ that had previously been made only by transmutation of $BkBr_3$).

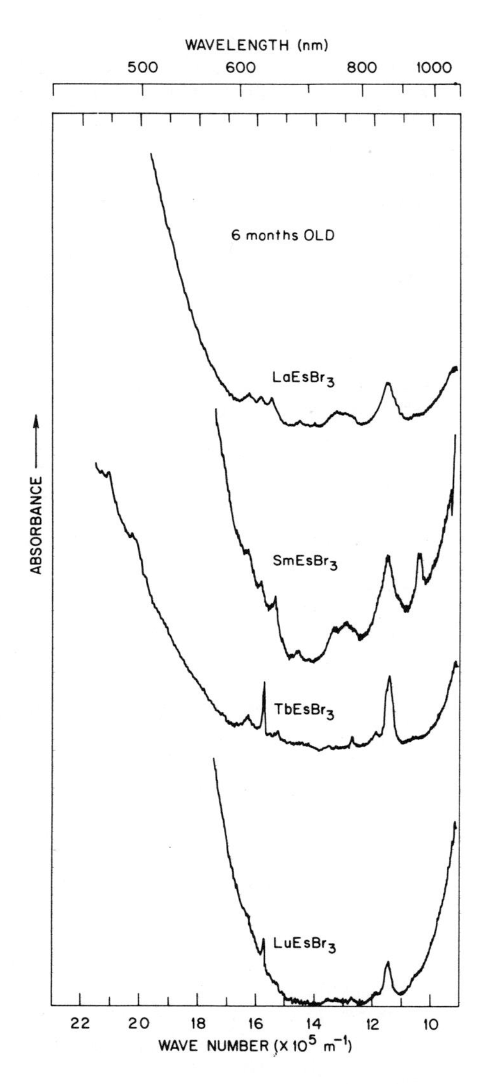

Figure 4. Spectra of 6-month-old mixtures of $EsBr_3$ (∿ 10 mole %) in several lanthanide bromides. Absorption peaks of trivalent californium are also present.

The spectrum of californium in $SmBr_3$ is not exactly like that of orthorhombic $CfBr_3$. The spectrum, rather, is like that of a mixture of orthorhombic and monoclinic $CfBr_3$. Although this seems to be an exception to what was predicted, the presence of such a mixture of two forms of $CfBr_3$ is probably to be expected since it is known that heat will convert orthorhombic $CfBr_3$ to the monoclinic form, and since the mixed crystal started out at some elevated temperature because of the presence of $EsBr_3$ (the decay energy of Es-253 is 1.5×10^4 kJ/mol-min). In the other three spectra, the absorption peaks of $CfBr_3$ in the hexagonal $LaBr_3$ are unique and differ from the similar spectra of $CfBr_3$ in rhombohedral $TbBr_3$ and $LuBr_3$. The spectra of californium (III) in the chloride hosts were weak in all cases, but the spectrum of californium in orthorhombic $TbCl_3$ is in line with what one would expect for orthorhombic $CfCl_3$. The absorption spectra of californium(III) growing into $LuCl_3$ is unlike any $CfCl_3$ spectrum we have previously seen. This last spectrum is similar to the spectrum attributable to californium in monoclinic $CfBr_3$. Taking into account the possible effects of heat, which is also an environmental effect, it appears that the spectra of californium progeny species tend to conform to that expected for the host lattice and not those of the pure californium halide structure.

The results of these and our previous studies of actinide compounds suggest that it is reasonable to assume that most of the progeny species are accounted for in our observations. The data also suggest that the oxidation state of progeny species is controlled by heredity and their structure by environment. We are not, however, invoking any severe chemical or structural instabilities in the model systems we have studied. In sharp contrast to our results, severe chemical instabilities could be created if oxidation state is preserved by progeny ions in α or β^- decay processes in elements or to elements other than those in the 4f or 5f transition element series. Many of these other progeny elements exhibit normal oxidation states that vary considerably from the (II), (III), or (IV) states more common for the f-transition series. It would seem that nature might correct these instabilities by appropriate redox reactions, thereby releasing or consuming electrons in the host storage material.

Acknowledgment

Research sponsored by the Office of Energy Research, U. S. Department of Energy under Contract W-7405-eng-26 with the Union Carbide Corporation and DE-AS05-76ER04447 with the University of Tennessee (Knoxville).

Literature Cited

1. Harbottle, G,; Maddock, A. G. "Chemical Effects of Nuclear Transformations in Inorganic Systems", North-Holland Publishing Co., Amsterdam, 1979.
2. Young, J. P.; Haire, R. G.; Fellows, R. L.; Peterson, J. R. J. Radioanal. Chem. 1978, 43, 479
3. Ensor, D. D.; Peterson, J. R.; Haire, R. G.; Young, J. P. J. Inorg. Nucl. Chem. 1981, 43, 2425.
4. Young, J. P.; Haire, R. G.; Peterson, J. R.; Ensor, D. D.; Fellows, R. L. Inorg. Chem. 1981, 20, 3979.
5. Peterson, J. R.; Ensor, D. D.; Fellows, R. L.; Haire, R. G.; Young, J. P. J. Phys. (Orsay, Fr.) 1979, 40, C4:111.
6. Ensor, D. D.; Peterson, J. R.; Haire, R. G.; Young, J. P. J. Inorg. Nucl. Chem. 1981, 43, 1001.
7. Young, J. P.; Haire, R. G.; Peterson, J. R.; Ensor, D. D.; Fellows, R. L. Inorg. Chem. 1980, 19, 2209.
8. Young, J. P.; Vander Sluis, K. L.; Werner, G. K.; Peterson, J. R.; Noé, M. J. Inorg. Nucl. Chem. 1975, 37, 2497.
9. Wild, J. F.; Hulet, E. K.; Lougheed, R. W.; Hayes, W. N.; Peterson, J. R.; Fellows, R. L.; Young, J. P. J. Inorg. Nucl. Chem. 1978, 40, 811.
10. Peterson, J. R.; Fellows, R. L.; Young, J. P.; Haire, R. G. Radiochem. Radioanal. Lett. 1977, 31, 277.
11. Haire, R. G. and others, unpublished results.
12. Haire, R. G.; Peterson, J. R. Adv. X-Ray Anal. 1979, 22, 101.
13. Hahn, R. L.; Garrett, D. A. in "Physical-Chemical Studies of Transuranium Elements"; Peterson, J. R.; Haire, R. G., Eds.; U. S. Department of Energy Document No. DOE/ER/04447-137, 1982, pp. 38-42.

RECEIVED October 20, 1983

WASTE FORM LEACHING BEHAVIOR

Effects of Water Flow Rates on Leaching

C. PESCATORE[1]
University of Illinois at Urbana-Champaign, Urbana, IL 61801
A. J. MACHIELS
University of Illinois at Urbana-Champaign, Nuclear Engineering Program, Electric Power Research Institute, Palo Alto, CA 94303

Waste form leach rates in a geologic repository will be affected by unknown water flow rates and by extensive cracking of the waste form monolith. An understanding of these effects is important in predicting the geochemical behavior of disposed radioactive waste forms over the full range of possible scenarios. The dependence of the waste form source term on the rate of renewal of aqueous solution is first established for the simple but important case of solubility-limited network dissolution control. Next, the case of selective leaching control is investigated. Leaching in the presence of both mechanisms is then explored by using the mechanistic leaching code, LIX. The latter appears particularly well suited for deriving, from information obtained in static tests, the leach rates applicable to low-flow conditions which are difficult to simulate experimentally. It is suggested that, for glass waste forms, the leaching process is eventually controlled by selective leaching of modifier species at very low flow rates and by network dissolution of the glass structure at higher flow rates.

In glass leaching experiments, the rate of renewal of the corrosion solution is an important system parameter, the effect of which has been investigated in a number of recent studies focusing on complex, simulated nuclear waste glasses (1-4). In particular, changes in the leachant renewal frequency have been found to strongly affect elemental releases of both network formers and modifers as well as the pH of the solution. The results parallel those for the effects of the sample surface area-to-solution volume ratio (5,6). Namely, higher mass losses are obtained the more dilute the contacting aqueous solution (i.e., the higher the solution flow rate).

[1]Current address: Brookhaven National Laboratory, Upton, N.Y. 11973.

0097-6156/84/0246-0349$06.00/0

In a geologic repository, waste form leaching will be affected by unknown flow rates both under normal repository operation and accidental conditions. An understanding of these effects is then necessary in order to predict the geochemical behavior of disposed radioactive waste over the full range of possible scenarios.

In the present paper, flow rate effects on leaching are analyzed from a theoretical point of view and are rationalized in a consistent, generic leaching model incorporating the dependence of the mechanisms of selective leaching and network dissolution on solution feedback effects.

Flow rate effects on the rate of dissolution of the glass network and on the rate of selective leaching of glass modifiers are discussed first separately, in order to address expected leaching behaviors under network dissolution control and selective leaching control, respectively. Flow rate effects are then analyzed in the presence of both leaching mechanisms. It is concluded that the distinguishing feature of flow rate effects on leaching is that they determine the long-term rate-controlling leaching mechanisms.

Effects on Network Dissolution

A distinguishing effect of flow rate is that saturation of the solution is never achieved under dynamic leaching conditions. Thus, network dissolution, a solubility limited process, is never allowed to halt. A steady-state condition is eventually established whereby leach rates of network formers equalize the rate of removal of species from the leachant due to the flowing solution. This can be illustrated by formulating a network dissolution model which incorporates the dependence of leaching of network formers on solubility limits and water flow rates.

With reference to the silica dissolution reactions:

$$[SiO_2](gl) + 2\,[H_2O](aq) = [H_4SiO_4](aq) \tag{1}$$

$$[H_4SiO_4](aq) = [H_3SiO_4](aq) + [H^+](aq) \qquad pK=9.8 \text{ at } 25°\text{ C} \tag{2}$$

the instantaneous silicon dissolution leach rate per unit area of the solid, L(t), can be expressed as: (4,7)

$$L(t) = L_o\ (1 - C/C_{sat}) \tag{3}$$

where C denotes the concentration of orthosilicic acid in solution. L_o is a kinetic parameter representing the rate of forward reaction (1) when the solution is uncontaminated with silica. C_{sat} is the saturation value of orthosilicic acid in solution. Both L_o and C_{sat} exhibit an Arrhenius dependence on temperature, but negligible dependence on pH in water (7,8).

Considering, then, an example of leaching under moderate pH conditions (pH 9, say), such that ionization of orthosilicic acid can be neglected, the balance of silicon species in a well-mixed solution can be written as follows:

$$\frac{dC}{dt} = \beta\ (9t) - \Phi(C - C^*) \qquad t>0 \tag{4}$$

β - ratio between the sample "true" surface area, A, and the volume of the contacting solution V;

Φ - ratio between the volumetric flow rate of the leachant, F, and the volume of the contacting solution, V, i.e., the leachant renewal frequency;

C^* - silicon volumetric concentration in the incoming solution;

t - time.

Indicating by C_o the initial concentration of silicon in solution, Equation (4) predicts it to evolve according to the expression:

$$C\ (t) = (C_o - C_\infty)e^{-t/\tau} + C_\infty \tag{5}$$

where:

$$C_\infty = \frac{\beta L_o + \Phi C^*}{\beta L_o + \Phi C_{sat}}\ C_{sat} \tag{6}$$

and

$$\tau = C_{sat} / \ (\beta L_o + \Phi C_{sat}) \tag{7}$$

Accordingly, a flow-rate limited steady-state characterized by a constant concentration of silicon in solution, C_∞, and a constant leach rate

$$L_\infty = L_o\ (1 - C^* / C_{sat}) \frac{\Phi C_{sat}}{\beta\ L_o + \Phi C_{sat}} \tag{8}$$

is eventually achieved. The time needed to reach 1% departure from steady-state, T, is approximately:

$$T = 5\tau = \frac{5}{\beta L_o\ /\ C_{sat} + \Phi} \tag{9}$$

Therefore, if leaching is network dissolution controlled, the higher the flow rate the sooner steady-state is achieved and the higher the leach rate, which accords with the experimentsl evidence. In particular, because $\beta = A/V$ and Φ - F/V, Equations (6)-(8) indicate that the only parameter characterizing the system at steady state is the ratio F/A, i.e., in a repository environment, the specific flow of groundwater past the waste form specimen.

Effects on Selective Leaching

Selective leaching of the glass matrix takes place by ion exchange of modifier ions (alkalis or alkaline earths, essentially) from the glass for hydronium ions in solution. This process results in dealkalinization of the original glass. Glass modifiers exhibit high solubilities in water indicating that selective leaching is not solubility limited and that flow rates are not expected to influence this mechanism unless in the extreme case of a stagnant solution. This can be illustrated by formulating a selective leaching model which incorporates ion exchange processes taking place within the glass bulk and at the glass-solution interface as well as the flow condition of the solution.

With reference to sodium species and to Figure 1, the ion exchange reaction

$$[Na^+](gl) + [H_3O^+](aq) \rightleftharpoons [Na^+](aq) + [H_3O^+](gl) \qquad (10)$$

can be modeled separately for the glass bulk and surface phases (4). Ion exchange in the bulk phase results in an ionic counterdiffusion process characterized by a concentration-dependent diffusion coefficient, $\tilde{D}$ (4,9). Ion exchange at the glass-solution interface can be modeled in terms of the phenomenological constants k^+ and k^- yielding the following expression for the leach rate of sodium species per unit area of the solid:

$$L(t) = k^+ n = k^- C_{sol}$$

where n represents the surface concentration of sodium species; C_{sol} the concentration of sodium species in solution; k^+n the rate of sodium release into solution; k^-C_{sol} the rate of sodium resorption on the glass surface. The parameters k^+ and k^- depend in general on temperature, and may depend on the physico-chemical properties of the solution and the glass surface.

Considering an example of leaching under moderate pH excursion, such that the autoprotolysis of water can be neglected and k^+ and k^- are approximately constant, the balance of sodium species in the system obeys the following set of equations: (4)

$$\frac{\partial C}{\partial t} = \frac{\partial}{\partial x}\left[\tilde{D}\,\frac{\partial C}{\partial x}\right] \qquad x,t>0 \qquad (12)$$

$$\frac{dn}{dt} = -\,L(t) + \left(\tilde{D}\,\frac{\partial C}{\partial x}\right)_{x=0+} \qquad t>0 \qquad (13)$$

$$\frac{dC_{sol}}{dt} = \beta\,L(t) - \Phi(C_{sol}-C^*) \qquad t>0 \qquad (14)$$

where C* represents the sodium concentration in the incoming leachant, henceforth assumed equal to zero. The two parameters β and Φ are the usual surface area-to-solution volume ratio and the leachant renewal frequency, respectively. The coefficient $\tilde{D}$ can be expressed in terms of the molar fraction of sodium in the glass, x_{Na}, and the self-diffusion coefficients of Na^+ and H_3O^+ species:

$$D = \frac{D_{Na} D_H}{x_{Na} D_{Na} + (1-x_{Na}) D_H} \tag{15}$$

In practice $D_{Na} >> D_H$, (10) and counterdiffusion is controlled by the slower hydronium ion. This suggests assuming that:

$$\tilde{D} \sim D_H$$

which linearizes the system of Equations (12)-(14). Assuming further chemical equilibrium between sodium species on the glass surface and sodium species immediately adjacent in the bulk phase, the law of mass action requires that:

$$\frac{C(0+,t)}{n(t)} = K \qquad t \geq 0 \tag{16}$$

where K is a true equilibrium constant.

Within the context of the above approximations, the original system of Equations (12)-(14) can be solved with the usual boundary and initial conditions:

$$C(\infty,t) = C_o \tag{17}$$

$$C(x,0) = C_o \tag{18}$$

$$n(0) = C_o/K \tag{19}$$

$$C_{sol}(0) = 0 \tag{20}$$

The predicted asymptotic behaviors for the leach rate are as follows: (4,11)

$$L_\infty(t) \sim Co\sqrt{\frac{D_H}{\pi t}} \cdot \frac{[k^+]^2}{2 K^2 D (k^- \beta)^2 t} \qquad \text{t large enough} \quad \Phi = 0 \tag{21}$$

and

$$L_\infty(t) \sim co\sqrt{\frac{D_H}{\pi t}} \qquad \text{t large enough} \quad \Phi \neq 0 \tag{22}$$

Equations (21) and (22) both suggest that leaching will eventually slow down to zero under selective leaching control. Under the conditions of Equation (21), leaching tends to zero as a build-up of glass modifiers in solution allows the solid surface to come to equilibrium with the leachant. Under the conditions of Equation (22), species build-up effects do not play a role, regardless of how small the flow rate, and leaching tends to zero as the glass matrix gets depleted of network modifiers. This accords with physical intuition.

Equation (22) is the classical expression for the leach rate per unit area under diffusion controlled conditions.

The General Case

While it proved convenient to consider flow rate effects on leaching under network dissolution control and selective leaching control separately for illustration purposes, these mechanisms are not independent of each other. Indeed, as ion exchange reactions of the kind (10) deplete the solution of hydronium ions, the solution becomes more alkaline causing orthosilicic acid to ionize and, consequently, additional dissolution of the glass matrix according to reactions of the kind (1). It is, however, the flow condition of the solution that determines the long-term rate-controlling leaching mechanisms. If the solution is stagnant, saturation of the solution is reached much sooner with respect to the network formers than the alkalis; thus leaching will eventually take place at a rate controlled by selective leaching of the alkalis. If the solution is flowing, saturation is never achieved, and selective leaching, a process which otherwise would slow down to zero, is controlled eventually by the constant network dissolution rate of the glass matrix. In this case, all elements eventually leach out at the same rate dictated by the dissolution kinetics of the network forming element with highest solubility in solution and by the flow conditions of the solution. This can be illustrated for the case of sodium and silicon leaching from nuclear waste glasses by combining together to the modeling approach outlined earlier for network dissolution and selective leaching controls, and by taking into account the dissociation of the orthosilicic acid, Reaction (2), and the autoprotolysis of water:

$$H_2O \rightleftharpoons OH^- + H^+ \tag{23}$$

The mathematical formulation of the model complicates somewhat (4), and it is not reported here for the sake of simplicity. In particular, the model results in a system of coupled, non-linear ordinary and partial differential equations which have been fully implemented in a computer code, named LIX, to predict elemental releases of silicon and sodium from borosilicate glass.

Figure 2 shows the predicted, normalized cumulative mass losses based on the behaviors of silicon and sodium for three different values of the leachant renewal frequency. The physical parameters used refer to the leaching of PNL 76-68 borosilicate glass in deionized water at 90°C, (4) and reference is made to the geometric surface area, SA, of the sample. In particular, the curves corresponding to silicon and sodium tend to have the same, constant slope with increasing flow rate. In particular, the curves corresponding to $\Phi = 1\ \text{day}^{-1}$ practically coincide, indicative of network dissolution control.

Figure 3 shows the predicted behavior of the pH of the solution as a function of leachant renewal frequency for the same system parameters. As can be seen, the higher the flow rate, the sooner steady state is achieved, and the closer the leachant composition to that of the original solution. In particular, the pH curve for the static case ($\Phi = 0$) shows that the solution pH has not reached steady state yet after 28-days leaching. Approach to steady state under the static leaching conditions can be a very lengthy process. However, an equilibrium pH value can be estimated by use of the solution electroneutrality condition as applied to the reactions modeled. Indeed, at all times:

$$[H^+] + [Na^+] - [OH^+] - [H_3SiO_4^-] =$$
$$\left\{ [H^+] + [Na^+] - [OH^-] - [H_3SiO_4^-] \right\}_{t=0} \qquad (24)$$

which yields an equilibrium pH of ~12 at 90°C based on the equilibrium relation between sodium solution and surface species:

$$[Na^+]_{eq} = \frac{k+C_o}{k^- K} \qquad (25)$$

and on values of the physical parameters from Reference 4.

Figure 4 shows a plot of the long term, normalized leach rate, $L^*(t)=L_\infty(t)/L_{Si}(t=0)$, based on the behaviors of sodium and silicon as a function of Φ, where leaching follows a law of the type (21). This lower range of Φ values may encompass flow rates in a nuclear waste repository under normal operational conditions. Network dissolution control operates in the higher ranges of Φ according to a law of the type (8). Higher ranges of Φ values may encompass flow rates in a nuclear waste repository under accident conditions. If species are released, whose saturation concentration is lower than the orthosilicic acid saturation concentration, new solid phases may form both in solution and at the solution-glass interface (12). Formation of new phases in solution may accelerate leaching; new phases at the solution-glass interface may stabilize the dealkalized layer and slow down both leaching mechanisms.

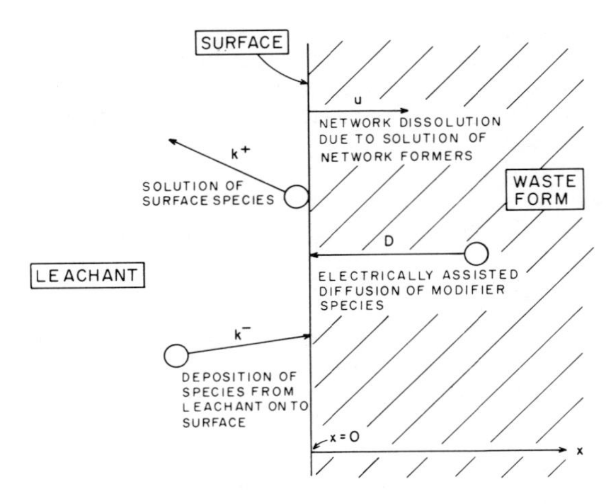

Figure 1. Graphical Representation of the Waste Form-Leachant System.

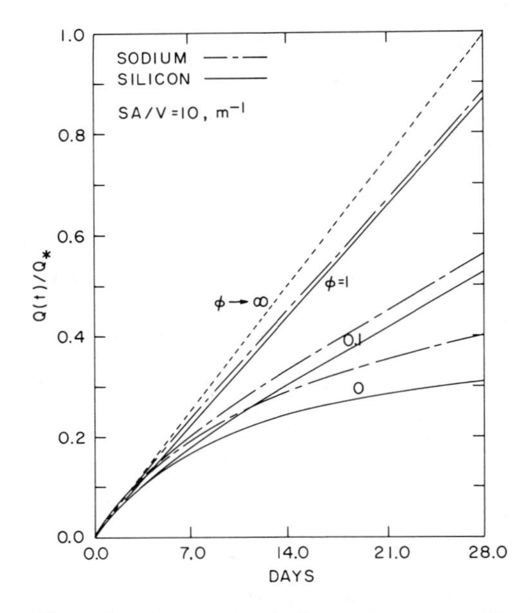

Figure 2. Predicted, Normalized Mass Loss vs. Time for Various Values of Φ. $[\Phi] = [day^{-1}]$; Q_* - grams of Glass Leached After 28 Days Assuming Congruent Dissolution at the Initial Silicon Leach Rate.

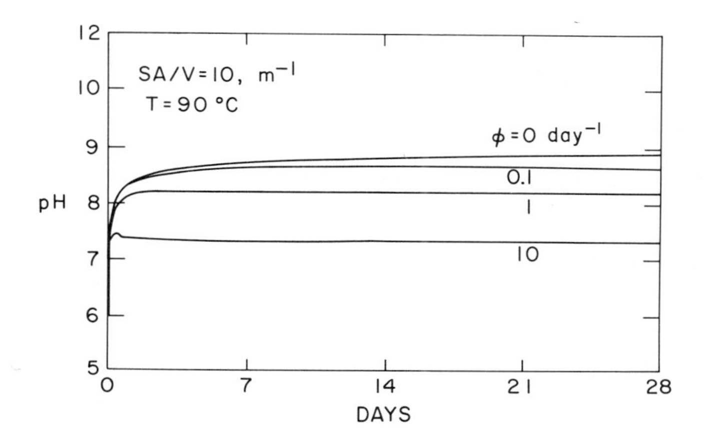

Figure 3. Predicted Behavior of Solution pH vs. Time for Various Values of Φ.

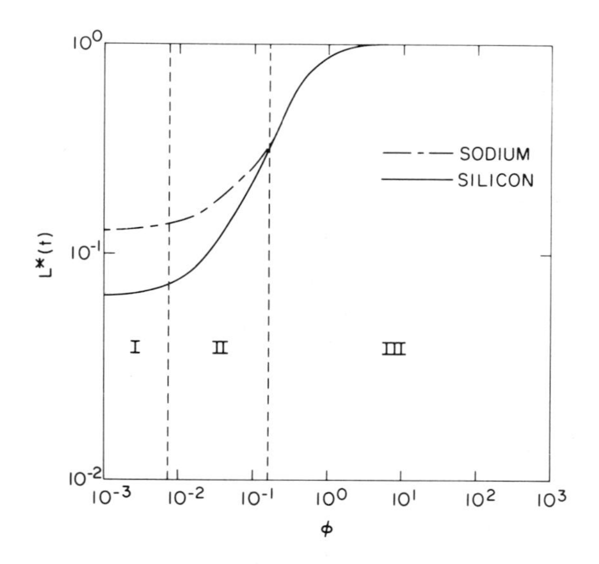

Figure 4. Long-Term Normalized Leach Rates vs. Time.

Conclusions

The distinguishing feature of flow rate effects on leaching is that they determine the long-term rate-controlling leaching mechanisms. Under dynamic leaching conditions saturation of the solution is never achieved; steady-state elemental releases are thus controlled by dissolution of the glass network. The higher the flow rate, the faster steady-state is approached, and the higher the resulting leach rates. Under low groundwater flow rates, leaching is controlled eventually by the selective release from the glass of modifier species which have high solubilities in water. The release of species with lower solubilities than the network former controlling the dissolution process may or may not slow down leaching of both glass modifiers and network formers. In general, the mechanistic leaching code LIX appears particularly well-suited for deriving from information obtained in static tests the leach rates applicable to any flow conditions, including the low flow conditions which are difficult to simulate experimentally.

Literature Cited

1. Coles, D. G. et al., UCID-19492-Rev. 1, Lawrence Livermore Laboratory, Berkeley, California, 1982.
2. Macedo, P. B.; Barkatt, A.; Simmons, J. H., Nuclear and Chemical Waste Management, 1982, 3, 13-21.
3. Macedo, P. B.; Barkatt, A.; Simmons, J. H., in "Scientific Basis for Radioactive Waste Management-V," Lutze, W., Ed.; North-Holland: New York, 1982; p. 57.
4. Pescatore, C., Ph.D. Thesis, University of Illinois, Urbana, Illinois, 1983.
5. Pederson, L. R., et al., PNL-SA-10841, Pacific Northwest Laboratory, Richland, Washington, 1982.
6. Machiels, A. J.; Pescatore, C. in "Scientific Basis for Nuclear Waste Management-VI"; Brookins, D. G., Ed.; North-Holland: New York, 1983; p. 209.
7. Rimstidt, J. D.; Barnes, H. L., Geochimica et Cosmochimica Acta, 1980, 44, 1683-1699.
8. Alexander, G. B.; Heston, W. M.; Iler, R. K., J. Phys. Chem., 1954, 58, 453-455.
9. Doremus, R. H., "Treatise on Material Science and Technology"; Tomozawa, M., Ed.; Academic: New York, 1979; Vol. 17, p. 41.
10. Frischat, G. H., "Ionic Diffusion in Oxide Glasses," Trans. Tech. Publications, Bay Village OH., 1975.
11. Pescatore, C.; Machiels, A. J., Proc. 10th IMACS World Congress on Systems Simulation and Scientific Computation, 1975, Vol. 2, p. 283.
12. Grambow, B., in "Scientific Basis for Radioactive Waste Management-V Lutze, W., Ed.; North-Holland; New York, 1982; p. 93

RECEIVED October 20, 1983

23

Characterization of Borosilicate Glass-Containing Savannah River Plant Radioactive Waste

MCC-1 Tests and Durability in Geologic Repository Groundwaters

NED E. BIBLER

E. I. du Pont de Nemours & Co., Savannah River Laboratory, Aiken, SC 29808

Results are presented from static leach tests on borosilicate glass containing high-level radioactive nuclear waste from the Savannah River Plant. Tests were performed in shielded facilities closely following MCC-1 procedures. Leachants were deionized water, MCC brine, or silicate water, all at 40°C. Normalized mass losses (g/m^2) based on ^{137}Cs, ^{90}Sr, and ^{238}Pu were calculated. Results of leach times of 3, 7, 14, 28, and 300 days are reported for deionized water. Results for 28 and 200 days are reported for silicate water and brine. Normalized mass losses and pH changes indicate that glass containing radioactive waste leaches similarly to glass containing nonradioactive, simulated waste. Release rates in the two simulated groundwaters were slightly less than in deionized water. Also, radiolysis of the leachant by alpha, beta, and gamma rays from the glass did not significantly affect the normalized mass losses or the pH changes due to leaching. Results of the long-term tests suggest that equilibrium concentrations of radionuclides will be achieved. Based on tests with different ratios of glass surface area to leachant volume, these concentrations are controlled more by solubility and surface layer effects than by the surface area of the glass.

An environmental assessment(1) concluding that borosilicate glass is suitable for the immobilization of Savannah River Plant (SRP) waste has recently been published by the Department of Energy. This assessment included results of leach tests with glass containing either actual(2,3) or simulated(4) radioactive waste in leachants that simulated groundwaters of possible geologic repositories: salt, basalt, tuff, and granite. In support of the environmental

This is Part 3 in a series.

0097-6156/84/0246-0359$06.00/0

assessment, a detailed report summarizing the extensive body of glass characterization data was prepared(5). That report concentrated on the performance of SRP borosilicate glass in generic repository environments. A major conclusion was that SRP waste glass could effectively limit radionuclide release as part of a waste package in these generic repository environments.

This paper presents further data on the leaching of SRP borosilicate glass containing actual radioactive waste. Two previous papers describing this glass are currently being published. One deals with general methods of preparation and characterization(2). The second is a study of devitrification of the glass(6). The results of that study indicate that devitrification (up to 19 vol %) does not significantly affect the leach rate. The present paper presents the results of MCC-1 leach tests(7) and of leach tests in which the leachant was periodically changed. Leaching was at 40°C in deionized water or in the MCC-1 brine or silicate solutions. This temperature approximates that expected at the onset of leaching in a salt or granite repository(5). The brine simulates groundwater in a salt repository. The silicate solution simulates groundwater in a hard rock repository. Normalized mass losses were calculated based on the amounts of ^{137}Cs, ^{90}Sr, or ^{238}Pu activity leached from the glass. Because of the intense radioactivity of the glass (surface dose rate of approximately 10^4 rad/hr), all the tests had to be performed remotely in shielded facilities. Results indicate that this radioactivity did not affect the leaching process. Also, in the long-term tests (approximately 200 days), radioactivity measurements indicate that the radionuclides are approaching equilibrium concentrations.

Glass Preparation and Composition

The glass was prepared by feeding an aqueous slurry of glass-forming frit (SRP Frit 131) and waste (approximately 70 wt % frit on a dry basis) to a joule-heated melter in the shielded cells. The waste was from SRP storage Tank 11. Major components of the waste were nonradioactive Fe, Mn, and Al resulting from chemical operations at SRP. These materials were highly contaminated with fission products and transuranic radionuclides. The composition of the glass (Table I) was determined by elemental analysis of solutions of dissolved glass. Determination of the specific activity of the glass was performed on the dissolved glass solutions following normal counting procedures. Sources of radioactivity were beta from ^{90}Sr, gamma from ^{137}Cs, and alpha from ^{238}Pu Specific activities (Table I) were very high. Also, the surface dose rate of the glass was estimated with an ion chamber to be 10^4 rad/hr. This was due primarily to ^{90}Sr beta particles. Quantitative metallography indicated that the glass contained <1 vol % crystalline material(6). These crystals were ferrite spinels.

Table I. Composition of SRP Radioactive Borosilicate Glass Containing Tank 11 Waste*

Nonradioactive Composition				
Component	Wt %	Component	Wt %	
SiO2	45.5	NiO	0.9	
Fe_2O_3	5.7	MgO	1.4	
Na_2O	15.8	ZnO	0.3	1.1
B_2O_3	9.1	Cr_2O_3	4	
Li_2O	2.3	U_3O_8	0.	
Al_2O	10.2	TiO_2	1.5	
MnO_2	1.7			
CaO	4.0			

Radioactive Composition**	
Radionuclide	Activity, mCi/g Glass
^{137}Cs	1.0
^{90}Sr	2.7
^{238}Pu	0.59

* Prepared by melting a mixture of waste and glass-forming frit at 1150°C.
** Other radionuclides were present but at much lower levels.

Leaching Procedures

Because of the intense radioactivity of the glass, four procedures prescribed in MCC-1 were modified. First, polypropylene leach containers rather than Teflon (Du Pont) were used because of the relatively low radiation stability of Teflon and the possibility of radiolysis producing HF from the Teflon. Second, the radioactive glass was placed in stainless steel baskets for leaching. Platinum baskets were used for brine leaching to avoid chloride induced corrosion of stainless steel. Plastic could not be used adjacent to the glass because it would quickly degrade from the intense beta radiation from the glass. Third, the vigorous cleaning procedure to remove contaminants from the leach vessels was not necessary because only radioactivity was used to measure leach rates. The vessels were soaked for 16 hours at room temperature in 1M HNO_3 followed by rinsing in deionized water, followed by two successive soakings at 90°C for 16 hours in deionized water. Lastly, glass samples for leaching were obtained by fracturing the glass rather than by core drilling. The fractured surfaces were not polished before leaching. Because fractured samples were used, the surface area had to be estimated. This was done by placing each surface against a calibrated grid and estimating its area. The error introduced by this procedure was <20%. As prescribed in MCC-1, the glass was washed in an ultrasonic cleaner in water and alcohol prior to leaching. Control experiments indicated that radioactive contamination was not a problem in the shielded cells. The tests contained 3-10 g glass and 200 ml leachant. During leaching, the sample bottles were placed in a 40 $\pm$2°C oven. After the test, the glass was removed from the leachant. The leachate in the container was then acidified and sampled after being left to stand one day. The ^{137}Cs, ^{90}Sr, and ^{238}Pu activities of the leachate were determined by standard counting techniques. The vessel was then rinsed with 4M HNO_3 to determine the amount of radioactivity sorbed on the container. Except for ^{238}Pu this amount of activity was negligible. For ^{238}Pu, 10 to 25% of the activity had sorbed on the walls of the leach vessel.

Leach test results were calculated in terms of normalized mass loss of the glass (mass loss/unit area) based on ^{137}Cs, ^{90}Sr, or ^{238}Pu. The appropriate equation is:

$$(NL)_{i,t} = \frac{V_l \cdot C_{i,l}}{SA \cdot C_{i,g}}$$

where $(NL)_{i,t}$, is the grams of glass leached per unit surface area in total time t based on species i. V_l is the volume of leachant, SA is the surface area of the glass, and $C_{i,g}$ is the concentration of species i in the glass (mCi/gram glass). $C_{i,l}$ is the concentration of i in the leachant (mCi/ml). This concentration was calculated to include the amounts of radionuclides that had sorbed on the walls of the leach vessel. In some of the leach tests, the leachant was changed periodically. The total normalized mass loss was then determined by summing the mass losses over all the previous time increments. These tests were performed so that both short and long term data could be obtained with the same sample of glass.

Leaching Results

Tests in Deionized Water. Results in deionized water for MCC-1 tests up to 28 days and for tests where the leachant was periodically replaced up to 300 days appear in Figure 1. For the latter type of tests, the glass was periodically transferred to a new vessel containing fresh deionized water. The glass was not allowed to dry during these transfers. The error brackets in Figure 1 indicate the deviation of the results of duplicate tests. Only single data points were available for the longer tests measuring ^{238}Pu. Data for the 3 day tests for each respective radionuclide are in good agreement. For the 7, 14, and 28 day results, data for the MCC-1 tests are lower than the results for the tests where the deionized water was changed at those respective days. A possible explanation for this is that saturation effects in the leachate are not as effective in the tests in which the leachate was changed. In all the tests, most of the leaching occurs in the first 28 days. The long-term tests indicate, especially for ^{137}Cs and ^{238}Pu, that leaching has nearly stopped. During this time, leaching is inhibited by surface layers being formed on the glass and by leachate saturation by various species from the glass(4,8). These results are consistent with other long-term leach tests of SRL 131 waste glass(9) and of Pacific Northwest Laboratory 76-68 waste glass(8) both of which contained nonradioactive simulated waste. On this basis, glasses containing actual and simulated waste leach similarly.

28 Day Tests in Deionized Water, Silicate Water, and Brine. Normalized mass losses for these tests are shown in Table II. The results for deionized water and silicate water are similar with results at 40°C based on leaching Cs, Sr, and U from SRL 131 glass containing nonradioactive simulated waste(9). Comparable data at 40°C for Cs and Sr in brine are not available. For U in brine, a value of 0.07 g/m^2 has been reported(9). Data in Table II indicate that silicate water is less aggressive than deionized water in agreement with results of Wicks(4) and Lokken(9). For brine, the results in Table II indicate that the leachability of ^{90}Sr is much higher than in the other leachants. This has been observed by Strachan(8) and has been attributed to the fact that the pH does not increase during leaching in brine as much as it

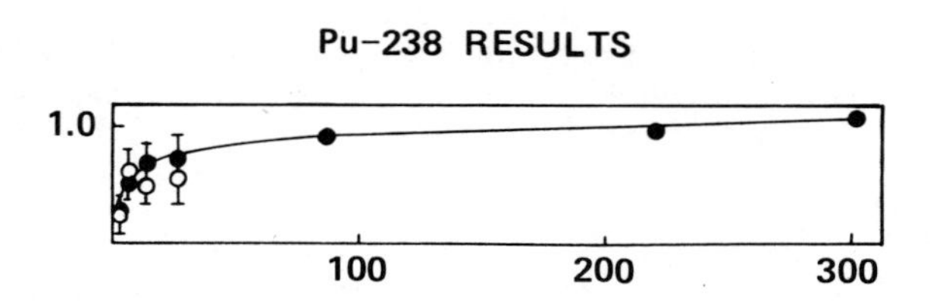

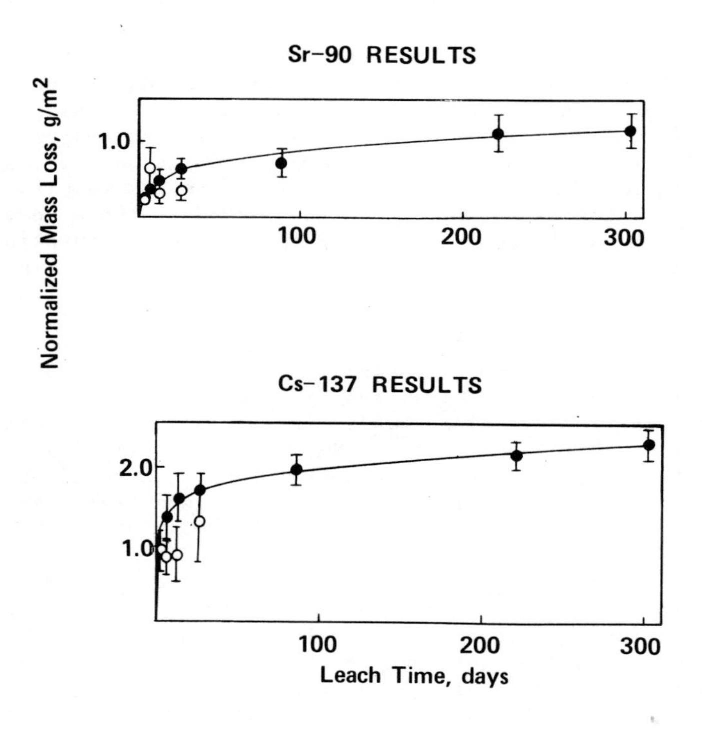

Figure 1. Time dependence of normalized mass losses in deionized water at 40°C based on Cs-137, Sr-90, and Pu-238 from SRP borosilicate glass containing actual radioactive waste. O, MCC-1 tests; ●, tests where leachant was replaced periodically. Error bars indicate the precision of duplicate tests.

does during leaching in deionized water or silicate water(8). Such a smaller increase in pH for brine was also observed in this study (Table III). Another reason for the higher normalized mass loss for ^{90}Sr (and possibly also ^{238}Pu) may be the low glass surface area to leachant volume ratio (SA/V) in this test compared to the tests with deionized and silicate water. This effect of SA/V on normalized mass loss is discussed later. The agreement of the results for normalized mass losses in Table II in the presence of stainless steel baskets with results where steel was absent,(9) suggest that the steel is not affecting the leaching process. Data indicating this have also been obtained in tests with nonradioactive glass(4). Thus, it appears that the stainless steel is not affecting the leaching chemistry.

The pH changes in these tests (Table III) are consistent with H^+ ions in solution exchanging with Na^+ ions in the glass. These changes are in agreement with results with nonradioactive glass(4,8); thus radiation from the glass containing actual waste does not affect the pH.

200 Day Leach Tests in Silicate Water and Brine. Normalized mass losses for four long-term tests in silicate water and brine are given in Table IV. With each leachant, the tests were performed at two different values of SA/V by increasing the amount of glass in the second tests while holding the leachant volume constant. A direct comparison of the results obtained in the simulated groundwaters with those obtained in deionized water at long times (Figure 1) cannot be made because the tests were performed differently. The long term normalized mass losses in deionized water are for tests where the leachant was changed periodically. Results for the simulated groundwaters (Table IV) are for single 200 day tests. Because of the limited availability of radioactive glass when these studies were initiated, the glasses used for the long-term tests in brine and silicate water had previously been leached in scouting tests for 28 days. Thus, the normalized mass losses in Table IV do not contain the contributions from these shorter times. However, the data in Table IV for each groundwater can be compared to each other. They indicate the effect of SA/V on the normalized mass loss. For smaller SA/V values, larger normalized mass losses are obtained. Based on these limited data, there is nearly quantitative agreement with the change in SA/V. A 2X decrease in SA (V was held constant) leads to a 2X increase in normalized mass loss. These results imply that in each respective simulated groundwater, each glass has lost the same amount of ^{137}Cs, ^{90}Sr, or ^{238}Pu. Data in Table V confirm this. The concentrations of gamma, beta, and alpha activity in the respective simulated groundwaters are equal regardless of the SA/V in the test. This dependence of normalized mass loss on SV/V in long-term tests has also been observed by Oversby(10) and Pederson, et al.(11).

Table II. 28 Day MCC-1 Leach Results for Radioactive SRP 131/Tank 11 Glass*

Leachant	SA/V cm^{-1}	Normalized Mass Loss, g/m^2		
		^{137}Cs	^{90}Sr	^{238}Pu
Deionized H_2O	0.1	1.3 **	0.32 **	0.37 **
Brine†	0.03	0.68	1.5	0.7
Silicate††	0.1	0.25	0.19	0.13

* T = 40°C.
** Standard deviation in three separate tests was: ^{137}Cs, 0.4; ^{90}Sr, 0.08; ^{238}Pu, 0.02
† Composition of brine, g/l; KCl, 48.2, NaCl, 90.0, and $MgCl_2$, 116.0. pH = 6.5.
†† Composition of silicate water, g/l; $NaHCO_3$, 0.179; SiO_2, 0.058. pH = 7.5.

Table III. pH Changes for 28 Day MCC-1 Leach Tests for Radioactive SRP 131/Tank 11 Glass*

Leachant	pH					
	Test 1		Test 2		Test 3	
	Initial	Final	Initial	Final	Initial	Final
Deionized H_2O	6.8	7.6	6.9	7.9	6.2	8.4
Brine***	6.5	6.6	6.5	**	6.5	**
Silicate H_2O***	7.5	8.4	7.5	6.3	7.5	9.3

* T = 40°C.
** Data for final pH not available.
*** Compositions are listed in Table II.

Table IV. Long-Term Leach Results for Radioactive SRP 131/Tank 11 Glass*

Leachant	Time, days	SA/V cm^{-1}	Normalized Mass Loss, g/m^2		
			^{137}Cs	^{90}Sr	^{238}Pu
Brine**	207	0.02	2.2 $\pm$0.5***	2.8 $\pm$0.7***	1.4 $\pm$0.3***
	197	0.04	1.2	1.7	0.83
Silicate H_2O**	192	0.1	0.30	0.31	0.18
	211	0.2	0.09 $\pm$0.01***	0.12 $\pm$0.03***	0.12 $\pm$0.05***

* T = 40°C.
** Compositions are listed in Table II.
*** Average value determined in duplicate tests.

Table V. Leachant Activity for Long-Term Leach Tests of Radioactive SRP 131/Tank 11 Glass*

Leachant	Time, days	SA/V cm^{-1}	Leachate Activity, dpm/ml ^{137}Cs	^{90}Sr	Alpha
Brine**	217	0.02	(9.6 $\pm$1.5)E03***	(2.9 $\pm$0.8)E04***	(3.4 $\pm$0.5)E02***
	197	0.04	9.7E03	3.8E04	3.8E02
Silicate H_2O**	192	0.1	6.5E03	1.8E04	2.2E02
	211	0.2	(4.0 $\pm$0.1)E03***	(1.3 $\pm$0.3)E04***	(2.8 $\pm$1.0)E02***

* T = 40°C, results for normalized mass losses are given in Table IV.
** Compositions are listed in Table II.
*** Average value determined in duplicate tests.

In regard to long-term geologic storage of borosilicate nuclear waste glass, these results imply that the surface area of the glass is not the main factor that will control the amount of radionuclides released. The main factors will be the volume of groundwater available, the solubility of radionuclides in that groundwater, and the formation of a gel layer on the glass that inhibits diffusion from the glass. Surface area of the glass will be an important factor in determining how fast these equilibrium concentrations will be reached. Considering that the residence time of water in an emplacement hole in a repository will probably be greater than 100 years,(5) saturation of this water and gel layer formation will easily be achieved. Recently, the use of solubilities has been employed to demonstrate the effectiveness of geologic storage for nuclear waste(12).

pH Changes in Long-Term Tests. The initial and final pH values in the long-term tests are shown in Table VI. Values for the longest test in deionized water (day 82 to day 220 in Figure 1) are also included. The values for deionized water and silicate water are consistent with pH changes caused by leaching, i.e. the pH increased. The small changes observed for the tests in brine are consistent with earlier data(8). During these long-term tests, the leachant was continuously being irradiated by gamma rays and alpha and beta particles from the glass. The dose to the leachant was ~10 Mrad. The data in Table VI clearly indicate that this dose is not affecting the pH change during leaching by producing HNO_3 from radiolysis of the N_2, O_2, and H_2O present. Production of HNO_3 and a lowering of the pH has been observed in leach tests where the entire system (leachate, glass, and air) was being simultaneously irradiated by an external ^{60}Co γ-ray source(13-14). Also, other work has shown that radiolysis of moist air produces HNO_3(15), while the gamma radiolysis of water and dissolved air in the absence of a vapor phase produces no HNO_3(16). In the tests described here with radioactive glass, the radiation dose to the vapor phase during the test was small because of the low gamma flux coming from the glass (~10 rads/hr based on ion chamber measurements); thus, the amount of HNO_3 formed in this manner would be small. Also, radiolysis of the leachant was primarily from fission product beta particles, which have the same effect as gamma rays(17). Thus, little HNO_3 would be formed by this process. Consequently, the HNO_3 production was insufficient to effect the pH change due to leaching. The above results and discussion are in complete accord with conclusions presented in a recent review on this subject by Burns, et al. (18).

Conclusions

The data presented in this paper support the following conclusions concerning the performance of SRP radioactive nuclear waste glass in deionized water and in two simulated groundwaters.

Table VI. pH Changes for 200 Day Leach Tests of Radioactive SRP 131/Tank 11*

	pH					
	Test 1		Test 2		Test 3	
Leachant	Initial	Final	Initial	Final	Initial	Final
Deionized H_2O	6.7	8.0	6.8	8.5	-	-
Brine**	6.5	5.9	6.5	6.3	6.5	5.7
Silicate H_2O**	7.5	8.4	7.5	8.2	7.5	8.5

* $T = 40^{\circ}C$, SA/V = 0.02 - 0.2.
** Compositions are listed in Table II.

1. Leach behavior of borosilicate glass containing actual radioactive SRP high-level waste is similar to glass containing nonradioactive simulated waste.
2. Release rates in MCC brine and silicate water are lower than in deionized water.
3. Radiolysis of the leachants by radiation from the glass does not affect the leaching process nor does it produce sufficient HNO_3 to cause a pH decrease.
4. In long-term tests (1 yr), the amount of ^{238}Pu, ^{90}Sr, and ^{137}Cs released are affected more by the radionuclide solubility in the leachant and layer formation on the glass than by the surface area of the glass.
5. Stainless steel does not affect the leaching chemistry.

Site-Specific Repository Testing Program at Savannah River Laboratory

The above conclusions are all based on data for a generic repository system. Each of these conclusions will now be tested under site-specific conditions. This is one of the objectives of the current program being developed at SRL. SRP borosilicate waste glass (both radioactive and simulated) will be leached under conditions expected in three site-specific repositories. These are salt, tuff, and basalt. SRL is currently working with developers of these repositories to estimate long-term storage conditions so that realistic testing parameters can be established. One unique feature of these tests will be that the actual repository rock will be used for the primary leaching vessel.

Acknowledgment

I wish to acknowledge D. Z. Nelson and F. A. Cheek, III for their technical support and their excellent abilities on the master-slave manipulators while performing the leach tests.

The information contained in this article was developed during the course of work under Contract No. DE-AC09-76SR00001 with the U.S. Department of Energy.

Literature Cited

1. Environmental Assessment. Waste Form Selection For SRP High-Level Waste, DOE/EA-0179, US Department of Energy, Washington, DC, 1982.

2. Bibler, N. E.; Smith, P. Kent, "Characterization of Borosilicate Glass Containing Savannah River Plant Radioactive Waste." To be published in Proceedings of the Materials Characterization Conference, August 16-18, 1982, Alfred, NY.
3. Walker, D. D.; Wiley, J. R.; Dukes, M. D.; Leroy, J. H. Nuclear and Chemical Waste Management, 1982, 3, 91.
4. Wicks, G. G.; Robnett, B. M.; Rankin, W. D., "Scientific Basis for Nuclear Waste Management"; Elsevier Publishing Co.: New York, 1982, Vol. V, p. 15.
5. Plodinec, M. J.; Wicks, G. G.; Bibler, N. E., "An Assessment of Savannah River Borosilicate Glass in the Repository Environment"; DP-1629, E. I. du Pont de Nemours & Co., Savannah River Laboratory, Aiken, South Carolina, 1982.
6. Bibler, N. E., "Characterization of Borosilicate Glass Containing Savannah River Plant Radioactive Waste. 2. Microstructure and Durability." To be published in the Proceedings of XIII International Congress on Glass, Hamburg, Germany, July 3-8, 1983.
7. Nuclear Waste Materials Handbook Waste Form Test Methods, MCC-1P Static Test, DOE/T1C-11400, Pacific Northwest Laboratory, Richland, Washington, 1981.
8. Strachan, D. M., in "Scientific Basis for Radioactive Waste Management," Lutze, W., Ed; Elsevier Science Publishing Co: New York, 1982; Vol. V., p. 182.
9. Lokken, R. O.; Strachan, D. M., "Long-Term Leaching of SRL Glass and SYNROC"; PNL-SA-10976, Pacific Northwest Laboratory, Richland, Washington, 1983.
10. Oversby, V. M., "Leach Testing of Waste Forms Interrelationship of ISO and MCC Type Tests"; UCRL-87621, Lawrence Livermore Laboratory, Livermore, California, 1982.
11. Pederson, L. R.; Buckwalter, C. Q.; McVay, G. L., "The Effects of Surface Area to Solution Volume on Waste Glass Leaching." PNL-SA-10306, Pacific Northwest Laboratory, Richland, Washington, 1982.
12. Chambre, P. L.; Pigford, T. H.; Zavoshy, S., "Solubility-Limited Dissolution Rate in Groundwater." UCB-NE-4016, University of California, Berkeley, California, 1982.
13. McVay, G. L.; Pederson, L. R. J. Amer. Cer. Soc. 1981, 64, 154.
14. Barkatt, A.; Barkatt, A.; Sousanpour, W. Nature, 1982, 300, 339.
15. Jones, A. R. Radiation Research, 1959, 10, 655.
16. Wright, J.; Linacre; J. K.; Marsh, W. R.; Bates, T. H. in Proc. Intern. Conf. Peaceful Uses Atomic Energy: UN-Geneva, 1955; Vol. 7, p. 560.
17. Saver, M. C.; Hart, E. J.; Flynn, K. F.; Girdler, I. E., "A Measurement of the Hydrogen Yield in the Radiolysis of Water by Dissolved Fission Products." ANL-76-46, Argonne National Laboratory, Argonne, Illinois, 1976.
18. Burn, W. G.; Hughes, A. E.; Marples, J. A. C.; Nelson, R. S.; Stoneham, A. M. J. Nucl. Materials 1982, 107, 245.

RECEIVED October 20, 1983

24

Leach Resistance of Iodine Compounds in Portland Cement

R. D. SCHEELE, L. L. BURGER, and K. D. WIEMERS

Pacific Northwest Laboratory, Richland, WA 99352

Several radioactive iodine isotopes arise from the production of electricity by nuclear power. Most of these are short-lived with half-lives less than a few days and will have decayed to negligible levels by the time the spent nuclear fuel is reprocessed; however, ^{129}I has a half life of 16 million years. Table I presents its radiological properties, environmental inventory and nuclear production rates. Because of the biological significance of iodine (it accumulates in the thyroid gland) and the amount of ^{129}I arising from the generation of electricity of nuclear power, ^{129}I is a potential hazard if released uncontrolled to the environment.

Table I. ^{129}I Properties, Inventory and Nuclear Production Rate

Properties	
• Radiation	β^-, 0.156 MeV max (0.06 MeV Ave), γ, 0.04 MeV
• Half Life	16 My
• Specific Activity	1.7×10^{-4} Ci/g
Environmental Inventory	
• Natural	40 Ci
• Total	300-1000 Ci
Nuclear Production Rate (33,000 MWd/tonne U	
• 0.037 Ci/tonne U	
• 1.25 Ci/GWe-yr	
• 10^4 Ci estimated by yr 2000 (1)	
• 56 Ci/yr for a 1500 MTU/yr reprocessing plant	

The ultimate objective of any disposal form development project is to find an immobilization system which is prepared easily from the waste products of the recovery processes and will not allow hazardous material release during transport, interim storage, normal disposal, and in the event of a credible

0097-6156/84/0246-0373$06.00/0

occurrence such as groundwater intrusion into the disposal repository. For ^{129}I the time required for radioactive decay makes achievement of this goal very difficult. In our laboratory we have examined methods of iodine removal from gas streams, techniques for fixing the iodine in stable solid forms, and methods of disposal. One of the techniques which proved attractive was to incorporate an insoluble, stable iodine compound in a cement matrix. The present report describes the development and leach testing of this system using Portland III cement as the solidification matrix. Other aspects of the iodine work have been reported elsewhere.(2, 3)

Development Considerations

A satisfactory disposal form requires that the material may be easily made from the iodine capture technology in place, and that it be resistant to and stable at the conditions imposed by transportation and storage and disposal environments. Temporary storage does not appear to be a problem. Transportation requirements are minimum for ^{129}I requiring only proper labeling and transportation on sole-use vehicles.

Capture Technologies. When spent nuclear fuel is treated to recover the valuable fissile content at a fuels reprocessing facility (FRP), the preferred scheme is to cause the iodine to volatilize during fuel dissolution and be released to the process off-gas system (POG). It is thus separated from the bulk of the radioactive wastes. Several technologies have been developed for collecting gaseous iodine. These technologies and their primary products and, with additional treatment, their secondary products are presented in Table II. The organic traps, charcoal and other organic sorbers are used at nuclear power plants but cannot be used at FRP's because of their reactivity. Methods of iodine removal from gaseous streams have been reviewed by Holladay (4) and by Jubin (5). The overall technology requirements for ^{129}I management have been reviewed by Burger and Burns (6), in an IAEA document (7), and by McKay, Miquel, and White (8).

Inspection of Table II shows that a variety of iodine products are formed with the iodine in several oxidation states. Thus fixation materials with iodine as complexed elemental iodine, iodide, and iodate, must be evaluated, and the solid fixation matrix selected must be able to accept a variety of materials.

Disposal Strategy. Two basic strategies for radioactive waste disposal have been developed: (1) isolation and (2) dispersion. For isolation, the waste can be stored in a dry geologic respository. Space disposal has been examined but is not considered atrtractive at the present (3, 9). ^{129}I dispersion in the ocean is attractive because of the large volume of the ocean

Table II. Iodine Capture Technologies

Technology	Product(s)	Potential Products after Secondary Treatment
Caustic scrubbers	5 NaI:1 $NaIO_3$	Insoluble iodate or iodide
Mercuric nitrate-nitric acid (Mercurex)	HgI^+	$Hg(IO_3)_2$, $Ba(IO_3)_2$, $Sr(IO_3)_2$, CuI
20-22 M HNO_3 (Iodox)	Iodic acid	Insoluble iodate or iodide
Solid sorbents containing silver:		
Silver mordenite (AgZ)	AgZI,	Insoluble iodate or other metal zeolite
Silver zeolite Type X (AgX)	AgXI	
Silver nitrate impregnated silica	AgI	
Purex Silver reactors	AgI, $AgIO_3$	
Charcoal	Iodine loaded charcoal, I_2	
Organic sorbers	Iodine loaded material, I_2	

and its large content of nonradioactive iodine. The released ^{129}I could be diluted to harmless levels (2, 10).

If the waste is isolated in a geologic repository, the iodine form should be stable to at least 100°C and possibly at 250°C depending on the repository site. If the waste form satisfies the thermal stability requirement, the most likely release mechanism then becomes leaching in the event that groundwater contacts the immobilization form. Allard et al. (11) report log Kd values for silicate minerals ranging from -0.5 to -3.5. Fried et al. (12) found little retention of iodine (as iodide or iodate) by Los Alamos Tuff. Thus, once the iodine has been removed by leaching, it will potentially move at the same velocity as the groundwater.

Ocean disposal requires that the ^{129}I release rate (by leaching) must be less than the mixing rate of the ocean to insure adequate mixing of the ^{129}I with stable iodine. No thermal stability requirements are seen for this dispersion strategy.

Immobilization Form Selection

Matrix. A matrix material to encapsulate the chosen iodine compound for storage, transportation, or disposal purposes is desirable regardless of the management mode chosen; several alternatives exist: bitumen (asphalt), synthetic minerals, plastic, glasses, and Portland cement.

Bitumen is attractive because it is relatively impermeable to water and presents a reducing, nonhydrolytic environment for the fixation compounds. A disadvantage is combustibility, of particular concern during the packaging operations. Minerals have not received much attention although preliminary work with sodalite was promising (13, 14), and the leachability of samples was very low (3). Organic polymers could provide good short term protection. Disadvantages would include chemical and radiation instability over very long time periods and potential fire hazards. Glasses have been examined at Rennes University (15) but have not proved completely satisfactory.

Cement was chosen for the present study because of its low cost, simplicity of use, and stability in most environments. It is, however, rather porous. Also, in the presence of water, hydrolysis reactions promoted by the high pH of the cement can cause conversion reactions resulting in the release of iodine from many compounds. Type III Portland cement was chosen because of its early strength (curing time 8 days).

Iodine Compounds. Since contact with water is the most likely release mechanism in either of the terrestrial disposal operations, we initially screened iodide and iodate compounds on the basis of their solubilities. Tables III and IV present rankings of the low solubility iodides and iodates, respectively, with the least soluble at the top.

Since we selected cement as our solidification matrix, it is also desirable to use an iodine compound which is resistant to hydrolysis. Tables III and IV also include solubility data for the hydroxide analogs of the low solutility iodides and iodates, respectively.

The only iodides which are more insoluble than their hydroxide analogs are Hg(I), Ag, and Tl. It should be mentioned that even though a compound has a less soluble hydroxide analog, the rate of conversion may be slow enough for the iodine release rate to be acceptable.

Additional factors used to screen candidate iodine fixation compounds were thermal and chemical stability and volatility. Of the low solubility iodides, those of Ag, Cu(I), Pb, Pd, and Tl meet the arbitrary 250°C stability requirement. Several, such as those of Bi and Hg, have excessive vapor pressures. Many of the iodates show excellent thermal stability including those of the alkaline earths, rare earths, Ag, Cu, Pb, Zn, Hg, Th, and U. Several, including $AgIO_3$ and $Hg(IO_3)_2$ convert to the iodide on heating(19).

Also considered was the thermodynamic resistance of the iodides to reaction with oxygen. Of the low solubility iodides, only AgI and Tl(I) were stable.

TABLE III. Ranking of Iodide Compounds by Solubility

	Compound	Solubility Product(16)	$[I^-]$ or $[OH^-]$, M
1.	Hg_2I_2	4.5×10^{-29}	6.7×10^{-15}
	$Hg_2(OH)_2$	2.0×10^{-24}	1.4×10^{-12}
2.	AgI	8.3×10^{-17}	9.1×10^{-9}
	AgOH	2.0×10^{-8}	1.4×10^{-4}
3.	PdI_2	2.5×10^{-23}(17)	3.7×10^{-8}
	$Pd(OH)_2$	1.0×10^{-31}	5.8×10^{-11}
4.	CuI	1.1×10^{-12}	1.0×10^{-6}
	CuOH	1.0×10^{-14}	1.0×10^{-7}
5.	BiI_3	8.1×10^{-19}	3.9×10^{-5}
	$Bi(OH)_3$	4.0×10^{-31}	3.0×10^{-8}
6.	HgI_2	1.1×10^{-12}	1.3×10^{-4}
	$Hg(OH)_2$	3.0×10^{-26}	3.9×10^{-9}
7.	TlI	6.5×10^{-8}	2.5×10^{-4}
	TlOH	2.4	1.6
8.	PbI_2	7.1×10^{-9}	2.4×10^{-4}
	$Pb(OH)_2$	1.2×10^{-15}	1.3×10^{-5}

Oxidation may be a problem only prior to closure of the waste form canister since the earth's crust is predominantly reducing. However, reduction of iodates becomes possible after water intrusion and oxidation of iodides can occur on reaching atmospheric environment. Allard et al. (11) predict that any released iodine in the groundwater will be as iodide.

Based on this screening and consideration of the products of the iodine capture technologies, AgI, CuI, PbI_2, $Ba(IO_3)_2$, $Ca(IO_3)_2$, $Pb(IO_3)_2$, $Hg(IO_3)_2$, iodine-loaded silver mordenite (AgZI) and iodine-loaded lead zeolite X (PbXI) were selected for incorporation into cement and subsequent leach testing.

Leach Resistance

Previous Studies. Clark (20), Morgan et al. (21), Partridge and Bosuego (22), and the Pacific Northwest Laboratory (PNL) (23-27) have performed leaching studies on selected iodine compounds in cement, evaluating various cement additives or coatings.

Clark used a modified IAEA dynamic leach simulation test to evaluate $Ba(IO_3)_2$, $AgIO_3$, $Hg(IO_3)_2$, and $Pb(IO_3)_2$ in Portland type I cement as disposal forms for ^{129}I captured using the Iodox process. Of the iodates tested, $Ba(IO_3)_2$ in cement was generally superior and gave the most consistent results; some samples with $Hg(IO_3)_2$ and $AgIO_3$ loadings performed comparably. Of the coatings and additives tried, addition of butyl stearate proved the most effective and reduced the leach rate by a factor of three.

TABLE IV. Ranking of Iodates by Solubility

	Compound	Solubility Product(16)	$[IO_3^-]$ or $[OH^-]$, M
1.	$Hg_2(IO_3)_2$	2.0×10^{-14}	1.4×10^{-7}
	$Hg_2(OH)_2$	2.0×10^{-24}	1.4×10^{-12}
2a.	$Hg(IO_3)_2$	3.2×10^{-13}	8.6×10^{-5}
	$Hg(OH)_2$	3.0×10^{-26}	3.9×10^{-9}
2b.	$Pb(IO_3)_2$	3.2×10^{-13}	8.6×10^{-5}
	$Pb(OH)_2$	1.2×10^{-15}	1.3×10^{-5}
3.	$AgIO_3$	3.0×10^{-8}	1.7×10^{-4}
	$AgOH$	2.0×10^{-8}	1.4×10^{-4}
4.	$CuIO_3$	1.4×10^{-7}	3.7×10^{-4}
	$CuOH$	1.0×10^{-14}	1.0×10^{-7}
5.	$Fe(IO_3)_3$	1.6×10^{-15}	6.0×10^{-4}
	$Fe(OH)_3$	4.0×10^{-38} (18)	5.9×10^{-10}
6.	$Ce(IO_3)_4$	5.0×10^{-17}	7.0×10^{-4}
	CeO_2	8.0×10^{-37}	
7.	$Ba(IO_3)_2 \cdot 2H_2O$	1.5×10^{-9}	1.4×10^{-3}
	$Ba(OH)_2$	5.0×10^{-3}	2.1×10^{-1}
8.	$Th(IO_3)_4$	2.5×10^{-15}	1.6×10^{-3}
	$Th(OH)_4$	4.0×10^{-45}	1.7×10^{-9}
9.	$TlIO_3$	3.1×10^{-6}	1.8×10^{-3}
	$TlOH$	2.4	1.6
10.	$La(IO_3)_3$	6.1×10^{-12}	2.1×10^{-3}
	$La(OH)_3$	2.0×10^{-19}	2.8×10^{-5}
11.	$Ni(IO_3)_2$	1.4×10^{-8}	3.0×10^{-2}
	$Ni(OH)_2$	2.0×10^{-15}	1.6×10^{-5}
12.	$In(IO_3)_3$	1.2×10^{-12}	3.1×10^{-3}
	$In(OH)_3$	6.3×10^{-34}	6.6×10^{-9}
13.	$Zn(IO_3)_2$	2.0×10^{-8}	3.4×10^{-3}
	$Zn(OH)_2$	1.2×10^{-17}	2.9×10^{-6}
14.	$UO_2(IO_3)_2$	3.2×10^{-8}	4.0×10^{-3}
	$UO_2(OH)_2$	1.1×10^{-22} (18)	6.0×10^{-8}
15.	$Pu(IO_3)_4$	5.0×10^{-13}	5.0×10^{-3}
16.	$Cu(IO_3)_2$	7.4×10^{-8}	5.3×10^{-3}
	$Cu(OH)_2$	2.2×10^{-20}	3.5×10^{-7}
17.	$Ce(IO_3)_3$	3.2×10^{-10}	5.6×10^{-3}
	$Ce(OH)_3$	1.6×10^{-20}	1.5×10^{-5}
18.	$Sr(IO_3)_2$	3.3×10^{-7}	8.7×10^{-3}
	$Sr(OH)_2$	6.0×10^{-5} (18)	8.0×10^{-2}
19.	$Ca(IO_3)_2 \cdot 6H_2O$	7.1×10^{-7}	1.1×10^{-2}
	$Ca(OH)_2$	5.5×10^{-6}	2.2×10^{-2}
20.	$Co(IO_3)_2$	1.0×10^{-4}	5.8×10^{-2}
21.	$Bi(IO_3)_3$	Insoluble	
	$Bi(OH)_3$	4.0×10^{-31}	3.3×10^{-8}

Morgan et al. evaluated the effect of water-to-cement ratios, curing time, iodine content, gamma irradiation, and different leachant compositions on iodine leaching from 9-15 wt% $Ba(IO_3)_2$ cement using a modified IAEA leach test. Their studies showed that only the leachant composition caused a significant change in leach rates. Simulated seawater, tap water, and spring water showed 10^4 times lower leach rates than distilled water. This reduction was attributed to the formation of a protective surface film which slowed iodine loss.

Partridge and Bosuego studied the leaching of 26.9 wt% $Hg(IO_3)_2$ in Portland type I-II cement. The $Hg(IO_3)_2$ was produced by a process that uses HNO_3 to oxidize iodine in Mercurex solution. In addition to their test using recovered pure $Hg(IO_3)_2$, another test using 2.9 wt% $Hg(IO_3)_2$ and 24.8 wt% $Hg(NO_3)_2$ was performed. They used a simulated dynamic leach procedure in both cases and observed a leach rate a factor of 2 greater for the former test based on the fraction leached.

Preliminary studies at PNL (23-27) consisted of three screening tests to evaluate the effects of leaching method, environmental parameters, cement additives and coatings, and the nature of iodine compound on iodine leachability from Portland type III cement. Summarizing the most significant factors in these tests: 1) for pure iodine compounds in cement, approach to dynamic leach conditions increased leach rates, 2) higher temperature increased leach rates, 3) high carbonate leachant concentration decreased leach rates, 4) iodine loading in cement of >5 wt% increased leach rates, 5) low iodine compound solubility decreased leach rates, and 6) no additives or coatings tested had a significant effect.

Altomare et al. (28) assessed the status of waste management of ^{129}I. Using the data of Clark (20) and Morgan et al. (21) for $Ba(IO_3)_2$-cement, Altomare and coworkers estimated that under dynamic conditions all of the iodine would be leached from the waste form in 10^4 years, a short time relative to the lifetime of ^{129}I.

Experimental. Characteristics of a typical iodine-containing cement monolith are listed in Table V. The monoliths were suspended in glass beakers containing the leachant at ambient temperature (19-23°C), and the tests were peformed in air. The leachant-to-surface-area ratio was 10:1. The results are presented as the normalized fraction leached, which is calculated by multiplying the fraction leached by the sample's geometric volume-to-surface area ratio, normally 0.5 cm.

Two leach test procedures were used: a modified IAEA dynamic leach test and a static leach test. In the former test the leachant was gently agitated on a mechanical shaker and was changed at selected intervals. A typical sampling schedule was three times daily for the first week, twice per day for week 2, once per day for weeks 3 through 5, and once per week through week 9.

TABLE V - Typical Characteristics of Iodine-Cement Monoliths

Property	Value
Iodine concentration, mmol I/g dry cement	0.8
Cement type	Portland Type III
Water/cement weight ratio	0.3
Height, cm	4.5
Diameter, cm	2.54
Geometric volume, V, cm^3	23.0
Geometric surface area, S, cm^2	45.0
V/S, cm	0.5
Curing time	7 days

The leachate was distilled water. The iodine forms tested by the modified IAEA method were AgI, $Ba(IO_3)_2$, $Ca(IO_3)_2$, and AgZI, all in Portland type III cement. In the static leach test, the leachant was sampled on days 3, 7, 14, 56, 112, and 180. The sample was replaced with an equal volume of leachate preequilibrated with a pure cement sample. The iodine forms tested by the static method were AgI, CuI, PbI_2, $Ba(IO_3)_2$, $Ca(IO_3)_2$, $Hg(IO_3)_2$, PbXI, and AgZI all in Portland type III cement. Distilled water was the leachate. For $Ba(IO_3)_2$ and $Hg(IO_3)_2$, seawater and Columbia River water leachates were also tested.

An iodide-specific ion electrode was used to analyze the leachate. For iodates, the iodine was first reduced to I^- using hydroxylamine. The detection limit was $\sim 10^{-7}$ $\underline{M}$ I^-.

Results. The CuI and PbI_2 cements lost iodine very rapidly, e.g., for CuI, 20% in 15 days. Even though CuI and PbI_2 have low solubilities, they are susceptible to oxidation and hydrolysis in the alkaline environment of the cement. Also, there are no cations present in the cement to lower the solubility of the released iodine. For these forms, bitumen may be a preferred matrix.

The AgI-cement exhibited the best leach resistance of all forms tested. Figure 1 presents the results for the static and dynamic leach tests of AgI cements. The leach rate unit of cm/d may be converted to fraction leached per day by multiplying by the surface area-to-volume ratio. Using the data presented in Figure 1 for the dynamic leach test between days 40 and 100 and assuming a linear extrapolation, over 4000 years are required to leach 1% of the iodine from a 208 L (55 gal) cement monolith. A 208 L steel drum is the typical waste container used for disposal of low activity radioactive waste and is used in this report as the standard waste package.

The results from the static leach tests for Ba, Hg, and Ca iodates are presented in Figures 2, 3, and 4, respectively. Little difference is seen between these forms except for $Ba(IO_3)_2$

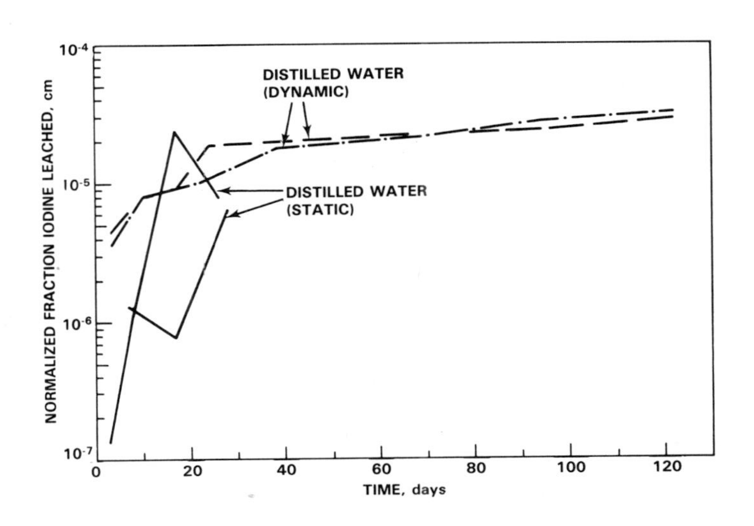

FIGURE 1. Static and Dynamic Leach Tests of AgI-Cement

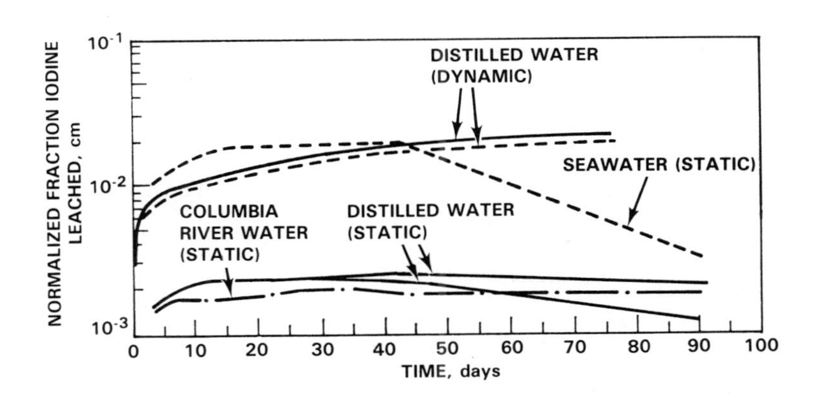

FIGURE 2. Static and Dynamic Leach Tests of $Ba(IO_3)_2$-Cement

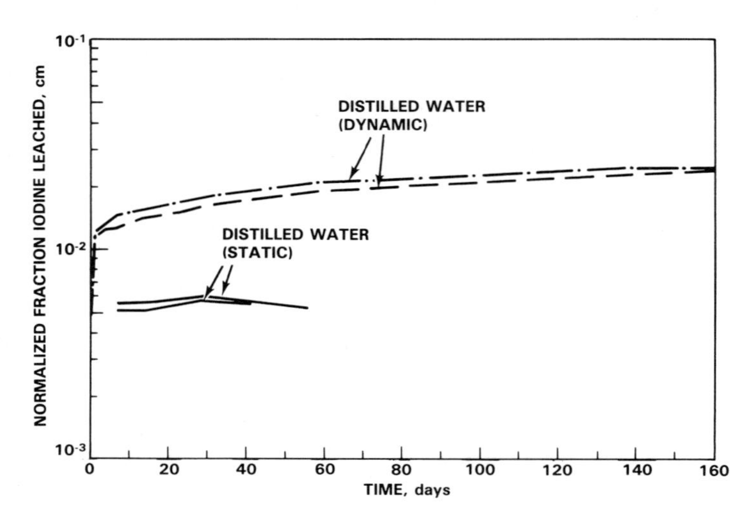

FIGURE 3. Static and Dynamic Leach Tests of $Ca(IO_3)_2$-Cements

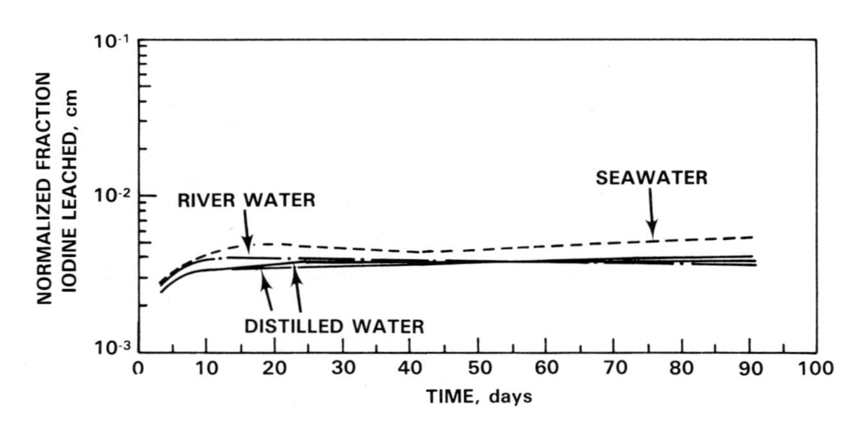

FIGURE 4. Static Leach Test of $Hg(IO_3)_2$-Cement

in seawater which exhibits a much higher leach rate than in other waters. No explanation has been determined for the latter result which is in contrast to that found by Morgan et al. (22) for $Ba(IO_3)_2$. Figure 2 also shows the dynamic leach results for $Ba(IO_3)_2$ in cement. Comparison of the dynamic leach test results shown in Figures 2 and 4 shows $Ca(IO_3)_2$ to behave similarly to $Ba(IO_3)_2$. Note also that the fraction leached is a factor of 10 greater than in the static tests. From the dynamic test results in Figure 2, using a linear extrapolation of the 40 to 70 day data, 5 years would be required to leach 1% of the iodine from a 208 L monolith. This estimated leach rate is comparable (factor of 3 less) to that found for 9 wt% iodine as $Ba(IO_3)_2$ by Clark (20).

In solubility tests of the pure compounds in distilled and cement-equilibrated water, iodate concentrations in both liquids were near the expected theoretical values for $Ba(IO_3)_2$ and $Ca(IO_3)_2$. For $Hg(IO_3)_2$ the concentration in distilled water corresponded to the expected value; however, the concentration in the cement-water was a factor of ~1000 higher. This higher concentration was attributed to hydrolysis due to the high pH (~12) of the cement-water. Hydrolysis would account for the release of iodine from the $Hg(IO_3)_2$-cement. The iodate concentrations found in the leachates of the cements containing $Ba(IO_3)_2$ and $Ca(IO_3)_2$ were a factor of 10 lower than expected. No metallic ion concentration was sufficiently high to cause precipitation of the iodate.

In the static leach test of AgZI in cement, presented in Figure 5, iodine leached from the AgZI cement at nearly the same rate as from the iodate cements. In the dynamic leach test, also presented in Figure 5, the leach rate was a factor of 10 faster than from AgI cement, and a factor of 10 slower than from $Ba(IO_3)_2$ cement. Based on dynamic leach test sample 1 in Figure 5, 100 years would be required to leach 1% of the iodine from a 208 L cement monolith. The iodine release rate from AgZI may depend on the iodine loading of the AgZ since a fraction of the iodine is apparently held as a less stable molecular complex with this fraction increasing with increasing iodine loading (2).

In contrast, iodine is readily lost from PbXI-cement (see Figure 6). When the concrete was prepared, the cement mixture turned purple suggesting release of elemental iodine, which in the alkaline environment of the cement, disproportionates to iodide and iodate.

Conclusion

Our studies have shown that there is a considerable difference in stability and leach resistance among the various iodine compounds in Portland III cement. Of the compounds examined, AgI had the best leach resistance followed by AgZI, $Ba(IO_3)_2$, $Ca(IO_3)_2$,

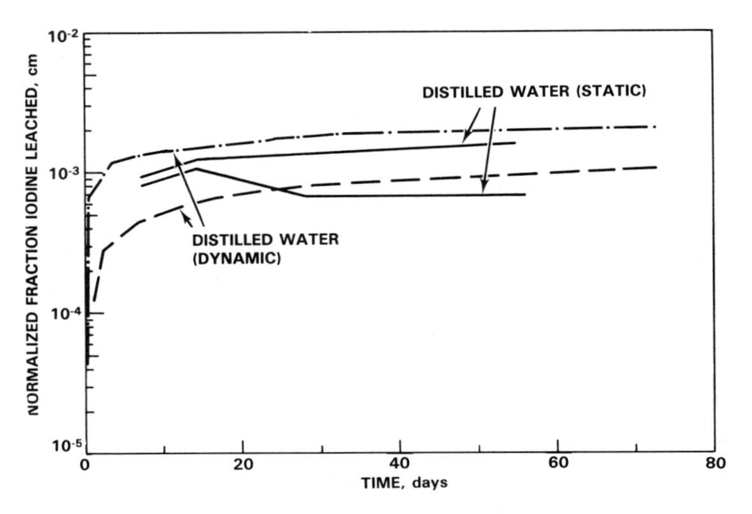

FIGURE 5. Static and Dynamic Leach Tests of AgZI-Cement

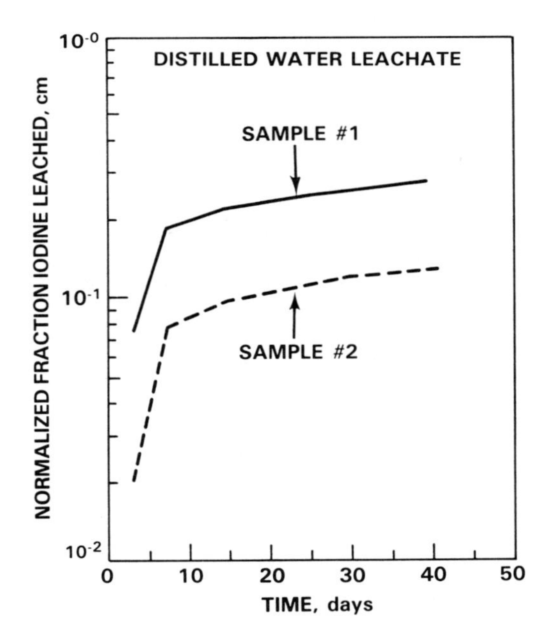

FIGURE 6. Static Leach Test of ~370 mgI/g PbX in Cement

$Hg(IO_3)_2$, and $Sr(IO_3)_2$. Long time extrapolations of leach rates are uncertain to one or two orders of magnitude, but the data indicate that if the disposal form comes in continuous contact with the water the ^{129}I will be released before it has decayed. Thus, good disposal strategy includes a mechanism for dispersal such that dilution to harmless levels is achieved. For interim storage, where a dry environment can be assured, and for transportation, the forms appear to be adequate.

Literature Cited

1. Russell, J. L.;Hahn, P. B. "Public Health Aspects of Iodine-129 from the Nuclear Power Industry." Radiol. Health Data Rep. 12:189, 1971.
2. Burger, L. L.; Scheele, R. D. "Iodine Fixation Studies at the Pacific Northwest Laboratory." In Management Modes for Iodine-129, W. Hebel and G. Cottone, eds. Harwood Academic Publishers, New York, 1982.
3. Burger, L. L.; Scheele, R. D.; Wiemers, K. D. Selection of a Form for Fixation of Iodine-129. PNL-4045, Pacific Northwest Laboratory, Richland, Washington, 1982.
4. Holladay, D. W. A Literature Survey: Methods for the Removal of Iodine Species from Off-Gases and Liquid Waste Streams of Nuclear Power and Fuel Reprocessing Plants with Emphasis on Solid Sorbents. ORNL/TM-6350, Oak Ridge National Laboratory, Oak Ridge, Tennessee, 1979.
5. Jubin, R. T. A Literature Survey of Methods to Remove iodine from Off-Gas Streams Using Solid Sorbents. ORNL/TM-6607, Oak Ridge National Laboratory, Oak Ridge, Tennessee, 1979.
6. Burger, L. L.; Burns, R. E. Technical Requirements for the Control of ^{129}I in a Nuclear Fuels Reprocessing Plant, PNL-3186, Pacific Northwest Laboratory, Richland, Washington, 1979.
7. IAEA. Radioiodine Removal in Nuclear Facilities. IAEA Technical Reports Series 201, International Atomic Energy Agency, Vienna, 1980.
8. McKay, H. A. C.; Miquel, P.; White, I. F. "Management Modes for Iodine-129." In Management Modes for Iodine-129, W. Hebel and G. Cottone, eds. Harwood Academic Publishers, New York, 1982.
9. Burns, R. E.; Defield, J. G. Disposal of Radioactive Iodine in Space. NASA Technical Paper 1313, Marshall Space Flight Center, Alabama, 1978.
10. Burger, L. L. Determining Criteria for the Disposal of Iodine-129, PNL-3496, Pacific Northwest Laboratory, Richland, Washington, 1980.
11. Allard, B.; Torstenfeld, B.; Andersson, K.; Rydberg, J. "Possible Retention of Iodine in the Ground." In Scientific Basis for Nuclear Waste Management, Vol. 2, p. 673, C. J. M. Northrup, ed., Plenum Press, New York, 1980.

12. Fried, S.; Friedman, A. M.; Cohen, D.; Hines, J. J.; Strickert, R. G. The Migration of Long-Lived Radioactive Processing Wastes in Selected Rocks. Ann. Rpt. Project AN0115A, FY 1977, ANL-78-46, Argonne National Laboratory, Argonne, Illinois, 1978.
13. Winters, W. I. The Effects of Hot-Pressing Conditions on the Properties of Iodine Sodalite. RHO-LD-153, Rockwell Hanford Operations, Richland, Washington, 1980.
14. Strachan, D. M.; Babab, H. Iodide and Iodate Sodalites for the Long-Term Storage of Iodine-129. RHO-SA-83, Rockwell Hanford Operations, Richland, Washington, 1979.
15. Malagani, J. P.; Wasniewski, A.; Doreau, M.; Robert, G.; Mercier, R. "Conductivite Electrique et Spectre de Diffusion Raman des Verres Mixtes $AgPO_3$-MI_3 avec M = Cd, Hg, Pb. Correlation entre Conductivite et Structure." Mat. Res. Bull. 13:1009, 1978.
16. Dean, J. A. Lange's Handbook of Chemistry. Eleventh Ed., McGraw-Hill Book Company, New York, New York, 1973.
17. Horner, D. E.; Mailen, J. C.; Bigelow, H. R. "Solubility of PdI_2 in Nitrate and Perchlorate Solutions.", CONF-761002-4, pp. 1-24, 1973.
18. Sidell, A. Solubilities of Inorganic and Metal Organic Compounds, Vol. 1. Third edition. D. Van Nostrand Company, Inc., New York, New York, 1953.
19. Duval, C. Inorganic Thermogravimetric Analysis. Trans. R. E. Oesper. Elsevier Publishing company, New York, 1963.
20. Clark, W. E. "The Isolation of Radioiodine with Portland Cement. Part 1: Scoping Leach Studies," Nuc. Tech. 36:215-221, 1977.
21. Morgan, M. T.; Moore, J. G.; Devaney, H. E.; Rogers, Williams, C.; Newman, E. "The Disposal of I-129." Paper presented at the Symposium on Science Underlying Radioactive Waste Management, November 28 - December 1, 1978, Boston, Massachusetts, 1978.
22. Partridge, J. A; Bosuego, G. P. Waste Management of Mercuric Nitrate Off-Gas Scrubber Solution. A Laboratory Study. HEDL-TME 79-11, Hanford Engineering Development Laboratory, Richland, Washington, 1979.
23. Wiemers, K. D. "Carbon-14 and Iodine-129 Fixation." Nuclear Waste Management Quarterly Progress Report October through December 1977. Eds. A. M. Platt, J. A. Powell. PNL-2377-4, Pacific Northwest Laboratory, Richland, Washington, 1978.
24. Wiemers, K. D. "Carbon-14 and Iodine-129 Fixation." Nuclear Waste Management Quarterly Progress Report January through March 1978. Eds. A. M. Platt , J. A. Powell. PNL-2378-1, Pacific Northwest Laboratory, Richland, Washington, 1978.
25. Wiemers, K. D. "Carbon-14 and Iodine-129 Fixation." Nuclear Waste Management Quarterly Progress Report January through March 1979. Eds. A. M. Platt, J. A. Powell. PNL-3000-1, Pacific Northwest Laboratory, Richland, Washington., 1979.

26. Wiemers, K. D. "Carbon-14 and Iodine-129 Fixation." Nuclear Waste Management Quarterly Progress Report April through June 1979. Eds. A. M. Platt, J. A. Powell. PNL-3000-2, Pacific Northwest Laboratory, Richland, Washington, 1979.
27. Wiemers, K. D. "Carbon-14 and Iodine-129 Fixation." Nuclear Waste Management Quarterly Progress Report July through September 1979. Eds. A. M. Platt, J. A. Powell. PNL-3000-3, Pacific Northwest Laboratory, Richland, Washington, 1979.
28. Altomare, P. M.; Barbier, M.; Lord, N.; Nainan, D. Assessment of Waste Management of Volatile Radionuclides. PB80-147754. The Mitre Corporation, McLean, VA, 1979.

RECEIVED November 3, 1983

LEGAL BACKGROUND

Nuclear Waste—A View from Washington, D.C.

MIKE MCCORMACK

McCormack Associates, Inc., 508 A Street, SE, Washington, DC 20003

If the American people were aware of the tremendous amount of scientific research--to say nothing of the outstanding quality of engineering development--that has gone into management policies and regulations for handling, transporting, and storing high-level and low-level radioactive waste, much of the emotionalism and fear associated with this subject would be eliminated.

It should be observed at this point that one of the responsibilities that falls on the shoulders of all scientists in this country is the scientists' obligation to help their fellow citizens, who have not had the advantage of technical or scientific training, to understand the realities of such subjects as nuclear waste, thus making it possible to legislate more intelligently and realistically in subject areas wherein technical and scientific facts are so important.

Last December the Congress enacted landmark legislation entitled, "The Nuclear Waste Policy Act of 1982." It establishes a number of federal programs for the management and storage of high-level nuclear waste and spent nuclear fuel. The U.S. Department of Energy (DOE) is presently developing programs for implementing that law. At the same time, the various state legislatures are preparing interstate compacts and submitting them to the Congress for approval, as called for in the Low-Level Radioactive Waste Policy Act of 1980. In addition, the new budget requests for FY 84 for nuclear waste research and development are now under consideration by the Congress.

Before reviewing these subjects in greater detail, it may be in order to call to mind several facts about nuclear waste that must be understood before any rational consideration of laws, programs, regulations, or research can proceed.

1. The first fact is that there are no scientific, technical, or economic obstacles that delay--let alone prevent--the safe handling of all nuclear wastes. A number of completely acceptable methods exist today for the removal of all high-level wastes from the biosphere, essentially forever, and certainly far beyond the time when they will have decayed away to activity levels below the natural uranium from which they came.

0097-6156/84/0246-0391$06.00/0

For chemists who are accustomed to severe peer review, the words "all" and "forever" have a slightly different meaning than they do for the average non-technical person. It is important that scientists recognize the discontinuity that exists between their professional language and the language of the average citizen. In the real world, it is an accurate statement of fact that the technology exists to totally remove all high-level nuclear waste from the biosphere, essentially forever; that inexpensive and simple technologies do exist to accomplish this. It is necessary to help others outside the scientific community to understand this fact.

2. The second fact is that there are a number of satisfactory methods for solidifying high- or low-level nuclear wastes for permanent disposal. One system or another may be preferred for a specific situation, and there will undoubtedly be several schools of thought as to which one is best in each case.

3. The third fact is that spent nuclear fuel is not waste. Spent nuclear fuel contains 2% to 3% waste, but is about 97% recoverable uranium and plutonium. Each bundle has the potential electric energy equivalent of more than 10 million barrels of oil. High-level nuclear wastes consist of fission products and actinides that are extracted from spent fuel, but not saved for commercial use or research. Spent fuel may be temporarily stored until it is reprocessed to separate the waste from the valuable plutonium and uranium. The remaining glassified waste will then be permanently entombed.

4. The fourth fact is that there will be no high-level waste extracted from spent commercial nuclear fuel until there is a reprocessing program with a chemical separation facility to isolate the waste from the reuseable fuel. Since this will not be accomplished during this decade, there will not be any high-level waste available from commercial nuclear fuel for geologic storage before 1990.

5. The fifth fact is that the most dangerous ingredients of high-level waste do not (contrary to some popular misconceptions) last for geologic periods of time. Highly active radio-isotopes cannot be long-lived. This is a contradiction in terms. The activity level of the most radioactive isotopes in waste, such as cesium-137 and strontium-90, decay in 1,000 yr by a factor of 10 billion, and glassified waste of that age would be scarcely more hazardous than the natural uranium ore mined today in the Western United States.

6. The sixth fact is that extremely demanding and rigid regulations, supported by exhaustive research and the highest standards of engineering, have been established for the management, transportation, and storage of all radioactive wastes--from the least dangerous low-level wastes originating in hospitals and research laboratories to the most dangerous high-level wastes from spent nuclear fuel. Strict compliance with every detail of every regulation reduces the threat to the environment or to public health and safety to a level substantially below that from any one of a multitude of non-nuclear sources. Accordingly, all such regulations must be observed religiously, thus insuring that any threat to the environment or to human health and safety shall continue to decline to levels below its present low level, even as the people of this country and the world depend more and more on nuclear medicine, research, industrial applications, and power production.

Another piece of legislation, The Low-Level Radioactive Waste Policy Act of 1980, establishes the policy that each state is responsible for disposal of the commercial low-level waste generated within its borders and that such waste can be most safely and efficiently managed on a regional basis. The law encourages states to enter into interstate compacts to establish and operate regional disposal facilities for low-level waste. This law requires subsequent approval of the compacts by the Congress.

Perhaps the most significant provision of the act was the authority given to Congressionally approved, compact states to exclude waste generated outside their compact region after January 1, 1986. Today, individual states with waste sites cannot refuse waste without being in violation of the commerce clause. Thus, the right of a region with an operating compact to refuse to accept waste from an outside state provides an incentive for the states of each region to form a compact and select a site within their region to store low-level waste. Furthermore, a state that refuses to join a compact will be obligated, starting in 1986, to take care of its own waste. This "stick" was inserted to encourage state legislators, many of whom did not realize that most low-level waste is generated in hospital and other health facilities, to face up to their responsibility for the benefit of the people they represent.

Passage of the policy act was a major step toward addressing a problem that had previously received far less attention than the issues of high-level waste disposal and interim storage of commercial spent fuel. This lack of interest was easy to understand. As long as there were three commercial sites open and providing adequate storage space, there was little incentive for other states to consider opening additional sites--to become embroiled in the troublesome issues involved in siting, licensing, constructing, and operating new sites. This situation was clearly an untenable one. It placed three states--South Carolina, Washington, and Nevada--in the position of shouldering the entire responsibility for disposing of waste generated in all fifty states. The fundamental unfairness and inefficiency of this system was brought to light in the summer of 1979 when the Governors of Nevada and Washington temporarily closed those sites, and the Governor of South Carolina announced that the amount of waste accepted at the Barnwell Site would be significantly reduced in the future. By these actions, South Carolina, Washington, and Nevada sent a clear warning signal to other states and to the federal government that action needed to be taken. An immediate crisis was averted, however, when the Washington and Nevada sites were reopened in late 1979.

In response to the 1980 enactment of the low-level waste law, compacts have been drafted and adopted or are under consideration in most of the states. In the northwest, a proposed compact has been ratified by Hawaii, Idaho, Oregon, Utah, and Washington; it is now before the legislatures of Alaska and Montana.

The Northwest Interstate Compact on Low-Level Radioactive Waste Management recognizes that low-level wastes are generated in constructive and essential activities, that they must be managed properly within the region with a minimum of handling and transportation, and that the economies of the area will be enhanced by sharing the responsibilities of such management. Each party state has agreed to adopt practices that will require low-level waste shipments originating within its borders and destined for a facility within another state to conform to applicable package and transportation requirements and regulations of the host state.

The compact provides for appropriate record keeping, inspections, penalties, and the imposition of fees by the host state upon those generating wastes and creates a Compact Committee to administer the business of the compact. The agreement of two-thirds of the member states and an affirmative vote by the host state must be obtained before waste from outside the compact area may be accepted.

The consent legislation for the Northwest Compact is S. 247 in the Senate. It has been referred to the Judiciary Committee and will be held at full Committee level. In the House, H.R. 1012 has been referred to the Energy and Commerce Committee, chaired by Congressman John Dingell of Michigan, and to the Interior Committee, chaired by Congressman Morris Udall of Arizona.

The Senate Judiciary Committee, under Senator Thurmond, held hearings on S. 247 during the week of March 2, 1983, along with an oversight hearing on the status of all the compacts and on issues that have surfaced in connection with them and the low-level waste law. Senator Thurmond is expected to wait for several more compacts to be introduced so that some comparison may be made as to how different regions are approaching the compacting process and handling the issues involved in disposal of low-level waste before bringing them to the full Senate for approval. On the other hand, he will not wait for all compacts to be introduced in Congress if that takes a long period of time. The 1986 date, after which Congressionally approved compacts may begin to exclude out-of-region waste, would appear to require approval within a year of all compacts submitted during this session of the Congress. As it is, there is likely to be a gap between the 1986 exclusionary date and the date by which some new disposal sites will be ready for operation. Senator Thurmond does not, however, support an extension of the exclusionary date as a means of resolving this problem. The Northwest Compact calls for the exclusion of waste from outside the region starting July 1, 1983. However, this will not be accepted by the Congress, and, in a compromise with those regions wishing further delay, the effective exclusionary date for all compacts will probably be held to January 1, 1986, as originally provided in the law.

Congress has the option of approving a compact in its entirety, rejecting the compact, or approving it subject to certain conditions. Should Congress choose to condition its approval of a particular compact, such conditions would not necessarily require subsequent action by state legislatures unless the alteration were substantial and went to the essence of the compact. This is not expected, even though Congress will move the Northwest exclusion date from July, 1983 to January 1, 1986.

There are options available to the states to provide for interim storage or disposal until new sites begin operation. One option is to establish interim storage facilities within the region. A second solution is to negotiate with sited regions for continued access to existing sites as long as those regions without sites have a compact in place and are making a good faith effort to designate a host state and bring a new disposal site on-line.

The Northwestern and Southeastern regions of the country are moving rapidly, but there has been less progress in other parts of the country. Some states, such as Texas and California, may "go it alone" as is allowed under the law. However, a quirk in the enabling legislation does not give them the right to refuse waste from outside their respective state boundaries, as it does for compact states. The question arises as to whether or not two states such as California and Texas, even though they are not contiguous, and even though each may plan to handle its own waste within its boundaries, may enter into a compact with each other for the sole purpose of preventing the importation of waste from outside their respective states. This quirk in the law was unintentional and came about through an oversight.

On January 7, 1983, President Reagan signed into law The Nuclear Waste Policy Act of 1982 for the management and storage of high-level, commercial nuclear waste and spent fuel. As a result, the Department of Energy has established a Nuclear Waste Policy Act Project Office. Its new director, Robert L. Morgan, has initiated a coordinated effort to meet the elaborate set of near-term actions that are required by this complex law.

The act requires major efforts in two primary areas, disposal and storage of spent fuel and high-level waste. This program is to be financed by a fee of 1 mil/KWhr of nuclear power produced, collected from utilities. The act provides for cooperative research, development, and demonstration activities at utility sites and federal sites.
The act finds that:

"1. The federal government has the responsibility to provide for the permanent disposal of high-level waste and spent nuclear fuel.

2. The generators and owners of high-level waste and spent fuel have the primary responsibility to provide for, and the responsibility to pay the costs of, the interim storage of such waste and spent fuel until such waste and spent fuel is accepted by DOE.

3. State and public participation in the planning and development of repositories is essential in order to promote public confidence in the safety of disposal of such waste and spent fuel.

4. High-level radioactive waste and spent nuclear fuel have become major subjects of public concern, and appropriate precautions must be taken to ensure that such waste and spent fuel do not adversely affect the public health and safety and the environment for this or future generations."

Under the act, DOE, with the concurrence of the U.S. Nuclear Regulatory Commission (NRC), has 180 days to issue guidelines for repository site recommendations. These guidelines must include consideration of geologic criteria; location of valuable natural resources; hydrology; geophysics; seismic activities; atomic energy defense activities; proximities to water supplies and population; the effect upon the rights of users of water; proximity to national parks, forests, etc.; proximity to sites where high-level radioactive waste and spent nuclear fuel is generated or stored; transportation and safety factors involved in moving such waste to a repository; the cost and impact of such transportation; and the advantages of regional distribution in the siting of repositories. To the extent practicable, DOE must recommend sites in different geologic media.

A site is automatically disqualified if any surface facility of the repository is located in a highly populated area or adjacent to an area 1 mile by 1 mile having a population of not less than 1,000 individuals.

The act establishes a step-by-step process by which the President, the Congress, the states, affected Indian tribes, DOE, and other federal agencies can work together in the siting, construction, and operation of a high-level nuclear waste repository. One of the priorities is to strengthen consultation and cooperation between DOE and affected states and affected Indian tribes.

Based on the information acquired thus far, DOE believes that the states of Louisiana, Texas, Utah, Mississippi, Nevada, and Washington each contains a potentially acceptable site for a repository. Accordingly, on February 2, 1983, DOE Secretary Hodel sent letters to the governors and legislative leaders of those states formally notifying them to these findings.

As an early step, DOE guidelines for the recommendation of sites for a repository were published in the Federal Register for public review and comment on February 7, 1983. These guidelines will be used to identify and nominate sites for characterization and, eventually, to determine the suitability of a site for development as a repository. Through consultation with the interested governors, the Council on Environmental Quality, the Environmental Protection Agency, the U.S. Geological Survey, and with the concurrence of the NRC, DOE had plans to finalize these guidelines no later than July 7, 1983, as required by the act. To facilitate public involvement, DOE was holding public hearings in March in Seattle, Chicago, New Orleans, Washington, D.C., and Salt Lake City.

The next step under the act is DOE's initial nomination of five locations for site characterization, based on a consideration of the guidelines and the environmental assessments to be prepared for each of the five nominated sites, followed by a recommendation of three candidate sites to the President for his approval. The DOE had intended to complete these actions in the summer of 1983 in order to permit the conduct of a sufficiently thorough site characterization program at each site to support the presidential recommendation of a site for the first repository by March 31, 1987, as required by the act. The act requires that for each site under consideration, DOE conduct public hearings to solicit recommendations on issues to be addressed in the environmental assessment and in any site characterization plan to be used if the site is approved by the President. The specific date and location for these hearings will be established after consultation with state representatives.

In FY 1984, the program begins a transition from the research phase to the engineering development phase. In basalt, drilling of an exploratory shaft will be completed and repository design studies will continue. In addition, at-depth testing in the preferred repository region will be initiated. In tuff, the exploratory shaft mining will be inaugurated. Confirmatory site characterization will continue. Additionally, R&D will continue on waste package design, materials characterization, and transportation studies. The aim of this overall work is to develop the data to assist a presidential site selection and subsequent repository license application for a first repository in the mid to late 1980's.

By June 1, 1985, a detailed study of the need for and feasibility of construction of one or more Monitored Retrievable Storage (MRS) facilities is required. The study is to include a proposal for the construction of one or more MRS facilities and to contain at least five alternative site-specific designs at a minimum of three different sites.

Implementation of the spent fuel storage provisions of the act is predicated upon the primary responsibility for storage remaining with the utilities. Federal actions are limited to: (1) encouraging and expediting effective use of existing at-reactor storage capacity through research, development, and demonstration, along with consultative assistance and (2) providing up to 1,900 metric tons of federal storage for use by utilities who have been certified by the NRC as having exhausted other reasonable alternatives.

As directed by the act, DOE intends, within one year, to enter into a cooperative agreement with one to three utilities desiring DOE assistance in at-reactor demonstrations of spent fuel storage technologies. The DOE is authorized to do unlicensed demonstrations for dry storage at DOE sites using up to 300 metric tons of spent fuel.

The act assigns the responsibility to DOE for interim storage of spent nuclear fuel from those civilian nuclear power reactors that cannot reasonably provide storage needed to assure their continued, orderly operation. The capacity provided by this program shall not exceed a total of 1,900 metric tons. Following an NRC determination of a utility's eligibility for interim storage, DOE will enter into contracts with the utility, take title to its fuel, provide the storage capacity for this fuel at a DOE site, ship the fuel to the storage site (using private industry under contract to provide these services), and store the fuel pending availability of disposal or other storage facilities.

The act stipulates that it shall be the policy of the United States to cooperate with and provide technical assistance to non-nuclear weapon states in the field of spent fuel storage and disposal.

These are only a small fraction of the requirements of that act, and they are greatly simplified in this review. Aside from making the federal government responsible for the ultimate storage of high-level commercial waste, and aside from the precedent-setting provision that the DOE will take title to spent fuel, the outstanding feature of the act is its meticulous attention to procedural details, especially in involving the public, the states where repositories may be located, and the agencies of government responsible for protection of the environment and human health and safety.

It is obvious that the DOE, Secretary Don Hodel, and Project Director Robert Morgan are committed to carrying out the most demanding requirements of the law even more rigorously than required. For instance, the DOE was to hold a public hearing in

Richland, Washington on March 25 on the environmental assessment it has prepared with respect to the vertical exploratory shaft it is planning to excavate in basalt at Hanford. This hearing was not required. However, the DOE is reaching out to obtain public involvement in all steps it takes in implementing the waste program.

Thus, the view from Washington is an optimistic one. With support of the vast amount of research, development and engineering that scientists and engineers have completed and will do in the future, this country certainly is now headed towards a successful and orderly program for handling all radioactive wastes safely and inexpensively. The high-level waste program, for instance, will require only about 2% of the cost of electricity produced from nuclear power.

There is one additional subject that deserves attention at this time. It is the need for the federal government to create, by law, a public corporation for handling the fuel cycle for commercial nuclear fuel. Such a Federal Nuclear Fuel Cycle Corporation (FNFCC) would handle almost all of the nuclear power fuel cycle in a manner similar to the DOE management of the fuel cycle for the weapons program. Except for the mining and milling of uranium and the fabrication of uranium (only) fuel elements, the federal government would pre-empt ownership of all facilities used in the fuel cycle and operate them by contract with private industry (as with the weapons fuel cycle). Ownership of all fissile and fertile material (and all existing and future fuel elements) would be pre-empted by law and vested in the FNFCC.

The FNFCC would lease fuel elements to any utility, foreign or domestic, (with IAEA supervision for foreign utilities) on condition that the fuel elements are returned for reprocessing.

The United States will probably need three fuel cycle centers. Each would include a reprocessing plant, an advanced fuel fabrication facility, and a waste glassification and storage facility.

The separations plants would produce a stream of blended uranium and plutonium, and a stream of waste for immediate glassification. The uranium-plutonium blend would be in such ratios as required for fuel but not for weapons production.

The FNFCC would probably be controlled by a Board of Directors, nominated by the President and confirmed by the Senate. Initial financing could come from assessments against utilities for reprocessing of existing spent fuel, as for the new high-level waste program. An assessment of 3 mils per kilowatt hour of nuclear electricity produced would probably fund all operations of the FNFCC, and all of the nuclear power research, development, and demonstraton presently funded in the DOE budget.

The FNFCC would absorb the high-level waste program, Barnwell, and all enrichment facilities. Thus, all these items would be "off-budget" in the future but still approved annually by the Congress.

Private industry would perform all operations on cost-plus contracts, thus eliminating potential conflicts between safety and profits.

Small nations with a few nuclear power plants would find it cheaper to lease fuel from the FNFCC than to try to reprocess it themselves. This would constitute a major, realistic, non-proliferation initiative. Candidate nations include the Republic of China, South Korea, the Philippines, Argentina, and Brazil.

Reprocesing fuel to extract and glassify waste for permanent geologic disposal is the most attractive method for handling spent fuel from a safety and environmental perspective. Also recycling fuel is probably more environmentally attractive than mining more uranium, especially from lower grade ores.

This concept of an FNFCC has been suggested to the Adminstration and to some Congressional leaders. There has been no known opposition expressed, but there is some indication of a reluctance to undertake such a major problem-solving initiative in one step.

RECEIVED December 12, 1983

INDEXES

Author Index

Subject Index

F

G

H

U

W

X

Z

Production by Anne Riesberg
Indexing by Deborah Corson
Jacket design by Anne G. Bigler

Elements typeset by Hot Type Ltd., Washington, D.C.
Printed and bound by Maple Press Co., York, Pa.

RECENT ACS BOOKS

"Industrial-Academic Interfacing"
Edited by Dennis J. Runser
ACS SYMPOSIUM SERIES 244; 176 pp.; ISBN 0-8412-0825-5

"Characterization of Highly Cross-Linked Polymers"
Edited by S. S. Labana and Ray A. Dickie
ACS SYMPOSIUM SERIES 243; 324 pp.; ISBN 0-8412-0824-9

"Polymers in Electronics"
Edited by Theodore Davidson
ACS SYMPOSIUM SERIES 242; 584 pp.; ISBN 0-8412-0823-9

"Radionuclide Generators: New Systems for Nuclear Medicine Applications"
Edited by F. F. Knapp, Jr., and Thomas A. Butler
ACS SYMPOSIUM SERIES 241; 240 pp.; ISBN 0-8412-0822-0

"Polymer Adsorption and Dispersion Stability"
Edited by E. D. Goddard and B. Vincent
ACS SYMPOSIUM SERIES 240; 477 pp.; ISBN 0-8412-0820-4

"Assessment and Management of Chemical Risks"
Edited by Joseph V. Rodricks and Robert C. Tardiff
ACS SYMPOSIUM SERIES 239; 192 pp.; ISBN 0-8412-0821-2

"Chemical and Biological Controls in Forestry"
Edited by Willa Y. Garner and John Harvey, Jr.
ACS SYMPOSIUM SERIES 238; 406 pp.; ISBN 0-8412-0818-2

"Chemical and Catalytic Reactor Modeling"
Edited by Milorad P. Dudukovic and Patrick L. Mills
ACS SYMPOSIUM SERIES 237; 240 pp.; ISBN 0-8412-0815-8

"Multichannel Image Detectors Volume 2"
Edited by Yair Talmi
ACS SYMPOSIUM SERIES 236; 333 pp.; ISBN 0-8412-0814-X

"Efficiency and Costing: Second Law Analysis of Processes"
Edited by Richard Gaggioli
ACS SYMPOSIUM SERIES 235; 262 pp.; ISBN 0-8412-0811-5

"Archaeological Chemistry--III"
Edited by Joseph B. Lambert
ADVANCES IN CHEMISTRY SERIES 205; 324 pp.; ISBN 0-8412-0767-4

"Molecular-Based Study of Fluids"
Edited by J. M. Haile and G. A. Mansoori
ADVANCES IN CHEMISTRY SERIES 204; 524 pp.; ISBN 0-8412-0720-8